Chelating Agents for Land Decontamination Technologies

SPONSORED BY
Hazardous Waste Committee of the Environmental Council
Environmental and Water Resources Institute (EWRI)
of the American Society of Civil Engineers

EDITED BY
Daniel C. W. Tsang
Irene M. C. Lo
Rao Y. Surampalli

ASCE

EWRI
ENVIRONMENTAL &
WATER RESOURCES
INSTITUTE

Published by the American Society of Civil Engineers

Cataloging-in-Publication Data on file with the Library of Congress.

Published by American Society of Civil Engineers
1801 Alexander Bell Drive
Reston, Virginia, 20191-4400
www.asce.org/pubs

ISBN 978-0-7844-1218-3 (paper)
ISBN 978-0-7844-7683-3 (e-book)
Manufactured in the United States of America.

Contents

Preface

Chelating agents (or chelants) refer to ligands that can occupy multiple positions in the inner coordination sphere of the central metal ion, leading to the formation of multidentate metal-chelant complexes (or chelates). The original Greek meaning of the word "chele" means a horse's hoof and, by extension, a crab's or scorpion's claw, which is symbolically illustrated on the book cover. Because of their strong interaction with metals, chelating agents are widely used in numerous applications, such as detergents, industrial cleaning, pulp and paper, photo industry, textiles, agriculture, cooling water, oil production, personal care, medicine, and food and beverage industry.

Contaminated land remediation has been a widespread and costly problem. Traditional excavation-and-disposal method is no longer regarded as a sustainable solution, while contaminant removal is often difficult to accomplish by means of physical separation methods only. Therefore, the proposed use of chelating agents for enhancing soil remediation has received extensive attention over the last two decades. There has been a significant increase in soil related research on the effects of chelating agents on metal solubility, mobility and bioavailability in soil, as well as degradability and plant uptake of metal-chelant complexes in the natural environment.

Chelating agents are able to enhance metal extraction from contaminated soil/sediment and facilitate metal mobility in the subsurface, making them potentially promising enhancement reagents for remediation technologies. This book focuses on the engineering applications of chelating agents for soil washing, soil flushing, phytoremediation, and electrokinetic remediation. The goal of this book is to provide environmental engineers and scientists with some practical considerations about the design and implementation of chelant-enhanced remediation technologies.

This book comprises two sections. Section 1 focuses on the application of chelating agents for ex-situ washing processes. Chapter 1 reviews the design and implementation of traditional soil washing and chelant-enhanced washing. It also discusses the economic and societal considerations associated with soil washing technology. Chapter 2 elaborates the application of chelant-enhanced washing for heavy metal-contaminated sediment. The chemistry of chelating agents in washing solution is explained with latest research findings. Chapter 3 focuses on the significance of operational conditions for effective chelant-enhanced soil washing. This chapter also reviews the results of a recent field demonstration case study. Chapter 4 focuses on the recovery of chelating agents from used washing solution. It suggests the state-of-the-art use of electrochemical treatment. Chapter 5 reviews the effectiveness of fresh and recovered chelating agents for treatment of spent catalysis.

This chapter also provides extensive characterization of recovered chelants and reviews available mathematical models for describing metal extraction kinetics.

Section 2 of this book discusses the application of chelating agents for in-situ soil remediation technologies. Chapter 6 gives an overview of chemical-enhanced soil flushing technology. It discusses the latest findings of metal removal by complexation with different chemical reagents. Chapter 7 evaluates heavy metal leaching during percolation of chelating agents. This chapter develops a comprehensive transport model for simulating the breakthrough of heavy metals and chelating agents under different flushing conditions. Chapter 8 examines the roles of iron, aluminum, and manganese hydroxides in chelant-enhanced phytoextraction. This chapter elaborates the complexation chemistry at the hydroxide surface with aid of latest research findings. Chapter 9 provides an extensive review of the use of chelating agents in electrokinetic remediation. This chapter compares in detail the use of different chelating agents and discusses comprehensive experimental results of electrochemical remediation.

This book provides a compilation of engineering applications and latest research findings for different chelating agent-enhanced remediation technologies. The contents of this book will be useful for engineers, scientists, and decision-makers in the area of contaminated land remediation.

The editors acknowledge the hard work and patience of all authors who have contributed to this book. The views or opinions expressed in each chapter of this book are those of the authors and should not be construed as opinions of the organizations they work for.

DT, IML, RYS

Contributing Authors

Alain Bourg, *University of Pau, Pau Cedex, France*

Fabrice Compère, *ERM, Faculté des Sciences, Poitiers, France*

Alain Dupuy, *Université of Bordeaux, Pessac Cedex, France*

Alok Gautam, *Indian Institute of Technology, New Delhi, India*

Shina Goel, *University of Newcastle, Callaghan, Australia*

Ying-Ying Gu, *University of Petroleum (East China), Qingdao, China*

Monika Kedziorek, *University of Bordeaux, Pessac Cedex, France*

Michael Komárek, *Czech University of Life Sciences, Prague, Czech Republic*

Domen Lestan, *University of Ljubljana, Ljubljana, Slovenia*

Irene M. C. Lo, *Hong Kong University of Science and Technology, Hong Kong, China*

Catherine Mulligan, *Concordia University, Québec, Canada*

Alessandra Polettini, *University of Rome "La Sapienza", Rome, Italy*

Raffaella Pomi, *University of Rome "La Sapienza", Rome, Italy*

Rongliang Qiu, *Sun Yat-Sen University, Guangzhou, China*

Rao Y. Surampalli, *U.S. Environmental Protection Agency, Kansas City, KS, USA*

Daniel Tsang, *University of Canterbury, Christchurch, New Zealand*

Albert Yeung, *University of Hong Kong, Hong Kong, China*

Weihua Zhang, *Sun Yat-Sen University, Guangzhou, China*

CHAPTER 1

Design, Implementation, and Economic/Societal Considerations of Chelant-Enhanced Soil Washing

Daniel C.W. Tsang, Irene M.C. Lo, and Rao Y. Surampalli

1.1 Introduction

1.1.1 Overview of Site Remediation

The proper management of heavy metal-contaminated soils has been a widespread and costly issue. Land contamination of hazardous wastes is a widespread problem. It has been estimated that the total number of contaminated sites are about 294,000 in the United States (U.S. EPA, 2004) and over 1,400,000 in Western Europe (ETCS, 1998). Despite substantial progress over the past two decades, a considerable amount of cleanup work remains and is expected to take more than 30 years (U.S. EPA, 2004). The prevalent contamination of heavy metals is of high concern. Heavy metals are, for instance, present in 77% of the U.S. EPA National Priority List sites, of which the most commonly occurring are arsenic, chromium, lead, zinc, nickel, and cadmium (U.S. EPA, 2004). Their anthropogenic sources include industrial activities (e.g., battery recycling, mine tailings, atmospheric deposition of smelter dust), agricultural activities (e.g., land application of fertilizers and sewage sludge), and military operation (e.g., shooting range). Heavy metals are highly persistent in the environment and excessive accumulation can have deleterious effects on soil fertility and ecosystem functions as well as impose health risks to animals and humans. Therefore, in order to reduce the potential risk to human health and the environment to an acceptable level, i.e., risk-based land management, managerial and/or remedial actions on contaminated land (Figure 1.1) are required to restrict the access of receptors to the site, block the exposure pathways, and/or remove the source contaminants. Remediation technologies that permanently remove the contaminants from the site are preferred where possible because they can eliminate the potential of long-term liability and allow flexible land use in the future.

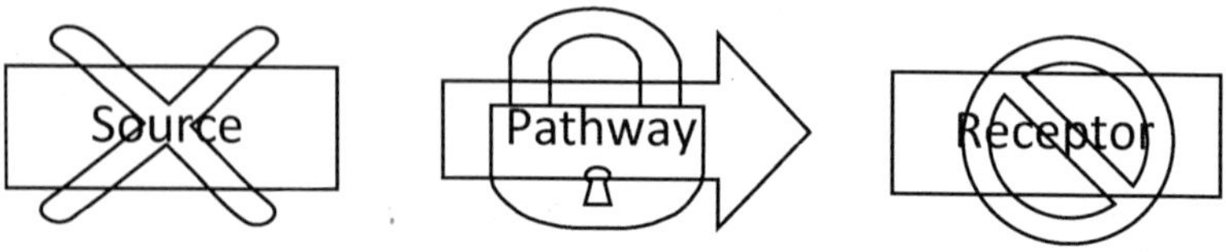

Figure 1.1 Management of contaminated land: restricting the access of receptors to the site (right) by managerial actions, blocking the exposure pathway (middle, e.g., containment) or removing the source contaminants (right, e.g., soil washing) by remedial actions.

The excavation of contaminated soils, followed by disposal at landfills (also known as "dig and dump"), was a simple and conventional source control method. However, there is a general consensus that this is not sustainable and is the least preferred compared with process-based remediation technologies. The position of the regulators and recent legislation are the key drivers for changing the "dig and dump" culture. In the United States, Section 121 (b) of the Comprehensive Environmental Response, Compensation, and Liability Act (CERCLA) mandates the U.S. EPA to select remedies that "utilize permanent solutions and alternative treatment technologies or resource recovery technologies to the maximum extent practicable" and to prefer remedial actions in which treatment that "permanently and significantly reduces the volume, toxicity, or mobility of the hazardous substances, pollutants, and contaminants is a principal element." There is a clear preference for permanent treatment over containment or removal and disposal in the remediation of Superfund sites (U.S. EPA, 2007).

On the other hand, the European Union (EU) Landfill Directive (EU, 1999) has a profound effect on the land remediation industries. When the waste is classified as hazardous, the Landfill Directive gives a hierarchy of treatment objectives to reduce the volume, reduce the hazardous nature, facilitate the handling, or enhance the recovery of the hazardous waste. Its implementation in 2004 reduced the available number of landfills for all forms of waste disposal and, for the first time, banned co-disposal of soils from contaminated land and non-hazardous wastes. The rules were tightened further in 2005 in that no wastes can be sent to hazardous waste landfills in excess of 6% organic matter and the new Waste Acceptance Criteria require that all wastes sent to hazardous waste landfills have to be pre-treated (Harbottle et al., 2007). Prices for waste disposal to hazardous landfills have risen dramatically and the use of excavation and disposal at landfills is therefore becoming less attractive as a method for contaminated land remediation.

As a result, there is a great need for remedial alternatives that ensure effective cleanup of heavy metal-contaminated soils. Soil washing is an ex-situ, water-based remediation technology that is one of the most promising options for soil treatment.

1.1.2 Scope and Purpose

Soil washing technology is a physical–chemical approach based on well-established mining and mineral processing principles and techniques. The design of soil washing systems involves a series of treatment operations and is flexible in terms of the number, type, and order of treatment processes such that the risk-based cleanup goals can be cost-effectively achieved, after considering the site-specific conditions and requirements. Figure 1.2 illustrates a schematic diagram of various types of soil washing. Some process units are used to separate contaminated fines from cleaner coarse fractions with the aid of scrubbing and washing the soils with water, while some units aim to chemically extract sorbed contaminants, either from entire soil matrix or contaminated sand/fines fractions, using washing (extracting) solutions that contain chemical agents such as acids/alkalis, chelating agents, solvents, surfactants, and reducing/oxidizing agents.

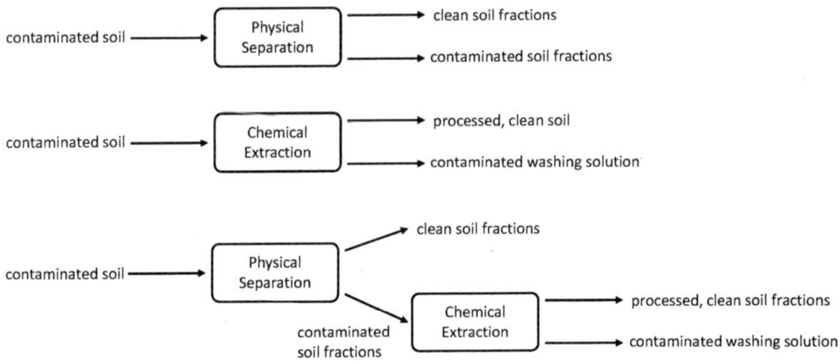

Figure 1.2 Schematic diagram of various types of soil washing.

Therefore, when employed primarily as a process for particle-size separation, soil washing is also known as physical separation or volume reduction (Griffiths, 1995; Mann and Groenendijk, 1996; Anderson et al., 1999; Richardson et al., 1999), which significantly lowers the remediation costs by reducing the expenditure for transportation and disposal at landfills. On the other hand, the term "soil washing" is also used in the literature for describing processes that involve the use of chemical extraction units and chemical agents in washing solution (Griffiths, 1995; Reed et al., 1996; Neale et al., 1997; Davis and Hotha, 1998), which facilitate a more complete removal of heavy metals from the soils, including the fines fraction. It should be noted that the latter type of soil washing is sometimes classified as chemical extraction (which is often preceded by physical separation) (FRTR, 2007).

Since there are many definitions of soil washing, it is important to clearly understand the technology details for implementation on a specific site (Mann and Groenendijk, 1996). Therefore, in this chapter the term "soil washing" refers to washing systems that primarily rely on physical separation, and "chemical-enhanced soil washing" (or "chelant-enhanced soil washing") to washing systems that exploit chemical (chelating agent) extraction besides physical separation. In general, physical separation is capable of removing particulate forms or surficial contamination of heavy metals, whereas chemical extraction is needed to remove sorbed or less labile heavy metals from the soils. Soil washing based on physical separation contributes to a significant reduction in the volume of soils that require further treatment or disposal at landfills. Chelant-enhanced soil washing enables a more complete extraction of heavy metals from different soil fractions.

This chapter provides an overview of generic soil washing technology, its applicability, advantages and limitations, approach to implementation, and system equipment. It discusses chelant-enhanced soil washing with respect to removal enhancement by chemical agents, advantages and limitations, use of chelating agents, and operational conditions. This chapter also provides a review of economic and societal considerations as well as status of soil washing technology.

1.2 Soil Washing Technology

1.2.1 Technology Description and Applicability

The concept of remediation by soil washing is based on the findings that in most cases heavy metals are bound to (as sorbed phase) or associated with (as particulate phases or surficial contamination) the fines fraction (smaller than 63 μm) of contaminated soils, i.e., clay, silt, and organic matter, while low concentrations of contaminants may exist in the oversize materials (larger than 2 mm) and sands (between 63 μm and 2 mm). The fines, in turn, are attached to sand and gravel particles by physical processes, primarily compaction and adhesion. Therefore, soil washing processes (Figure 1.3) are operated to disintegrate the soil aggregates, remove surficial contamination, and separate discrete metal precipitates by attrition scrubbing, and then to physically separate the soils into specific size fractions by exploiting the differences in size, density, hydrophobic surface properties and magnetism, using various separation units. The use of water is necessary for soil disaggregation and effective particle-size separation.

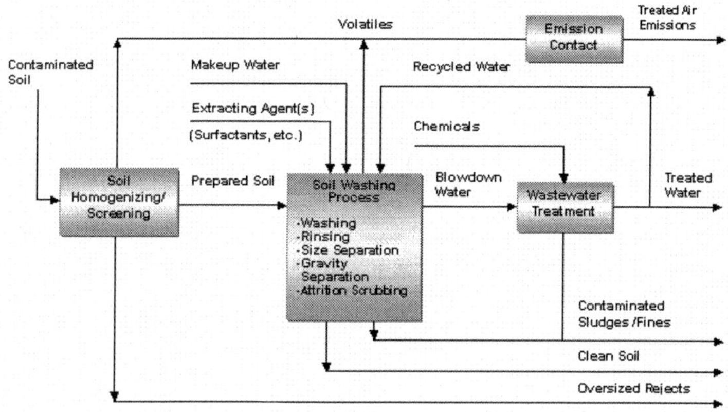

Figure 1.3 Typical soil washing process (FRTR, 2007).

Through these processes, large metallic particles with a distinctive particle-size range can be removed and metal-laden silt, clay, and humic materials can be separated from clean, coarse soil particles. The volume of contaminated materials is, therefore, significantly reduced. The contaminated fines are consolidated, either dewatered into sludge cake and disposed of, or further treated by chemical or metallurgical processes. The clean oversize and sand fractions, after attainment of the treatment standards is confirmed, can be recombined, augmented with nutrients, if required, and returned to the site as clean backfill. If unacceptable levels of heavy metals remain, the soils should be stockpiled or fed directly to the next step for additional treatment. Process water is typically recycled for reuse within the system, after being treated with technologies such as chemical precipitation and ion exchange (Griffiths, 1995; Mann and Groenendijk, 1996; Anderson et al., 1999; Richardson et al., 1999). There are more sidestreams and equipment involved in the actual operation than is shown in Figure 1.3, and more details are given in later sections.

The duration of soil washing is typically short- to medium-term (FRTR, 2007). The performance of a soil washing system that primarily relies on physical separation is typically measured by the volume reduction attained, and calculated by weighing the clean products (i.e., oversize and sand fractions) that meet the specified cleanup standards (Mann, 1999).

$$volume\ \ reduction\ (\%) = \left(1 - \frac{feed\ \ soil\ (tons) - clean\ \ products\ (tons)}{feed\ \ soils\ (tons)}\right) \times 100\%$$

Soil washing can be used independently or in conjunction with other treatment technologies. In general, soil washing is applicable for the following situations (Griffiths, 1995; Mann and Groenendijk, 1996; Anderson et al., 1999; Mann, 1999; Richardson et al., 1999; Sharma and Reddy, 2004; FRTR, 2007):

- treatment of a wide range of inorganic and organic contaminants, including heavy metals, radionuclides, polyaromatic hydrocarbons, polychlorinated biphenyls, and pesticides;
- soils that consist of less than 30 percent fines (slit and clay) fraction, because otherwise the separation of the fine particles from the washing solution may be less effective and there is a low degree of volume reduction;
- soils in urban or industrial areas that are contaminated by anthropogenic source (e.g., shooting range and mining/smelting sites), where metals may be preferentially sorbed on fine particles or present as surface coatings and discrete precipitates;
- sites that contain at least 5,000 tonnes of contaminated soils, as on-site soil washing is typically not cost effective because of the economy of scale (i.e., small-scale treatment is more costly than large-scale treatment on a per-tonne basis); however, this depends on the efficiency of the soil washing as well as the costs of available remediation alternatives, that is, soil washing for less than 5,000 tonnes may be economically competitive at some sites;
- sites that have sufficient space for on-site treatment, as space requirements are variable based on the design of the soil washing system, throughput rate, staging for untreated and treated soils, and site logistics.

1.2.2 Advantages and Limitations

Soil washing presents a number of advantages (Mann, 1999; Sharma and Reddy, 2004; FRTR, 2007; Dermont et al., 2008), including:

- separation technologies are well established in the mining and mineral processing industries and the operational costs are usually low;
- full-scale modular treatment train systems and mobile unit systems are available;
- high throughput rate: about 70 tonnes per hour;
- processed coarse fractions of soils can be returned to the site as backfill or used as construction-grade material for other on-site development;
- volumes of soils that need further treatment or disposal are considerably reduced;
- potential to remove both organics and inorganics in the same treatment system and address a broad range of influent contaminant concentrations;
- metals may be recovered and recycled in some cases (e.g., smelting and shooting range sites);
- closed system permits good control of the operating conditions such as pH and temperature;
- relatively few permits are required for air emission (if volatile organics are present) and wastewater discharge, compared with other treatment systems.

On the other hand, there are factors that may limit the applicability and effectiveness of soil washing (Mann, 1999; Sharma and Reddy, 2004; FRTR, 2007; Dermont et al., 2008):

- treating soils with a silt/clay content in excess of 30 to 50 percent is generally not cost-effective due to difficulties in physical handling and solid/solution separation; it should be noted that a high percentage of fines does not necessarily preclude the use of the soil washing process (a combination of attrition scrubbing, wet screening and hydrocyclones can be used for fine-grained sediments), but results in a more costly treatment;
- high humic content in soils (e.g., agricultural soils) can inhibit soil washing as a result of strong sorption of heavy metals (particularly Pb, Cu, Cd, etc) on the soil organic matter;
- insignificant difference in density or surface properties between metal-bearing particles and soil matrix may result in insufficient separation;
- coarse fractions may not meet cleanup goals, despite attrition scrubbing, if metals are strongly bound in all particle size fractions;
- heterogeneities in soils and initial contaminant concentrations (due to variations at different depths and locations) complicate the operational settings and result in inconsistent performance, thus the use of additional techniques, such as blending of feed soils, may be needed to provide a more uniform feedstream;
- presence of heavy fuel oil that is of high viscosity, such as No. 6 fuel oil, present separation problems for soil washing systems;
- cost-effective and economically competitive soil washing operations usually require a large volume of soils to be treated (> 5,000 tonnes when conventional remediation alternatives are available);
- on-site space requirement for the soil washing system equipment and stockpiles of untreated and treated soils;
- as soil washing is an ex-situ technology, there are concerns about possible exposure of the contaminants to the personnel at the site and the neighbouring public during soil excavation and handling;
- process water needs to be treated prior to recycle or discharge.

The key to successful soil washing lies in an understanding of the soil matrix and contamination characteristics, thereby making appropriate arrangements and configurations of the unit processes, as elucidated in the following sections.

1.2.3 Approach to Implementation

To investigate whether soil washing is technically feasible for treating the contaminated soils at a particular site, a systematic tiered (or phased) approach should be implemented (U.S. ITRC, 1997; Mann, 1999; Richardson et al., 1999). Three tiers are suggested prior to the full-scale application.

1.2.3.1 Tier 1 (Characterization Studies)

Due to the inherently heterogeneous nature of soils and the variability between sites, it is important that every soil washing project should be qualified by screening characterization studies. Site characterization is conducted to determine the basic mineral and physical properties of the soils and contaminants as well as their relative associations. Representative samples need to be collected throughout the site. Sieving analysis of the soils is performed to construct the particle-size distribution curve. After separation, each particle-size fraction is then chemically analyzed (total concentrations, sequential extractions, scanning electron microscopy, etc) for the contaminants of concern. The mode of contamination in each fraction (particulate, coated, soluble, weakly/strongly bound, etc) is also important. The analytical results are then overlain on the particle-size distribution curve. Such information on the soil properties and the relationship between particle size and contamination characteristics, along with existing site information, provides significant insight into the possible treatment scenarios and allows an informed decision to be made regarding the feasibility of soil washing. Most importantly, substantial volume reductions can be obtained by understanding the particle-size/contaminant relationship and by screening and separating the contaminated soils for the most appropriate treatment (Mann, 1999).

1.2.3.2 Tier 2 (Treatability Studies)

Based on the characterization studies of the properties of the soil components and contaminants and their association mechanisms, bench-scale treatability studies are performed to evaluate the performance of specific process units and to design a treatment system that can accomplish contaminant separation to meet the cleanup goals. Differences in the intrinsic properties of coarse and fine particles (and discrete metal precipitates, if present), such as particle size, density, settling velocity, surface flotation characteristics, and magnetic properties, are exploited to achieve effective separation. Generally, the greater the property difference, the easier the separation. Screening, hydrocyclone, froth flotation, and filtration studies are conducted to select the treatment units, screening points, cut-off points, and to determine the polymer, flow rate, and throughput requirements. Treatability studies should also focus on the metal concentrations in the products of the process, specifically the oversize, sand, and fines fractions (Mann and Groenendijk, 1996).

Afterwards, computer simulations of various conceptual process designs (i.e., probable combinations of selected technologies) are performed. Flowsheet simulation systems assist in comparing the process alternatives and optimizing the design of the process flow diagrams (Toebermann et al., 1999). The simulation input values include the soil characteristics in terms of particle size, particle density, and contaminant distribution as well as the design parameters and separation performance of each process unit. The system performance can then be predicted in terms of separation efficiency, volume reduction, and concentrations in the oversize, sand, and fines fraction.

In addition to technical effectiveness, there are a number of factors that need to be taken into account in the design selection, such as economic evaluation (e.g., net savings realized from soil washing compared with available remediation alternatives), treatment facility (e.g., maintenance, operational control and safety, long-term reliability), regulatory concerns, potential environmental impact, and public acceptance. The report of the treatability studies provides the confirmed process flow diagram, general specifications for the actual facility, unit treatment price, and any particular contractual qualifications.

1.2.3.3 Tier 3 (Pilot-Scale Study)

Pilot-scale studies are normally used for testing and revising the full-scale treatment design. The pilot-scale facilities consist of the full range of required treatment units, yet run at a lower capacity (e.g., 5–15 ton/h). The scope and location of the pilot-scale facility depends directly on the size and complexity of the project. Adequate process control should be built into the plant to enable personnel to verify that the operating parameters established during bench-scale testing can be met in the field. Field sampling and analytical programs should also be defined to allow quality control and assurance of the processes and products.

It may not be necessary to conduct a pilot-level study if bench-scale studies can provide enough information and the site situation closely matches current experience, but it is extremely helpful to demonstrate the capability of the specified treatment process at the actual site prior to the full-scale implementation. Based on the lessons learned from previous field applications (Mann and Groenendijk, 1996; U.S. DoD, 1997; Mann, 1999), it should be recognized that no site can ever be fully characterized prior to site construction, and that the plant design should be flexible enough to handle the expected variability in the texture and metal content of the soils. Pilot-scale studies provide opportunities to identify potential errors in the preceding bench-scale testing or operational problems in the field (e.g., inadequate process control), thereby allowing modifications of the engineering design and equipment, which are crucial for ensuring subsequent full-scale remediation to be reliable and successful. Nevertheless, it is important to emphasize that, during the course of full-scale application, flexibility in the field program should be allowed for revision based on field analytical results of the process and products.

1.2.4 System Equipment

The original concepts of soil washing root in the mining and mineral processing industries, in which separation techniques have been used for years to concentrate metal particles for extraction. Most of the equipment for soil washing systems, such as trommels, screw classifiers, attrition mills, hydrocyclones, etc, has been in common use in the mining and mineral processing industries with proven effectiveness and reliability. Table 1.1 summarizes the equipment and particle size range, according to underlying separation principles. As to the most commonly used devices, wet vibrating screens and trommels are used to separate gross oversize materials (e.g., cobbles and debris; > 5 cm) and process oversize materials (e.g., gravel; 2 mm – 5 cm) (Mann, 1999). Most of the hydrodynamic classifiers and gravity concentrators have a good applicability for sand fraction (63 μm – 2 mm) but are not suitable for fine particles (< 63 μm), while the MGS-Mozley, hydrocyclones, and froth flotation (or dissolved air flotation) are effective for separating fine particles (10 – 63 μm) (Dermont et al., 2008).

Table 1.1 Physical Separation Units

Exploitable Feature	Operation Unit	Typical Technologies	Particle Size Range [#]
Difference in particle size	Mechanical Screening	Barrel trommel	6-60 mm
		Vibrating screens	150 μm – 20 cm
		Gyratory screens	60□ μm – 1 cm
Difference in settling velocity	Hydrodynamic Classification	Hydrocyclones	5-200 μm
		Screw classifier	50-1000 μm
		Hydraulic classifier	100-1000 μm
Difference in particle density	Gravity Concentration	Mineral jig	0.8-6 mm
		Shaking table	63-2000 μm
		Spiral concentrator	63-2000 μm
		MGS-Mozley	1-1000 μm
		Dense media separation cyclone	1-10 mm
Difference in hydrophobic property of particle surface	Froth Flotation	Flotation cell	10-300 μm
Difference in magnetic property	Magnetic Separation	High intensity magnetic separation	1-1000 μm
		Low intensity magnetic separation	0.1-20 mm

[#] (Dermont et al., 2008)

The purpose and use of commonly used process units in previous pilot/field-scale studies (U.S. ITRC, 1997; Anderson et al., 1999; Richardson et al., 1999; Mann, 1999) are briefly summarized as follows. The vibrating screen is used to remove gross oversize materials that are not suitable for soil washing. Natural materials such as roots and rocks can be returned to the site after meeting specified testing requirements, whereas man-made materials such as demolition debris and fill materials should be separated for disposal as solid waste. The maximum size of particles allowed in the feedstock varies with the equipment used, ranging from 1 to 5 cm. The soils are then fed by belt conveyor to the trommel or wet screening module, where high-pressure water sprays help in the deagglomeration of the soil particles and process oversize materials (> 2 mm) are separated. The process oversize fraction is staged for confirmation of meeting the cleanup goals and return to the site as clean backfill.

The soil/water slurry is subsequently pumped to hydrocyclone combinations or screw classifiers that are able to separate coarse/fine sand (63 μm – 2 mm) from fines (< 63 μm). Attrition scrubbing/milling, which causes intensive interparticle abrasion, can be used to remove adhered fines and surficial contamination from sand particles. Afterwards, froth flotation (where a surfactant is added to enhance the fine/contaminant flotation) or spiral concentrators are employed to separate the sand and fines again. The sand fraction is dewatered by means of a dewatering screen, analyzed according to site-specific protocols, and returned to the site as clean backfill, after attainment of the treatment standard is confirmed. Concurrently, the fines are consolidated in lamella clarifiers (where polymer flocculant is needed for coagulation/flocculation), and dewatered by a pressurized belt filter press into sludge cake that is disposed of with or without further treatment (e.g., stabilization/solidification).

It should be noted that there is no single, universal soil washing system. The systems employed at different sites, and by various vendors, vary in the selection and sequence of the unit processes. The full-scale soil washing systems usually employ the previously described equipment in four major subsystems: mechanical and wet screening, separation using hydrocyclones, sand handling and treatment, and fines handling and treatment.

1.3 Chelant-Enhanced Soil Washing

1.3.1 Removal Enhancement by Chemical Agents

Because soil washing with water is ineffective for removing sorbed contaminants from soil particles, chemical agents (also referred to as extracting chemicals or extractants) can be added to the washing solution in order to promote the overall efficiency of contaminant removal. The purposes of chemical-enhanced soil washing are to: (1) produce a cleaner sand fraction that would otherwise fail to meet the specified cleanup goals by soil washing with water; or (2) treat the entire soil matrix, including the most contaminated fines fraction, to an extent that is compatible with the specified cleanup goals so that no further treatment or disposal of the soil is required. Chemical-enhanced soil washing is a physical-chemical technology for transferring contaminants from soil, sludge, and sediment into the washing solution, which is then recycled and reused after on-site wastewater treatment processes. Efficient water quality management is very important, because it not only reduces the overall amount of water used, but also ensures that contaminants and chemical agents in the washing solution are effectively recovered or removed. In addition, an effective containment system may be required to minimize the environmental impact associated with spills.

The system design varies with the primary goal of chemical-enhanced soil washing. For the purpose of producing clean sand fraction, for example, at the

Vineland Chemical Company Superfund Site in New Jersey (U.S. ACE, 2005), oversize materials (> 2 mm) are removed by a trommel and vibrating wet screens, and fine particles (< 100 μm in this project) are removed by hydrocyclones, respectively, prior to chemical extraction. Only the sand fraction is mixed with the chemical agents and ground in a rotating ball mill that provides aggressive mixing. The resulting product is clean sand that can be returned to the site as backfill. The contaminated washing solution is further processed using pH adjustment, flocculation and sedimentation. The sludge generated, as well as fines initially separated by the hydrocyclones, is consolidated into a highly concentrated form and disposed of at an off-site hazardous waste landfill.

By contrast, to meet the aim of returning all size fractions of soils to the site as clean backfill, chemical extraction can take place before and after particle size separation. For example, as reported in Griffiths (1995), the washing solution that contains chemical agents is first mixed with the feed soil in a drum washer where water knives to promote breakup of the soil lumps. After an appropriate washing time (also called the residence or contact time), process oversize fraction (> 2 mm) is separated by a trommel. The sand and fines fraction (< 2 mm) are continuously mixed with chemical agents in stirring tanks. After an adequate residence time the soil slurry is pumped to hydrocyclones, which are operated primarily to perform solid/solution separation rather than particle-size separation (Griffiths, 1995). It is important that, when extraction is complete, the processed soils are rinsed with clean water to remove entrained chemicals and contaminants. The chemical agents are recovered by precipitation, or other appropriate methods, and reused. The process oversize, sand, and fines fraction can be returned to the site after verifying the attainment of cleanup goals. This setup was employed for soil remediation of a small-arms range site (an Army Base) in Louisiana (U.S. DoD, 1997).

The operation and maintenance for chemical-enhanced soil washing is typically medium-term (FRTR, 2007). If the overall goal of chemical-enhanced soil washing is to produce a cleaner sand fraction, the performance is measured by the volume reduction as described in the previous section. If the overall goal is to treat the entire soil matrix, the performance is determined by the extraction efficiency calculated by the initial and final metal concentrations in the soils, as below.

$$extraction\ \ efficiency\ (\%) = \left(1 - \frac{final\ \ concentraion\ \ in\ \ soil\ (mg/kg)}{initial\ \ concentraion\ \ in\ \ soil\ (mg/kg)}\right) \times 100\%$$

1.3.2 Advantages and Limitations

Despite the fact that the advantages and limitations diverge according to the chemical agents used, chemical-enhanced soil washing generally has the following advantages as compared with soil washing that merely relies on physical separation (U.S. DoD, 1997; Peters, 1999; FRTR, 2007; Dermont et al., 2008):

- chemical-enhanced extraction is, in principle, not constrained by the proportion of fines because contaminants can also be extracted from this fraction by chemical agents; in field applications, chemical-enhanced soil washing has been shown to be applicable to contaminated soil, sediment, and sludge that contain up to a 50% fines fraction;
- both sorbed metals and surface precipitates can be extracted or dissolved with the aid of chemical agents; thus, soil washing technology is no longer limited to performing volume reduction only;
- fine particles, the most contaminated fraction, can be treated to meet specified cleanup goals, allowing a complete and permanent treatment of contaminated soils such that off-site disposal is not necessary;
- able to produce a cleaner sand fraction that otherwise is not qualified as clean backfill; therefore, the expectancy of success and the performance reliability of the soil washing system are promoted.

The large-scale application of chemical-enhanced soil washing may encounter certain limitations, which are highly dependent on the soil properties, contamination characteristics, and site conditions (U.S. DoD, 1997; Peters, 1999; FRTR, 2007; Dermont et al., 2008):

- high clay content (< 2 μm, the smallest particles of the fines fraction) may require long contact times, resulting in low throughput rates, e.g., 10 tonnes per hour;
- high calcite content and/or substantial amounts of amorphous iron hydroxides may interfere with the metal extraction due to non-selective consumption of the chemical agents;
- metals imbedded in the mineral lattices are usually non-extractable (nevertheless, these metals are probably non-bioavailable to microorganisms, plants, and humans);
- high humic content in soils may inhibit metal extraction as the soil organic matter has a high sorption affinity for many metals;
- complex mixtures of contaminants, e.g., co-presence of cationic and anionic (oxyanions such as arsenate and chromate) metals, require various chemical agents or operational conditions for best extraction;
- non-aqueous phase liquids, if co-exist, may physically isolate the particle surface and hamper the metal extraction;
- chemical agents need to be recovered or removed from the washing solution using commercially available methods;
- residual chemical agents in the treated soils, if present despite rinsing with water, may pose a threat to ecological systems;
- treated soils may need to be supplemented with nutrients for re-vegetation or mixed with clean background soils prior to being returned to the site, because of the possible effects of chemical agents on the physico-chemical and microbiological properties of the soils;
- higher treatment costs resulting from the chemical cost and additional treatment process for recovery or removal of chemical agents;

- treatment costs may be prohibitive for high levels of contamination coupled with stringent cleanup standards.

1.3.3 Use of Chelating Agents

Chemical agents are selected in consideration of the contaminant characteristics. Chelating agents, acids, and reducing/oxidizing (redox) agents are most applicable to enhance metal extraction. Chelating agents extract metals from soils primarily by forming highly stable and soluble metal complexes that tend to dissociate from the sorption sites on the soil surfaces (Nowack, 2002; Lestan et al., 2008). Strong acids (e.g., hydrochloric acid, sulphuric acid, phosphoric acid, and nitric acid) extract metals by dissolving discrete/surface metal precipitates and soil minerals such as iron hydroxides, on which metals are strongly sorbed, and via ion exchange, that is, proton (H^+) competition for cationic metals or oxyanion (SO_4^{2-} or PO_4^{3-}) competition for anionic metals (Peters, 1999; Dermont et al., 2008). Redox manipulation promotes metal desorption through converting the metals into more soluble forms by a valence change. Reducing agents such as sodium bisulphite ($Na_2S_2O_5$) and hydroxylamine hydrochloride (NH_2OH-HCl) lead to dissolution of Fe/Mn oxides, thus enhancing the extraction of metals bound to these oxides. Oxidizing agents such as sodium hypochlorite ($NaClO$) and potassium permanganate ($KMnO_4$) facilitate chromium removal by a conversion of trivalent chromium, which is readily sorbed or precipitated, to hexavalent chromium, which is soluble (Peters, 1999; Dermont et al., 2008). Redox agents are usually used as a complement to chelating agents.

It is suggested that an ideal chemical agent would: (i) interact very weakly with the soil matrix as compared to the target contaminants, (ii) increase the solubility and mobility of the target contaminants, and (iii) be generally non-toxic and biodegradable (Vulava et al., 2000). It is, however, also recognized that it is nearly impossible to have a single chemical agent that possesses all these desirable characteristics. Despite the proven efficiency of acid extraction in full-scale applications for non-calcareous soils, strong acids result in significant dissolution of soil minerals and organic matter (up to 50%), destruction of the basic nature and soil structure, together with an increase in the soil acidity. Severe damage to physical, chemical, and biological properties of soils by acid extraction limits the suitability of the treated soils for being returned to the site (Reed et al., 1996; Neale et al., 1997; Davis and Hotha, 1998; Peters, 1999; Dermont et al., 2008). On the other hand, chelating agents are capable of extracting metals with much less impact on soils. Therefore, soil-related research on chelating agents has increased dramatically since the mid 1990s due to the proposed use of chelating agents for soil remediation (Nowack and VanBriesen, 2005).

Chelating agents have been widely used in many applications (total worldwide use was 200,000 tonnes in 2000), e.g., industrial cleaning, detergents, photos, pulp and paper, textiles, and agrochemicals (Nowack and VanBriesen, 2005). The group of aminocarboxylates are most commonly used, containing one or more tertiary or secondary amines and two or more carboxylic acid groups, as illustrated in Figure 1.4.

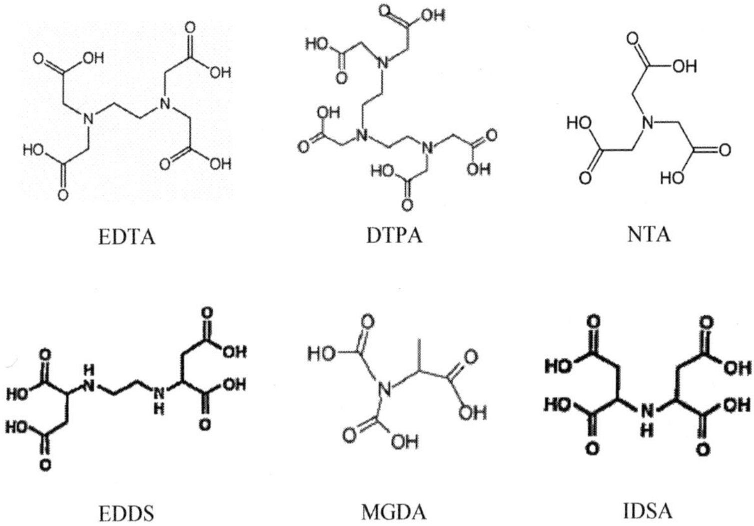

Figure 1.4 Structure of commonly used aminocarboxylate chelating agents.

The best known chelating agents are EDTA (ethylenediaminetetraacetic acid), NTA (nitrilotriacetic acid), and DTPA (diethylenetriaminepentaacetic acid) (Nowack and VanBriesen, 2005). Most studies reported in the literature investigated the effectiveness and impacts of EDTA for soil washing and other soil remediation technologies (e.g., heap leaching, phytoextraction, and soil flushing), due to its low cost and high efficacy of metal extraction (Nowack, 2002; Nowack and VanBriesen, 2005; Nowack et al., 2006; Lestan et al., 2008). EDTA offers the best cost/performance ratio of all chelating agents; however, its low biodegradability, and thus its high persistence in the natural environment, has been a recent concern. The transport of metal-EDTA complexes under pulsed and continuous flushing could be simulated using advection-dispersion transport equations modified with kinetic terms for metal extraction and mineral dissolution (Kedziorek et al., 1998; Friedly et al., 2002; Tsang et al., 2007a; Kent et al., 2008). Residual metal–EDTA complexes would travel in the subsurface with high mobility and possibly lead to adverse health and environmental effects.

The European authorities have conducted extensive evaluation of EDTA and the risk assessment report indicated that EDTA has a low toxicity profile for humans, and the environmental risks are limited to some localized cases involving high emissions at concentrations above 2.2 mg/L (about 10 μM) (Nowack et al., 2006). This is a value hardly ever reached in natural waters, but is very likely to be exceeded, by 100-10,000 times for a typical range of 1-100 mM, in chelating agent-enhanced remediation technologies (Nowack et al., 2006). Although the overall loss of soil weight due to EDTA-enhanced soil dissolution may be insignificant (< 13 g kg^{-1}), dissolution of amorphous Fe and Al oxides would reduce the shear strength, destabilize the soil aggregate stability, and mobilize colloids and fine particles (soil dispersion) (Tsang et al., 2007b). Moreover, the metal mobility, bioavailability, and fractionation could be modified after EDTA applications and require further investigations (Udovic ad Lestan, 2009; Zhang et al., 2010)

Therefore, risk reduction measures should be considered to ensure that no residual EDTA is allowed to remain in the treated soils or to leach into the environment after soil remediation. At the end of soil washing, the processed soils should be rinsed with clean water to remove residual EDTA, and the washing solution should be treated to remove or recover EDTA. Additions of ferric chloride, sodium phosphate, calcium hydroxide, and pH adjustment have been shown to effectively recover both metals and EDTA by precipitation and sedimentation (Lo and Zhang, 2005). On the other hand, metal-EDTA complexes can be separated from the washing solution by membrane separation and electrochemical treatment, or degraded by advanced oxidation processes (Finzgar and Lestan, 2006, 2008; Pociecha and Lestan, 2009), although these treatments are more costly than chemical precipitation.

Moreover, biodegradable chelating agents such as EDDS ([S,S]-ethylenediaminedisuccinic acid, which is a stereoisomer of EDTA), IDSA (iminodisuccinic acid), and MGDA (methylglycinediacetic acid) have received increasing attention in recent years (Nowack and VanBriesen, 2005). In particular, EDDS has been considered a promising substitute for EDTA in soil remediation technologies. EDDS can be fully degraded in wastewater treatment (downward flow biological aerated filter reactor) (Vandevivere et al., 2001a), and in soils after an initial lag phase that is necessary for the population growth or adaptation of microbes (Tandy et al., 2006; Wang et al., 2007; Meers et al., 2008). The length of the lag phase and the biodegradation rate rely on the metal type, the extent of metal contamination, and the soil type. A major limitation in the application of EDDS, as well as other biodegradable chelating agents, is that EDDS cannot be recovered and reused. There is an inevitable increase in the chemical cost and whether this can be offset with a possible decrease in the cost of required treatment process needs to be justified.

1.3.4 Operational Conditions

It should be recognized in the first place that chelant-enhanced metal extraction depends on: (1) soil characteristics (e.g., soil texture, types and content of mineral oxides, and organic matter content); (2) metal contamination characteristics (type, concentration, contamination age, operational distribution, and chemical speciation); and (3) operational conditions. It is particularly important that these unchangeable, site-specific factors should be prudently and extensively characterized at the outset of soil remediation, as outlined in previous section.

The key operational conditions of chelating agent-enhanced soil washing include the chelating agent-to-metal molar ratio, solution pH, solid-to-solution ratio, washing time, and temperature (Vandevivere et al., 2001b; Tandy et al., 2004; Hauser et al., 2005; Polettini et al., 2007; Zou et al., 2009; Yan et al., 2010). The molar ratio of the chelating agents in solution to the metals in the soils is of overriding importance, because one mole of an aminocarboxylate chelating agent forms complexes with one mole of metals. Under EDDS deficiency, initial extraction of Zn and Pb is followed by re-sorption as a result of metal exchange of newly formed $ZnEDDS^{2-}$ and $PbEDDS^{2-}$ with sorbed Cu on the soil surfaces (Tsang et al., 2009; Yip et al., 2009a; 2010; Lo et al., 2011a; 2011b), which is supported by speciation calculations using Visual MINTEQ (Gustafsson et al., 2008). Although a molar ratio of 1 is theoretically needed to extract all the target metals, a ratio of between 1 and 2 is usually required for optimum performance in consideration of the fact that a portion of chelating agent is consumed by non-selective complexation with mineral cations. While it has been shown that metal extraction is faster and more complete as the ratio increases to 10 or even 100 (Kim et al., 2003; Nowack et al., 2006), higher concentrations of chelating agents lead to higher chemical and wastewater treatment costs.

The pH dependence of metal extraction in the normal pH range between 5 and 8 is marginal for copper and zinc, but more pronounced for lead. This is a result of a number of factors that vary with solution pH: conditional strength of metal-chelating agent complexes (as indicated by stability constants), metal speciation in the solution (metal precipitates, metal-hydroxyl complexes, or free ions), speciation and sorption tendency of chelating agents, and mineral dissolution that competes for chelating agents (Nowack, 2002; Nowack et al., 2006; Tsang et al., 2009; Yip et al., 2009a; Zou et al., 2009; Yan et al., 2010). The pH effects on the metal extraction are generally less significant at higher chelating agent-to-metal molar ratios because more free (i.e., uncomplexed) chelating agents are available in the solution.

On the other hand, there are mixed effects on different metals due to the variation of solid-to-solution ratios (Vandevivere et al., 2001b; Tandy et al., 2004; Zou et al, 2009; Yan et al., 2010). At the same chelating agent-to-metal molar ratio, lead extraction is independent of the solid-to-solution ratio, whereas zinc extraction decreases and copper extraction increases with increasing solid-to-solution ratio, which may be due to a change in the concentration of counter ions in solution or organic matter dissolution. It is interesting to note that a higher solid-to-solution ratio results in a smaller volume of wastewater (washing solution) and a higher concentration of the chelating agent that requires treatment and recovery, which helps reduce the treatment and process operation costs. Yet, a lower solid-to-solution ratio, and thus a lower concentration of the chelating agent, is needed to achieve the same chelating agent-to-metal molar ratio. This has less impact on the soil structure, microorganisms, and plants, even if the residual chelating agent is present in the treated soils. Besides, the selection of the solid-to-solution ratio should also enable effective particle-size or solid/solution separation after chelating agent-enhanced extraction.

Metal extraction generally follows two-phase kinetics, that is, a rapid desorption within the first 1-4 hours followed by a subsequent gradual release (Whitworth et al., 1999; Polettini et al., 2006; Fangueiro et al., 2005; Wasay et al., 2007; Yip et al., 2009b). The required contact time in the extraction unit varies with the soil type, metal type, metal distribution, metal speciation, and metal loading. However, most bench-scale studies employ a 24-h reaction time, after which extraction can reach apparent equilibrium. This is undoubtedly important for result analysis and discussion, but it should be noted that the washing (residence) time is usually much shorter in large-scale applications at field sites (e.g., 10-60 min), for the purpose of maintaining a high throughput rate and low operation cost (U.S. DoD, 1997). If the soil texture and contamination level are highly variable at the site, some degree of overdesign is advisable to maintain the desired processing rate for the plant. Furthermore, if a long contact time is needed, successive (or multi-step) washing is probably more effective than a single extraction with a concentrated chelating agent (Finzgar and Lestan, 2007, 2008; Zou et al., 2009). Nevertheless, the capital, operational and maintenance costs for chemical extraction and solid/solution separation units inevitably increase. The process flow diagrams should be optimized in the treatability studies as described in the previous section.

In addition, there are possible operational amendments that can enhance the metal extraction efficiency, such as the addition of redox agents, increase in mixing intensity, ultrasonic treatment, and soil heating (Peters, 1999; Dermont et al., 2008; Zou et al., 2009). However, these alternatives are less likely to prove uneconomical in field-scale applications, in which large volumes of soils are to be treated.

1.4 Economic and Societal Considerations

In addition to the technical effectiveness, remedy selection is also influenced by the regulatory policies for considering land use, economic factors and societal impact (U.S. EPA, 1999, 2004). The European Union Contaminated Land Rehabilitation Network for Environmental Technologies (CLARINET) suggests the use of risk-based land management, that is, sustainable remediation should ensure that the site is fit for the designated future use, that the environment is protected, and that long-term care is an important factor (Vegter et al., 2002). In the United States, the use of innovative technologies has particularly been encouraged, as this may provide advantages in ensuring that remediation is implemented cost-effectively and expeditiously to enable redevelopment of contaminated sites (U.S. EPA, 2001). In general, the potential of long-term liability of the associated risk to human health and the value of flexible future land use are the key drivers for soil remediation projects.

Table 1.2 Elements of Costs for Soil Washing

Fixed Costs	**Variable Costs**
Permitting, Safety, and Regulatory	Site Excavation
Site Characterization	Equipment Lease and Depreciation
Characterization Studies	Labour (1/2/3 shifts)
Bench-Scale Treatability Tests	Personal Protective Equipment
Vendor Selection/Contracting	Fuel/Electricity
Process Design and Optimization	Water
Site Infrastructure Requirements and Preparation	Chemical agents (for chemical-enhanced soil washing)
Transport of Equipment to the Site	Sampling and Chemical Analysis
Plant Erection	Process Water Treatment
Decontamination and Decommissioning of Equipment	Disposal Cost of Contaminated Fines Fraction (optional in chemical-enhanced soil washing)
Transport of Equipment from the Site	Disposal Cost of Treatment Process Wastes (e.g., sludge cake)

Table 1.2 summarizes the cost elements of field-scale applications of soil washing technology (U.S. DoD, 1997; U.S. ITRC, 1997; CL:AIRE, 2007). The treatment costs vary widely with the project size, system design, and complexity. The key cost driver is the amounts of soils to be treated, i.e., the economy of scale, because the fixed costs contribute a large part of the overall treatment cost. For example, fixed costs amount to 59% of total cost for treating 835 tonnes of soils by chemical-enhanced soil washing, whereas it is only 41% of the total cost for treating 10,000 tonnes of soils (U.S. DoD, 1997). Variable costs depend on the volume of soils to be processed, soil type, contamination characteristics, throughput rate, cleanup goals, process design, chemical agents, project duration, etc. Some variable costs may be site-specific.

The costs are generally lower if treatment goals can be achieved merely by physical separation, and costs in Europe are also lower because of the more mature market, for example, the costs in 2001 were 20-45 euro per tonne in the Netherlands (Honders et al., 2003). According to the Federal Remediation Technologies Roundtable (2007), the estimated costs for soil washing based on physical separation are between USD 70 (large site) and 187 (small site) per cubic metre (USD 47 and 125 per tonne, respectively, if a soil bulk density of 1.5 tonne per cubic metre is assumed), while the costs for chemical-enhanced soil washing are between USD 358 (large site) and 1,717 (small site) per cubic metre (USD 239 and 1,145 per tonne, respectively, calculated using a soil bulk density of 1.5 tonne per cubic metre). Soil washing was deselected at several Superfund sites because of small soil volume and high costs (Dermont et al., 2008).

A demonstration project for site remediation of a small-arms range compared chemical-enhanced soil washing with off-site landfilling and in-situ solidification/stabilization, which are alternative technologies often considered when addressing metal contamination (U.S. DoD, 1997). It was noted that landfilling, if disposal is permitted and landfill is available in the vicinity, is the cheapest option at sites involving less than about 2,600 tonnes of soils. It was also noted that solidification/stabilization is often cheaper than chemical-enhanced soil washing, but the potential for liability remains. Chemical-enhanced soil washing eliminates long-term liability and allows greater flexibility for future land use. In general, compared with landfill disposal, soil washing offers advantages having much less social impact resulting from noise and dust due to the transportation requirements of landfilling (Harbottle et al., 2008).

Soil washing technology is well established and extensively employed in the Netherlands, Germany, Belgium, Sweden, Norway, and Switzerland, and more recently it has been applied in the United Kingdom, the United States, Canada, and Japan. Soil washing technology is employed independently or as an integral part of remediation plans for a broad range of soil remediation projects. Many fixed site facilities (stationary plants) for soil washing are present in Europe and Canada, and mobile soil washing units are also commercially available from many vendors (e.g., ART Engineering, 2010; A&G Milieutechniek, 2010; DEC Environmental, 2010; Grontmij UK, 2010; Heijmens Blackwell Remediation, 2010; Royal Boskalis Westminster, 2010). The soil washing applications in Europe have been more extensive than in the United States. For instance, in the Netherlands an average of 855,000 tonnes per year were treated by soil washing between 1991 and 2001 (Honers et al., 2003). In the United Kingdom, soil washing has been recently applied for soil remediation of 2,770 cubic metres at a former gasworks site in Elgin (Grontmij UK, 2010), 150,000 tonnes at Basford Gasworks in Nottingham (CL:AIRE, 2007), and over 500,000 tonnes in East London (DEC Environmental, 2010).

On the other hand, in the United States between 1982 and 2005, soil washing and physical separation were used for source treatment in 6 and 21, respectively, while chemical treatment was applied in 9 projects (U.S. EPA, 2007). Remediation projects at the King of Prussia Superfund site (19,200 tonnes of soils contaminated by chromium, nickel, and copper) (U.S. EPA, 1995; Mann, 1996) and the Vineland Chemical Company Superfund Site (410,000 tonnes of soils contaminated by arsenic) in New Jersey (U.S. ACE, 2005; ART Engineering, 2010) are two large-scale successful demonstration examples. A recent summary of 37 soil washing projects for metal-contaminated soil remediation (about one third undertaken after 2000), including full-scale operations and pilot/field demonstrations, indicated that 16 projects involved particle-size separation only, 18 projects involved both particle-size separation and chemical extraction, and 3 projects involved chemical extraction only (Dermont et al., 2008).

1.5 Conclusions

Soil washing offers a sustainable and cost-effective alternative to total landfill disposal of contaminated soils. Soil washing based on particle-size separation has proved successful for volume reduction in many full-scale applications. In recent years chemical-enhanced soil washing, especially for chelating agents, has been intensively investigated for metal extraction applications, and there is an increasing number of pilot- and full-scale demonstrations. In general, soil washing is more economically competitive in larger-scale projects. With prudent characterization and treatability studies, soil washing technology can provide permanent remedies, eliminate long-term liability, allow flexible future land use, and impose relatively less impact on the landscape and stakeholders.

1.6 References

Anderson, R., Rasor, E. and van Ryn, F. (1999). "Particle size separation via soil washing to obtain volume reduction." *Journal of Hazardous Materials*, 66, 89-98.

A&G Milieutechniek (2010). Decontamination – Soil Washing. http://www.aengbedrijven.nl/?page=text&id=14 (Accessed on May 30, 2010).

ART Engineering, LCC (2010). Soil Washing, Soil Decontamination, Soil Cleaning, Sediment Treatment, Soil and Groundwater Remediation Project Experience. http://www.art-engineering.com/index.htm (Accessed on May 30, 2010).

Contaminated Land: Applications in Real Environments (CL:AIRE) (2007). "Understanding Soil Washing." Technical Bulletin, TB 13, Contaminated Land: Applications in Real Environments, UK.

Council of the European Union (1999). Council Directive 1999/31/EC of 26th April 1999 on the Landfill of Waste.

Davis, A. P. and Hotha, B. (1998). "Washing of various lead compounds from a contaminated soil column." *Journal of Environmental Engineering*, 124, 1066-1075.

DEC Environmental Contractors (2010). Mobile Soil Washing Systems. http://www.decnv.com/EN/techniques/soil_and_groundwater (Accessed on May 30, 2010).

Dermont, G., Bergeron, M., Mercier, G. and Richer-Lafleche, M. (2008). "Soil washing for metal removal: A review of physical/chemical technologies and field applications." *Journal of Hazardous Materials*, 152, 1-31.

European Topic Centre Soil (ETCS) (1998). "Topic Report – Contaminated Sites." European Environment Agency, Copenhagen, Denmark.

Fangueiro, D., Bermond, A., Santos, E., Carapuca, H. and Duarte, A. (2005). "Kinetic approach to heavy metal mobilization assessment in sediments: Choose of kinetics equations and models to achieve maximum information." *Talanta*, 66, 844-857.

Federal Remediation Technologies Roundtable (FRTR) (2007). Remediation Technologies Screening Matrix and Reference Guides, Version 4.0. http://www.frtr.gov/matrix2/top-page.html (Accessed on May 30, 2010).

Finzgar, N. and Lestan, D. (2006). "Heap leaching of Pb and Zn contaminated soil using ozone/UV treatment of EDTA extractants." *Chemosphere*, 63, 1736–1743.

Finzgar, N. and Lestan, D. (2007). "Multi-step leaching of Pb and Zn contaminated soils with EDTA." *Chemosphere*, 66, 824–832.

Finzgar, N. and Lestan, D. (2008). "The two-phase leaching of Pb, Zn and Cd contaminated soil using EDTA and electrochemical treatment of the washing solution." *Chemosphere*, 73, 1484–1491.

Friedly, J. C., Kent, D. B., and Davis, J. A. (2002). "Simulation of the mobility of metal–EDTA complexes in groundwater: The influence of contaminant metals." *Environmental Science and Technology*, 36, 355–363.

Griffiths, R. A. (1995). "Soil-washing technology and practice." *Journal of Hazardous Materials*, 40, 175-189.

Grontmij UK (2010). Gasworks Remediation – Elgin, Moray. http://www.grontmij.co.uk/site/en-gb/Projects/Transportation/Articles+and+documents/Gasworks+Remediation+at+Elgin.htm (Accessed on May 30, 2010).

Gustafsson, J. P. (2008). Visual MINTEQ Version 2.53; Department of Land and Water Resources Engineering, Royal Institute of Technology: Stockholm, Sweden.

Harbottle, M. J., Al-Tabbaa, A. and Evans, C. W. (2007). "A comparison of the technical sustainability of in situ stabilization/solidification with disposal to landfill." *Journal of Hazardous Materials*, 141, 430-440.

Harbottle, M. J., Al-Tabbaa, A. and Evans, C. W. (2008). "Sustainability of land remediation. Part 1: Overall analysis." *Proceedings of the Institution of Civil Engineers–Geotechnical Engineering*, 161(2), 75-92.

Heijmens Blackwell Remediation Ltd (HBR) (2010). Soil Washing. http://www.hbr.eu.com/hbr/services/soils_washing (Accessed on May 30, 2010).

Honders, A., Maas, Th. And Gadella, J. M. (2003). Ex-situ treatment of contaminated soil—the Dutch Experience. Centre for Soil Treatment, Dutch Ministry of Spatial Planning and the Environment, The Netherlands.

Kedziorek, M. A. M., Dupuy, A., Bourg, A. C. M., and Compère, F. (1998). "Leaching of Cd and Pb from a polluted soil during the percolation of EDTA: Laboratory column experiments modeled with a non-equilibrium solubilization step." *Environmental Science and Technology*, 32, 1609-1614.

Kent, D. B., Davis, J. A., Joye, J. L. and Curtis, G. .P. (2008). "Influence of variable chemical conditions on EDTA-enhanced transport of metal ions in mildly acidic groundwater." *Environmental Pollution*, 153, 44-52.

Kim, C., Lee, Y. and Ong, S. K. (2003). "Factors affecting EDTA extraction of lead from lead-contaminated soils." *Chemosphere*, 51, 845-853.

Kirpichtchikova, T. A., Manceau, A., Spadini, L., Panfili, F., Marcus, M. A. and Jacquet, T. (2006). "Speciation and solubility of heavy metals in contaminated soil using X-ray microfluorescence, EXAFS spectroscopy, chemical extraction, and thermodynamic modeling." *Geochimica Cosmochimica Acta.*, 70, 2163-2190.

Lestan, D., Luo, C. L. and Li, X. D. (2008). "The use of chelating agents in the remediation of metal-contaminated soils: A review." *Environmental Pollution*, 153, 3-13.

Lo, I. M. C. and Zhang, W. (2005). "Study on the optimal conditions for EDTA recovery from soil washing effluents." *Journal of Environmental Engineering*, 131, 1507-1513.

Lo, I.M.C., Tsang, D.C.W., Yip, T.C.M., Wang, F. and Zhang, W. (2011a). "Influences of injection modes on EDDS-flushing of metal-contaminated soil." *Journal of Hazardous Materials*, 192, 667-675.

Lo, I.M.C., Tsang, D.C.W., Yip, T.C.M., Wang, F. and Zhang, W. (2011b). "Significance of metal exchange in EDDS-flushing column experiments." *Chemosphere*, 83, 7-13

Mann, J. M. and Groenendijk, E. (1996). "The first full-scale soil washing project in the USA." *Environmental Progress*, 15(2), 108-111.

Mann, J. M. (1999). "Full-scale and pilot-scale soil washing." *Journal of Hazardous Materials*, 66, 119-136.

Meers, E., Tack, F. M. G. and Verloo, M. G. (2008). "Degradability of ethylenediaminedisuccinic acid (EDDS) in metal contaminated soils: Implications for its use soil remediation." *Chemosphere*, 70, 358-363.

Neale, C. N., Bricka, R. M., and Chao, A. C. (1997). "Evaluating acids and chelating agents for removing heavy metals from contaminated soils." *Environmental Progress*, 16(4), 274-280.

Nowack, B. (2002). "Environmental chemistry of aminopolycarboxylate chelating agents." *Environmental Science and Technology*, 36, 4009-4016.

Nowack, B. and VanBriesen, J. M. (eds.) (2005). Biogeochemistry of Chelating Agents. ACS Symposium Series Vol. 910; American Chemical Society: Washington, DC.

Nowack, B., Schulin, R. and Robinson, B. H. (2006). "A critical assessment of chelant-enhanced metal phytoextraction." *Environmental Science and Technology*, 40, 5225-5232.

Peters, R. W. (1999). "Chelant extraction of heavy metals from contaminated soils." *Journal of Hazardous Materials*, 66, 151-210.

Pociecha, M., Lestan, D. (2009). "EDTA leaching of Cu contaminated soil using electrochemical treatment of the washing solution." *Journal of Hazardous Materials*, 165, 533–539.

Polettini, A., Pomi, R., Rolle, E., Ceremigna, D. (2006). "A kinetic study of chelant-assisted remediation of contaminated dredged sediment." *Journal of Hazardous Materials*, B137, 1458-1465.

Polettini, A., Pomi, R. and Rolle, E. (2007). "The effect of operating variables on chelant-assisted remediation of contaminated dredged sediment." *Chemosphere*, 66, 866-877.

Reed, B. E., Carriere, P. C. and Moore, R. (1996). "Flushing of a Pb (II) contaminated soil using HCl, EDTA, and $CaCl_2$." *Journal of Environmental Engineering*, 122, 48-50.

Richardson III, W. S., Phillips, C. R., Luttrell, J., Hicks, R. and Cox, C. (1999). "Application of remedy studies to the development of a soil washing pilot plant that uses mineral processing technology: A practical experience." *Journal of Hazardous Materials*, 66, 47-65.

Royal Boskalis Westminster N.V. (2010). Mobile Soil Washing Plant. http://www.boskalis.com/vervolg_vloot.php?pageID=455 (Accessed on May 30, 2010).

Sharma, H. D. and Reddy, K. R. (2004). Geoenvironmental Engineering: Site Remediation, Waste Containment and Emerging Waste Management Technologies. John Wiley & Sons, New York.

Tandy, S., Bossart, K., Mueller, R., Ritschel, J., Hauser, L., Schulin, R. and Nowack, B. (2004). "Extraction of heavy metals from soils using biodegradable chelating agents." *Environmental Science and Technology*, 38, 937-944.

Tandy, S., Ammann, A., Schulin, R. and Nowack, B. (2006). "Biodegradation and speciation of residual SS-ethylenediaminedisuccinic acid (EDDS) in soil solution left after soil washing." *Environmental Pollution*, 40, 2753-2758.

Toebermann, J. C., Werther, J., Rosenkranz, J. and Gruhn, G. (1999). "Flowsheet simulation of soil-washing processes." *Chemical Engineering & Technology*, 22(7), 558-567.

Tsang, D. C. W., Lo, I. M. C. and Chan, K. L. (2007a). "Modeling the transport of metals with rate-limited EDTA-promoted extraction and dissolution during EDTA-flushing of copper-contaminated soils." *Environmental Science and Technology*, 41, 3660-3667.

Tsang, D. C. W., Zhang, W. and Lo, I. M. C. (2007b). "Copper extraction effectiveness and soil dissolution issues of EDTA-flushing of artificially contaminated soils." *Chemosphere*, 68, 234-243.

Tsang, D. C. W., Yip, T. C. M. and Lo, I. M. C. (2009). "Kinetic interactions of EDDS with Soils: 2. Metal-EDDS complexes in uncontaminated and metal-contaminated soils." *Environmental Science and Technology*, 43, 837-842.

Udovic, M. and Lestan, D. (2009). "Pb, Zn and Cd mobility, availability and fractionation in aged soil remediated by EDTA leaching." *Chemosphere*, 74, 1367–1373.

U.S. Army Corps of Engineers (2005). "Soil washing proves innovation in remediation." The Corps Environment. Vol 6. no.2, U.S. Engineering and Support Center, Huntsville, AL.

U.S. Department of Defense (1997). ESTCP Cost and Performance Report: Joint Small-Arms Range Remediation. Environmental Security Technology Certification Program (ESTCP), U.S. Department of Defense, Arlington, Virginia.

U.S. Environmental Protection Agency (1995). Soil Washing at the King of Prussia Technical Corporation Superfund Site, Winslow Township, New Jersey. Office of Solid Waste and Emergency Response, Technology Innovation Office, U.S. Environmental Protection Agency.

U.S. Environmental Protection Agency (1999). A Sustainable Brownfields Model Framework. Office of Solid Waste and Emergency Response, U.S. Environmental Protection Agency Report No. EPA500-R-99-001.

U.S. Environmental Protection Agency (2001). Brownfields Technology Primer: Requesting and Evaluating Proposals That Encourage Innovative Technologies for Investigation and Cleanup. Office of Solid Waste and Emergency Response, U.S. Environmental Protection Agency Report No. EPA542-R-01-005.

U.S. Environmental Protection Agency (2004). Cleaning up the Nation's Waste Sites: Markets and Technology Trends: 2004 Edition. Office of Emergency and Remedial Response, U.S. Environmental Protection Agency Report No. EPA 542-R-04-015.

U.S. Environmental Protection Agency (2007). Treatment Technologies for Site Cleanup: Annual Status Report (Twelfth Edition). Office of Solid Waste and Emergency Response, U.S. Environmental Protection Agency Report No. EPA-542-R-07-012.

U.S. Interstate Technology and Regulatory Council (ITRC) (1997). Technical and Regulatory Guidelines for Soil Washing. Interstate Technology Regulatory Cooperation Work Group, Metals in Soils Work Team, Washington, DC.

Vandevivere, P. C., Saveyn, H., Verstraete, W., Feijtel, T. C. J. and Schowanek, D. R. (2001a). "Biodegradation of metal-[S.S]-EDDS complexes." *Environmental Science and Technology*, 35, 1765–1770.

Vandevivere, P. C., Hammes, F., Verstraete, W., Feijtel, T. C. J. and Schowanek, D. R. (2001b). "Metal decontamination of soil, sediment, and sewage sludge by means of transition metal chelant [S,S]-EDDS." *Journal of Environmental Engineering*, 127, 802-811.

Vegter, J., Lowe, J. and Kasamas, H. (eds.) (2002). Sustainable Management of Contaminated Land: An Overview. CLARINET Report, Umweltbundesamt, Vienna, Austria.

Vulava, V. M. and Seaman, J. C. (2000). "Mobilization of lead from highly weathered porous material by extracting agents." *Environmental Science and Technology*, 34, 4828-4834.

Wang, G., Koopmans, G. F., Song, J., Temminghoff, E. J. M., Luo, Y., Zhao, Q. and Japenga, J. (2007). "Mobilization of heavy metals from contaminated paddy soil by EDDS, EDTA, and elemental sulfur." *Environmental Geochemistry and Health*, 29, 221-235.

Wasay, S. A., Barrington, S., Tokunagal, S. and Prasher, S. (2007). "Kinetics of heavy metal desorption from three soils using citric acid, tartaric acid, and EDTA." *Journal of Environmental Engineering Science*, 6, 611-622.

Whitworth, D. J., Achterberg, E. P., Herzl, V., Nimmo, M., Gledhill, M. and Worsfold, P.J. (1999). "Development of a simple extraction procedure using ligand competition for biogeochemically available metals of estuarine suspended particulate matter." *Analytica Chimica Acta*, 392, 3-17.

Yan, D. Y. S., Yui, M. M. T., Yip, T. C. M., Tsang, D. C. W. and Lo, I. M. C. (2010). "Influence of EDDS-to-metal molar ratio, solution pH, and soil-to-solution ratio on metal extraction under EDDS deficiency." *Journal of Hazardous Materials*, 178, 890-894.

Yip, T. C. M., Tsang, D. C. W., Ng, K. T. W. and Lo, I. M. C. (2009a). "Kinetic interactions of EDDS with Soils: 1. Metal resorption and competition under EDDS deficiency." *Environmental Science and Technology*, 43, 831-836.

Yip, T. C. M., Tsang, D. C. W., Ng, K. T. W. and Lo, I. M. C. (2009b). "Empirical modeling of heavy metal extraction by EDDS from single-metal and multi-metal contaminated soils." *Chemosphere*, 74, 301-307.

Yip, T.C.M., Yan, D.Y.S., Yui, M.M.T., Tsang, D.C.W. and Lo, I.M.C. (2010). "Heavy metal extraction from an artificially contaminated sandy soil under EDDS deficiency: Significance of humic acid and chelant mixture." *Chemosphere*, 80, 416-421.

Zhang, W., Huang, H., Tan, F., Wang, H., and Qiu, R. (2010). "Influence of EDTA washing on the species and mobility of heavy metals residual in soils." *Journal of Hazardous Materials*, 173, 369-376.

Zou, Z., Qiu, R., Zhang, W., Dong, H., Zhao, Z., Zhang, T., Wei, X. and Cai, X. (2009). "The study of operating variables in soil washing with EDTA." *Environmental Pollution*, 157, 229-236.

CHAPTER 2

Remediation of Metal-Contaminated Sediments by Means of Chelant-Assisted Washing

Alessandra Polettini and Raffaella Pomi

2.1 Background on Contaminated Sediments

Sediments include all the solid materials that have settled down to the bottom of water bodies from a state of suspension in a liquid. Such solid materials originate from eroding soil and decomposing plants and animals, as well as from particles released by wastewater treatment plants (Bortone, 2007). Environmental agents including wind, water and ice may transport such particles to large distances from the point at which they are generated (Bortone, 2007). It is estimated that the annual sediment generation rate at the European level is about 1800×10^6 t, out of which $\sim 540 \times 10^6$ t are stored in river channels and on floodplains, $\sim 346 \times 10^6$ t accumulate in reservoirs, $\sim 200 \times 10^6$ t are mined from active fluvial areas, while the remaining $\sim 714 \times 10^6$ t are either deposited in lowland zones (estuaries, harbours, deltas) or discharged into seas and oceans (Bortone, 2007).

The relevance of sediments is recognized not only for ecological reasons, but also from a social and economic standpoint. From the ecological point of view, sediments represent an essential component of the aquatic ecosystem, providing the habitat and food resources for many organisms and contributing to the nutrient cycle to and from the overlying water.

Surface waters receive discharges from both point and diffuse sources of various liquid and solid wastes, which may potentially transfer contaminants to sediments. The pollutants found in sediments have both an anthropogenic and a natural origin. Potential sources of contaminants include: discharges from industrial installations, wastewater treatment plants, storm water discharges or combined sewer overflows; surface runoff or erosion of soils located in contaminant-bearing areas; wet and dry deposition of contaminants emitted into the atmosphere by different sources; upwelling or seepage of contaminated groundwater or non-aqueous phase liquids (NAPL) into a water body; direct disposal or accidental spill from ships, or release of contaminants from in-water and over-water structures or ship maintenance facilities (U.S. EPA, 2005).

The contaminants from the above mentioned sources may sorb onto sediment particles, to a degree which depends on their mineralogical composition (clay constituents, organic matter, hydrous iron and manganese oxides, etc.) and physico-chemical characteristics (particle size distribution, pH, redox conditions, salinity, etc.) (Andreottola et al., 2010), which all affect the relative proportion of contaminants between the solid particles and pore water (U.S. EPA, 2005). According to the U.S. EPA (2004), contaminated sediments are defined as aquatic sediments containing chemical substances in excess of appropriate geochemical, toxicological or sediment quality criteria or measures, or otherwise believed to pose a threat to human health or the environment. Many contaminants persist for years or decades due to the very slow or even null degradability in the aquatic environment. As a consequence, polluted sediments become a sink for contaminants, which, upon re-suspension or re-dissolution, have the potential of migrating into the water column, representing a possible threat to water quality (Mulligan et al., 2001; Barceló and Petrovic, 2007; Andreottola et al., 2010). Contaminated sediments may also be toxic to fish and other aquatic organisms through direct ingestion and absorption by dermal contact, or may enter the food chain through ingestion of a contaminated prey or direct exposure to water or sediment, thereby causing possible bioaccumulation effects (U.S. EPA, 2004). According to the U.S. EPA (1998), documented adverse ecological effects from contaminated sediments include fin rot, increased tumour frequency and reproductive toxicity in fish, as well as decreased biodiversity of the aquatic ecosystem, which may in turn result in altering the energy and nutrient flows, the productivity and the decomposition processes in the aquatic ecosystem (U.S. EPA, 2004).

As mentioned earlier, in addition to posing threats to human health and the environment, the presence of contaminated sediments may also exert potential economic impacts, limiting or hindering commercial and trade activities (shipping, fishing), tourism and recreational activities, as well as maintenance interventions (dredging).

It has been estimated that ~5% of watersheds in industrialized countries contain contaminated sediments and that 10% of marine and estuarine sediments are potentially hazardous to the aquatic environment (De Gioannis et al., 2008).

Dredging of sediments may be applied for different purposes. Maintenance (or navigational) dredging is required to develop or maintain existing waterways (ditches), to protect rivers e.g. against flooding and to ensure adequate shipping routes in waterways and harbours (Detzner et al., 2007). In such cases, dredging is typically done routinely, and the changes in the characteristics of sediments are generally limited to within a known range. For such reasons, it is common practice to relocate the largest part of the dredged material in the same water system or on the nearby embankment when the amounts of dredged material are small (Detzner et al., 2007). If these options are undesirable or impracticable for environmental, morphological or spatial reasons, alternative options may include reuse, treatment and/or confined disposal (Detzner et al., 2007). On the other hand, environmental dredging is required when the presence of contaminated sediments is identified and the need is recognized to remove sediments from the site to reduce health and environmental hazards, for attaining the quality standards required for the aquatic environment. It is estimated that ~500 millions m^3 of sediments are dredged each year for navigational purposes, of which ~1–4% requires dewatering and treatment prior to disposal (De Gioannis et al., 2008).

Sediments are a heterogeneous mixture of components, including organic matter, iron and manganese oxyhydroxides, carbonates, alumino-silicates (e.g., clay minerals) and sulphides, along with interstitial water (U.S. EPA, 1993b; Mulligan et al., 2001; McCready et al., 2003). Organic and inorganic contaminants are bound to the various geochemical phases through precipitation, ion exchange, sorption, complexation and inclusion in the crystalline structure. Factors affecting the retention of contaminants in sediments include physico-chemical parameters such as pH, particle size distribution, mineralogical composition, natural organic matter, redox potential, cation exchange capacity, specific surface area, nature and age of contamination, as well as presence of interfering species (Peters, 1999). As a result, the nature and strength of the interactions occurring between the sediment particles and the contaminants will determine the particular chemical speciation in sediments, which will in turn dictate their mobility and potential bioavailability.

For the reasons outlined above, it is commonly acknowledged that the information on the total contaminant content in sediment is inadequate to predict the potential environmental impacts posed by the material (U.S. EPA, 2005). Rather, this issue can be more accurately assessed from an understanding of the geochemistry of contaminants in sediments, indicating the existing chemical associations with the main mineral phases (e.g., oxides, hydroxides, carbonates, silicates) and other constituents (e.g., organic matter) of the solid matrix. Therefore, numerous sequential extraction procedures have been developed over the years, mainly as changes made to the original analytical method introduced by Tessier et al. (1979), on account of the main characteristics of the solid matrix of concern and the specific targets of the analysis (see e.g., the review by Hass and Fine (2010), and references therein). Sequential extraction procedures are assumed to be sufficiently accurate in selectively extracting the following components: soluble and/or exchangeable, bound to carbonates, occluded in Mn oxyhydroxides and amorphous Fe oxyhydroxides, sorbed or bound to organic matter, associated to well-crystallized Fe oxyhydroxides, and included in the aluminosilicates structure. Such geochemical fractions are ranked in an order of decreasing mobility, with contaminants associated to the soluble and/or exchangeable fractions being present in the most mobile (or labile) form, while those incorporated in the mineral lattice of the material being in the least mobile form (detrital fraction). The latter can only be extracted from the material provided that solid matrix dissolution is attained.

Most of the contaminants found in sediments have a low solubility in water, since it is expected that soluble compounds eventually dissolve in the water phase and therefore do not accumulate in sediments (Rulkens, 2005). The contamination found in sediments is typically of both organic and inorganic nature and comprises a wide spectrum of compounds. Important contaminants commonly include polycyclic aromatic hydrocarbons (PAHs), mineral oil compounds, metal (Cd, Cu, Cr, Pb, Hg, Ni, Zn, As) species, tributyl tin (TBT) compounds, polychlorinated biphenyls (PCBs), organochlorine pesticides, organophosphorous pesticides, dioxins and endocrine disruptors (Rulkens, 2005).

2.2 Management and Treatment of Contaminated Sediments

For the reasons outlined above, contaminated sediments deserve serious attention worldwide due to both the distinct contamination characteristics and the significant amounts to deal with. After dredging of contaminated sediments, major costs are involved in either proper disposal or remediation of the material.

Different approaches can be used for management of contaminated sediments, depending on site-specific conditions. When dredging is not required to ensure navigation in rivers and harbours, sediment management options include confined aquatic disposal and possibly in-situ treatment. Confined aquatic disposal involves underwater capping with clean sediments or sand, geotextiles or liners, either with or without a lateral confinement. In-situ treatment involves the use of reactive caps or the addition of additives to promote chemical or biological degradation of the contaminants (U.S. EPA, 2005). Ex-situ management options include either disposal in confined disposal facilities and upland landfills, or the application of remediation treatments for beneficial reuse or safe disposal.

Two different approaches can be adopted for the remediation of contaminated sediment, namely immobilization of the contaminants within the solid matrix and extraction/separation of the pollutants or contaminated sediment fractions (Peters, 1999). Although immobilization techniques are currently more reliable and generally involve lower remediation costs, extraction/separation methods have the advantage that contaminant concentrations may be reduced to below target regulatory limits and the treated material may be suitable for beneficial reuse (Meegoda and Perera, 2001), with positive environmental and economic outcomes.

Treatment processes applied to contaminated sediments have mostly been derived from soil remediation techniques and mineral processing methods. However, their applicability to sediments is strongly affected by the specific characteristics of the material, such as the high water and salt content and the presence of finely grained particles, which can adversely affect both the operation and the remediation efficiency (Andreottola et al., 2010). Despite decades of research, surprisingly the base of experience for remediation of contaminated sediment is still limited (U.S. EPA, 2005) and only little information is available on successful treatment of contaminated sediments (Colacicco et al., 2010).

Due to the variety of contaminants in dredged sediments, remediation is not usually a single process, but often requires a combination of techniques as the components of a treatment train to address multiple contaminant problems (Rulkens, 2005; U.S. EPA, 2005; Andreottola et al., 2010). Potential components of a sediment treatment train include pre-treatment, operational treatment, and/or effluent treatment/residual handling (U.S. EPA, 2005). The layout of a possible treatment train for dredged sediments is illustrated in Figure 2.1.

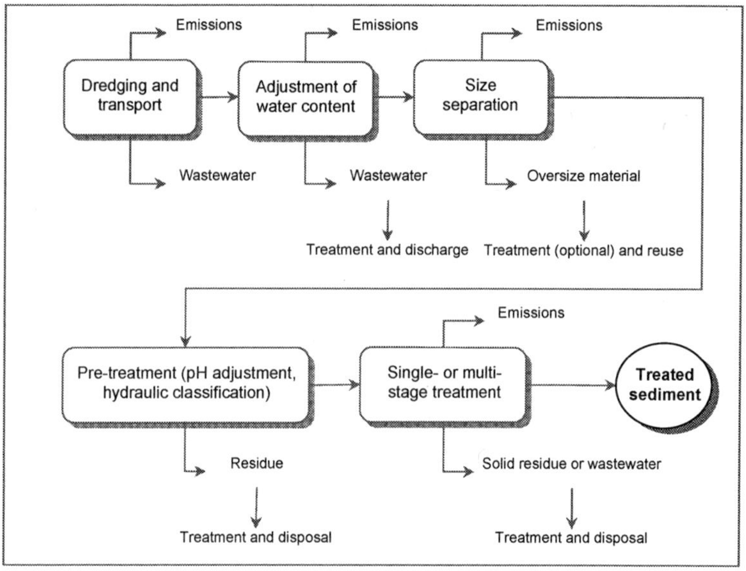

Figure 2.1 Schematic layout of a sediment treatment train.

Processing upstream of the core remediation treatment is generally performed to condition the material to meet the chemical and physical requirements for treatment or disposal, and/or to reduce the volume/weight of sediment that requires transport, treatment or restricted disposal (U.S. EPA, 2005). Pre-treatments typically include removal of debris, dewatering and physical separation. Dewatering is normally required to improve the handling characteristics of sediment and reduce the volume to be further treated. The extent of dewatering required is a function of the dredging method used (water contents are typically >50% and ~20% for mechanically and hydraulically dredged sediments, respectively) and the remediation technology to be applied (water contents ≤40% are commonly required for many processes) (Mulligan et al., 2001). The water generated during dewatering generally contains low levels of contaminants and requires treatment (U.S. EPA, 1993b). Physical separation aims at dividing sediment into separate fractions of different quality standards, on account of the fact that often the coarse materials (gravel and sand fractions) contain lower contaminant levels and may be suitable for unrestricted disposal and/or beneficial use (Mulligan et al., 2001; U.S. EPA, 2005; Detzner et al., 2007). Unit operations used for physical separation commonly include hydraulic classification systems (centrifugation, flocculation, floatation, hydrocycloning, sedimentation), although wet screening may also be used (Mulligan et al., 2001).

Different treatment processes can be applied to reduce the toxicity, mobility, bioavailability or concentration of the contaminants in sediments. The type of treatment to be adopted depends on the nature of contaminants, their concentrations, as well as on the physical, chemical and mineralogical characteristics of the sediment. Available disposal options and capacities may also affect the decision as to whether and how to treat sediments. Treatment processes are based upon different principles, including contaminant destruction or detoxification, extraction/separation of contaminants from sediment, reduction of sediment volume, or immobilization (U.S. EPA, 1993b, 2005; Meegoda and Perera, 2001). Treatment technologies can be classified as of physical, chemical, biological and thermal type (U.S. EPA, 1993b, 2005; Mulligan et al., 2001; Rulkens, 2005; Detzner et al., 2007). Processes involving combinations of treatments are also common.

In the following, the most typical treatment processes proposed for sediment remediation will be briefly reviewed, while a more detailed description of chemical washing using chelating agents will be provided in a dedicated section, since this forms the specific focus of this chapter.

Among the physico/chemical processes, solidification/stabilization is a well-known technology developed for hazardous wastes and contaminated soils as a means to reduce the mobility of contaminants. The process is based on the addition of binders, including different types of cements, lime, pozzolanic materials, thermoplastic resins and/or a variety of other chemicals (U.S. EPA, 1993b, 2005; Mulligan et al., 2001; Rulkens, 2005). Immobilization may occur due to physical encapsulation of contaminants within the solidified structure (resulting from a reduction in specific surface area and porosity), reduction in contaminant solubility (caused by changes in pH and alkalinity of the system) and/or chemical bonding/incorporation by the hydration phases formed upon treatment. The result of the treatment is the reduction in contaminant leachability from treated sediment to below the standards required for final disposal. Concerns about solidification/stabilization processes arise upon consideration of the following issues (Dermont et al., 2008): 1) contaminants are not removed from the contaminated material; 2) the treatment involves considerable volume increase; 3) assessing the process effectiveness requires constant monitoring of the treated site; 4) the long-term performance of the treated material is difficult to predict.

A variety of chemical processes can be applied to extract the contaminants from sediment or change their chemical form by either converting them into less harmful compounds or destroying them completely. Solvent extraction, washing and electrokinetic remediation belong to the former type of technologies, while neutralization, precipitation, chemical oxidation and chemical dechlorination belong to the latter. Both environmental and economic issues should be considered when a chemical treatment process is selected for sediment remediation. An important aspect involves in the first place an assessment of the fate of contaminants during the treatment process, in order to control the risks of formation of intermediates or by-products that are even more toxic or more mobile than the original pollutants (Rulkens, 2005). Another issue to be evaluated is the amount of chemicals used during treatment to ensure adequate remediation performances, with implications on both consumption of natural resources and treatment cost. With a view to beneficial reuse of treated sediments, the changes induced by the chemical treatment on the physical and chemical properties as well as on the mineralogy of the material should also be carefully controlled. Strong extracting agents are indeed known to lead to destruction of the solid matrix by dissolution of some mineral constituents (Peters, 1999; Dermont et al., 2008; Zhang et al., 2010; Zou et al., 2010), which may be detrimental for several sediment uses. When extracting agents are used for sediment remediation, concerns also arise as to whether the reagent remaining in the treated material may cause increased mobilization of the residual contaminants afterwards (Mulligan et al., 2001; Nowack et al., 2001; Zhang et al., 2010) promoting the migration of mobilized species in the surrounding environment (Peters, 1999; Dermont et al., 2008)0, or exerting ecotoxicological effects on the impacted ecosystems (Polettini et al., 2006). In this regard, recycling of the extracting agent is highly recommended, as described in more detail in Section 4. Furthermore, the need for appropriate processing and disposal of the liquid or solid residues is an additional aspect that must be taken into account (Dermont et al., 2008).

Biological remediation processes include bioleaching, aerobic or anaerobic degradation and phytoremediation (this being also enhanced using chemical agents) (U.S. EPA, 1993b; Mulligan et al., 2001; Rulkens, 2005). Bioremediation makes use of low-cost technologies, and therefore has the potential for widespread use. However, while it can effectively treat a wide range of organic contaminants, the effectiveness on inorganic contaminants is much lower or even null. A potential risk associated to biological remediation is due to the fact that partial degradation products may be more soluble or toxic than the original contaminants (U.S. EPA, 1993b). The efficiency of the degradation process can also be reduced by high organic concentrations, oxygen deficiency, lack of nutrients, and low temperature (U.S. EPA, 1993b). The use of slurry-phase systems for biological remediation may eliminate the need for preliminary dewatering of sediment, with obvious economic advantages.

Thermal treatment technologies are energy-intensive (Hakstege, 2007). They also typically require appropriate pre-treatments of sediments to remove excess water and, in some cases, coarse fractions and soluble salts. Among thermal processes, thermal desorption is a method of removing species with a high volatility and stability at high temperatures, namely volatile (VOC) and semi-volatile (SVOC) organic compounds (including mineral oil, mono-aromatic hydrocarbons, PAHs, PCBs, cyanides, chlorinated solvents and TBT), by application of appropriate temperature conditions (typically in the range 100–600 °C). Volatile metals may also be removed through thermal desorption. The volatilized pollutants can be removed and concentrated from the gas phase by absorption or condensation (Rulkens, 2005). Thermal desorption is basically a mass transfer process that concentrates the contaminants in a small volume. The final residue representing the concentrated contaminant stream must be treated afterwards.

Incineration is a high-temperature oxidation process in which organic contaminants are destroyed and converted into gaseous compounds. The process is effective in treating solid materials containing primarily organic contaminants such as halogenated and non-halogenated VOCs and SVOCs, PCBs, pesticides, dioxins/furans, and organic cyanides(U.S. EPA, 1993b).

Thermal stabilization includes a number of high-temperature processes aimed at immobilizing the contaminants within the treated sediment. Irrespective of the specific technology used, such processes are aimed at destroying organic pollutants and fixing inorganic contaminants in the mineral phase under high-temperature conditions (Detzner et al., 2007). Products of thermal immobilisation may include bricks, pellets to be used as lightweight aggregate, cement, pozzolanic materials or a slag/glassy solid (Rulkens, 2005); Detzner et al., 2007; Hakstege, 2007).

For thermal processes to be successfully applied, specific sediment characteristics must be obtained, including the absence of very fine particles that may pass through the system, a low moisture content to prevent costly vaporization of water, a good heating value, the absence of volatile metals, and elevated levels of halogenated organics (U.S. EPA, 1993b). Obviously, a major disadvantage of thermal processes is the considerable remediation cost resulting from their high energy demand.

2.3 Sediment Washing Using Chelating Agents

The principle of washing for sediment remediation is derived from soil washing techniques, which have been widely investigated and are well documented in the scientific literature (see, for example, the reviews by Peters, 1999; Mulligan et al., 2001a,b; Dermont et al., 2008 and related references therein). Notwithstanding the fair amount of research which has been conducted on soil washing, much less knowledge has been gained for sediment treatment. In this respect, only a limited number of literature references are specifically available on sediment remediation through washing (Meegoda and Perera, 2001; Fangueiro et al., 2002; Ceremigna et al., 2005; Di Palma and Mecozzi, 2007; McCready et al., 2003; Polettini et al., 2006, 2007, 2009; Yuan et al., 2010). Some field applications of the process are also documented (Amiran and Wilde, 1994; NJDEP, 2001; U.S. EPA, 1993a, 1994, 1995).

The term "washing" when applied to soil and sediment remediation is generally used to comprise both physical and chemical techniques to extract/separate contaminants. As mentioned before, physical separation through washing is aimed at concentrating the contaminants or the contaminated fraction(s) into a smaller volume by exploiting differences in specific physical characteristics (e.g., grain size, density) between the contaminant-bearing and uncontaminated particles (Dermont et al., 2008). The separated contaminated fraction is subsequently treated or disposed depending on the contamination level. Chemical washing aims at extracting contaminants bound to the solid particles of the material due to chemical interactions of various nature (co-precipitation, sorption, ion exchange, surface complexation). In such applications, the washing solution typically contains extracting agents of different types such as acids, bases, salts, oxidizing or reducing agents, chelating agents, surfactants and solvents (Peters, 1999; Polettini et al., 2009). Physical methods including sonication, abrasion or attrition scrubbing may also be used to enhance the chemical extraction process (Dermont et al., 2008).

To the authors' opinion, the term "washing" is more appropriate to indicate the chemical extraction process, while definitions such as "hydraulic separation" or "wet classification" should be used to indicate separation processes where the action of the processing fluid is only of physical nature. In the following, these distinct definitions will be adopted to refer to physical separation and chemical washing.

The washing process can be applied to treat soils and sediments contaminated with SVOCs, fuel hydrocarbons, and inorganic contaminants (mainly toxic metals and metalloids). It is less effective for VOCs and pesticides (Peters, 1999). The process has a poor efficacy for materials with high amounts of fine particles (silt and clay) and high humic contents, as well as for low-permeability materials (Peters, 1999).

The type of extraction agent to be used in chemical washing depends on a number of factors including the nature of contaminant(s) to be removed, the type and extent of their interactions with the solid matrix, the contamination level, the presence of interfering or competing species, as well as the operating conditions of the washing treatment.

Among the chemicals investigated for soil and sediment remediation, chelating agents have been widely assessed as potentially efficient extracting agents to enhance the performance of the extraction process (e.g., Peters, 1999; Dermont et al., 2008 and references therein). Chelant-assisted washing for the removal of metals from dredged sediments is the specific subject of the present chapter. Since the principle of chelant-based extraction of contaminants from sediments is exactly the same as that used in soil washing, the reader is referred to the pertinent chapter of this book for a more detailed description. Rather, here the effects of the main parameters of the chelant-assisted washing process will be reviewed, since this will make the basis for discussion and interpretation of the experimental results presented later in this chapter.

The most important factors which affect the performance of the washing process are associated to the properties of the solid matrix, the specific characteristics of the contaminants to be removed and the process parameters (Zou et al., 2010).

While the effects of the physico-chemical properties of sediment and the contaminants characteristics have been mentioned earlier in this chapter, the influence of the process parameters has not been discussed here so far. Among the process parameters, the most important in affecting the remediation efficiency in sediment washing include the extracting agent type and dosage, the solution pH, the liquid-to-solids (L/S) ratio, the solid/solution contact time, and the mode of extraction (Peters, 1999; Polettini et al., 2007; Zou et al., 2010).

The nature of the extracting agent is a factor of paramount importance in determining the metal removal yield of the washing process. Ideally, an optimal chelating agent should display a high extraction efficiency, a high selectivity for the target contaminants, a high solubility and thermodynamic stability of the formed metal complexes, a low tendency to adsorb onto the solid in both the complexed and the uncomplexed form, as well as low toxicity characteristics (Lim et al., 2004; Polettini et al., 2007). The complexing capacity of a given chelating agent towards a metal contaminant can be measured through the stability constant of the complexes formed. In this respect, it should be taken into account that different metal complexes can be formed depending on the solution conditions. In particular, the competition exerted by H^+ (towards the metal ion) and OH^- ions (towards the ligand) should be appropriately evaluated; at low pH values H^+ ions are more competitive than metals for complex formation, while at high pH values OH^- ions are more competitive than the ligand, so that metal hydroxo-complexes are preferentially formed (Stumm and Morgan, 1996). For instance, in the case of a generic divalent metal Me^{2+} and a tetravalent ligand L^{4-}, depending on pH other chelates, in addition to the MeL^{2-} complex, can be formed including the metal complexes $MeHL^-$, $MeH_2L_{(aq)}$ or MeH_3L^+ (which may have quite different stability constants) and the protonated ligand forms HL^{3-}, H_2L^{2-}, H_3L^-, $H_4L_{(aq)}$, H_5L^+, H_6L^{2+}; metal hydroxo-complexes including $Me(OH)_{2(aq)}$, $Me(OH)_3^-$, $Me_2(OH)^{3+}$, $Me_3(OH)_4^{2+}$, $Me_4(OH)_4^{4+}$ and others may be more stable than chelates when an excess of OH^- ions is present. The evaluation of which form is more stable under certain solution conditions requires specific speciation studies (Williams, 2005).

Another important aspect to be evaluated is the competition by major cations in solution towards chelate formation (Kim et al., 2003; Lim et al., 2004; Tandy et al., 2004; Polettini et al., 2007). Strong competitors in the case of soils and sediments are represented by Ca, Mg, Fe, Mn and others, which are typically present in the washing solution at much higher concentrations than metal contaminants, and may thus be favored over trace metals for the formation of chelates although the values of their stability constants are usually lower. Considering a generic divalent major metal cation Me_m^{2+} and a generic divalent trace metal cation Me_t^{2+} competing to each other for complex formation with a tetravalent ligand L^{4-}, the two following complex formation reactions can be written, each having a specific value of the stability constant K:

$$\mathrm{Me_t^{2+} + L^{4-} \Leftrightarrow Me_tL^{2-}} \qquad (K_t) \qquad (1)$$

$$\mathrm{Me_m^{2+} + L^{4-} \Leftrightarrow Me_mL^{2-}} \qquad (K_m) \qquad (2)$$

Subtracting reaction (2) from reaction (1), the following are obtained:

$$\mathrm{Me_t^{2+} + Me_mL^{2-} \Leftrightarrow Me_tL^{2-} + Me_m^{2+}} \qquad (K = K_t / K_m)$$

$$\frac{[\mathrm{Me_tL^{2-}}]}{[\mathrm{Me_mL^{2-}}]} = K\frac{[\mathrm{Me_t^{2+}}]}{[\mathrm{Me_m^{2+}}]} \qquad (3)$$

Equation (3) demonstrates that the relative abundance of the two complexes Me_tL^{2-} and Me_mL^{2-} in solution depends on the relative concentration of the two metal cations Me_t^{2+} and Me_m^{2+}, decreasing as the ratio $[Me_t^{2+}]/[Me_m^{2+}]$ increases. To account for competition phenomena by other cations, the conditional stability constant is used as a measure of the stability of a given complex. For the example taken above, the conditional stability constant would be defined as:

$$K_{cond} = \frac{[\mathrm{Me_tL^{2-}}]}{[\mathrm{Me_t^{2+}}][\mathrm{Me_mL^{2-}}]}$$

which, according to equation (3), would be expressed as:

$$K_{cond} = \frac{K_t}{K_m}\frac{1}{[\mathrm{Me_m^{2+}}]} = \frac{K}{[\mathrm{Me_m^{2+}}]}$$

Obviously, in the case of multiple competing cations, different stability constants need to be defined. This is the reason why, although many chelating agents such as aminopolycarboxylic acids usually form 1:1 complexes with most metal contaminant ions (Knepper, 2003; Stumm and Morgan, 1996), a >1 molar ratio between the chelating agent and the metal contaminants is typically required to yield adequate extraction efficiencies (Elliott and Brown, 1989; Polettini et al., 2006; Zou et al., 2010). An additional reason for a higher consumption of the chelating agent compared to the theoretical stoichiometric value may be found in the interactions with other constituents of the solid matrix. Indeed, it is known that organic ligands are capable of causing the dissolution of mineral phases of soils and sediments through surface complexation mechanisms (Nowack and Sigg., 1996; Nowack et al., 2001; Lim et al., 2004; Polettini et al., 2007; Zhang et al., 2010). The mobilization of organic matter has also been documented, showing that chelating agents are capable of causing the dispersion of humic aggregates into smaller colloidal particles, resulting in desorption and restructuring of organic matter in soils and sediments (Yang et al., 2001). This may also have a positive effect on the mobilization of hydrophobic organic contaminants such as PAHs, since in soils and sediments they are typically adsorbed to natural organic matter.

As mentioned before, solution pH is another very important factor in washing-based remediation. The effect of pH is due not only to the fact that it directly affects the type and speciation of chelates as described above in this section, but also to its influence on the solubility of chelating agents and on the degree of contaminant retention by surface functional groups of soil and sediment depending on the point of zero charge (pH_{zpc}) (Peters, 1999). For pH values below the pH_{zpc}, protonation of bridging oxo- or hydroxo- surface groups occurs, weakening surface metal atom bonds with the bulk solid matrix (Carbonaro et al., 2008) thereby favouring metal detachment (Peters, 1999; Zou et al., 2010). However, as already mentioned, too acidic pH values should be avoided to limit the negative effects caused on the chemical and physical structure of the solid matrix.

The remediation yield of the washing process is also known to be related to the L/S ratio adopted, increasing as the L/S ratio is increased. If the concentration of the chelating agent in the washing solution is maintained at the same value, the positive influence of increased L/S ratios is obvious since they are in turn associated to higher chelant dosages. However, when the chelant dosage is fixed, the L/S ratio may still positively affect metal mobilization for a number of indirect reasons, since in more diluted solutions: 1) the amount of chelating agent that can be dissolved is increased, 2) the equilibrium pH changes, and 3) lower dissolved metal concentrations may enhance metal detachment from the solid matrix. It should however be considered that adopting high L/S ratios results in generation of high volumes of residual solution, which may complicate the subsequent effluent treatment.

Since the extraction of metals and metalloids from soil is a kinetically-controlled process, the contact time is another key operating parameter in soil washing. According to Carbonaro et al. (2008), the extraction process occurs in subsequent steps, including: (1) diffusion of ligand to the solid surface; (2) formation of an initial adsorbed species or complex; (3) conversion to an adsorbed species capable of detaching surface-bound metal atoms; (4) detachment of the metal-chelant complex from the surface; and (5) diffusion of this complex into the liquid solution. For surface-bound metal atoms in the +II and +III oxidation states, steps (1), (2) and (5) are unlikely to be rate-limiting (Carbonaro et al., 2008). This implies that the total amount of chelating agent available to adsorb onto the solid surface and dissolve the adsorbed species decreases with time. For this reason, the increasing trend of metal concentration in the washing solution is usually characterized by a declining rate over time, attaining a final equilibrium value which depends, for given operating conditions, on the type and strength of the interactions between the contaminants and the solid surface. In the short term, extraction of metals is dominated by the most labile species, while in the long-term it is dictated by species more tightly bound to the solid matrix constituents (Zou et al., 2010).

The extraction conditions of the washing process can also be optimized in order to enhance metal removal. The use of multiple washing stages has been investigated as a means to improve the metal extraction capacity of chelating agents. The main reasons for improved efficiency are related to 1) the effect on removal of residual metal chelates trapped in the pore solution or re-adsorbed onto the sediment particles in the previous washing steps; 2) the reduced interference of major competitive cations, which are preferentially removed in the initial extraction steps; 3) the reduced re-adsorption phenomena of detached metals onto the solid surface; and 4) the possibility to avoid the approach to saturation conditions (Polettini et al., 2009; Zou et al., 2010). However, the incremental gain obtained in the extraction yield typically decreases as additional extraction steps are introduced, so that the overall removal has an upper threshold. This is due to the fact that, as the washing treatment proceeds, it becomes increasingly difficult to remove the residual contaminants, which are more strongly retained by the solid matrix and are thus not amenable to chelant extraction (Neaman et al., 2004; Andrade et al., 2007; Polettini et al., 2009).

Additional improvements in the washing conditions may be obtained with the aid of physical methods including ultrasonic processing (Peters, 1999; Meegoda and Perera, 2001; Polettini et al., 2009; Zou et al., 2010), attrition washing (BioGenesis Enterprises, 1999), and improved agitation (Zou et al., 2010).

2.4 Recovery of Chelating Agents from Waste Washing Solutions

As mentioned before, recovery and recycling of the chelating agent after the washing treatment are crucial in order to improve the overall environmental profile of the process. This is particularly important when potentially ecotoxic compounds are used as the extracting agents, which may negatively impact ecosystems if the treated sediment is reused after washing.

So far researchers have extensively addressed the recovery and recycle of EDTA, although data predicted on the basis of theoretical calculations are also available for different chelating agents (Chen et al., 1995). Two options can be applied, either individually or in combination, to recover chelating agents from spent washing solutions. The first one involves the addition of an appropriate electrolyte and pH adjustment to substitute a non-hazardous cation (e.g., Fe) for the toxic metal in the complex. In view of subsequent reuse, the neo-formed complex should exhibit an adequate extraction capacity towards contaminants. The second option relies on an electrolytic treatment to promote toxic metal migration and chelating agent recovery. According to the first mechanism, a metal complexed by a ligand can be removed by promoting the formation of a different complex with a non-hazardous cation (which may also require a shift in pH), and subsequently removing the metal from the solution through precipitation. This principle was applied by Chang et al. (2007) to recover Cu and chelating agents from solutions obtained from a washing treatment of industrial sludge. Zero-valent iron at Fe/Cu molar ratios of 0.5–40 and pHs of 2, 3 and 4 units was used to break down Cu-EDTA and Cu-DTPA chelates and precipitate Cu from the solution, according to the following reactions:

Fe^0 + Cu-EDTA → Fe(II)-EDTA + $Cu_{(s)}$
Fe^0 + Cu-DTPA → Fe(II)-DTPA + $Cu_{(s)}$.

An NaOH solution was afterwards used to shift pH to 13, a condition under which $Fe(OH)_3$ was precipitated and the free forms of the chelating agents were formed. The recovered solutions were also tested for use in washing treatments and found to display extraction capacities comparable to those obtained by applying pure chelating agent solutions.

A similar process was applied to recycle a synthetic Pb-EDTA wastewater (Kim and Ong, 1999)0. Pb substitution in the EDTA complex was attained by adding Fe(III) at a low pH, while subsequent Pb precipitation was obtained using potassium sulfate or monosodium phosphate. It was found that an Fe/Pb molar ratio of 1.5 was required to adequately remove Pb from the Pb-EDTA complex. After Pb precipitation, Fe(III) was then removed from the resulting Fe-EDTA solution by precipitation at high pH values. The recovered EDTA solution was found to exhibit the same extraction capacity as the original EDTA solution, and was recycled several times without any appreciable negative effect on the extraction yield. If Fe is not precipitated from the Fe-EDTA solution, this form could directly be used for metal removal at $pH > 7$, where the compound displays a chelation capacity towards other metals. The same study also demonstrated that Pb precipitation using phosphate gave a slightly better recycled EDTA solution than when using sulfate.

Ager and Marshall (2003) studied the feasibility of using Mg and Pd to substitute for toxic metals in EDTA complexes. Hong et al. (1999) and Zeng et al. (2005) evaluated the feasibility of using Na_2S and $Ca(OH)_2$ to precipitate toxic metals and recover EDTA. In particular, Zeng et al. (2005) observed that both metals and EDTA in the spent solutions could be separated almost completely by adding Na_2S and $Ca(OH)_2$. The authors found that Na_2S was able to separate Cd, Cu and Pb from EDTA, while Zn separation required the addition of $Ca(OH)_2$ to the Na_2S solution. A reagent loss of 19–23% was observed after seven reuse cycles of the recovered EDTA solution. Given the high metal content of the precipitated metal sulfides, they should be conveniently treated for metal recovery or final disposal. As both the recovered Ca-EDTA and Na-EDTA forms were found to be effective in metal complexation, the authors suggested that Ca-EDTA may be preferable in view of soil or sediment remediation, since it may also supply Ca ions and mitigate the dissolution of natural organic matter.

Hong and Jiang (2005) studied the feasibility of recovering and recycling chelating agents including EDTA, TMDTA, NTMP, GcG, SCMC and ADA after use for Pb extraction from a contaminated soil. The recovery and recycle process involved the addition of precipitating agents ($FeCl_3$, $Ca(OH)_2$ or sulfides) to the spent solution. The results showed that chelant recovery was effective even for the strongest chelating agents tested, namely EDTA and DTPA.

In order to reduce the overall amounts of spent solution to be treated, Di Palma et al. (2003) proposed a preliminary stage to reduce the volume of the spent solution. Evaporation and reverse osmosis were used for this purpose.

The electrolytic separation of metals and the chelating agent from the spent washing solution was proposed by Allen and Chen (1993), where a two-chamber cell separated by a cation exchange membrane was used to prevent migration to the anode and the oxidative destruction of negatively charged metal-EDTA complexes.

Gyliene et al. (2004) studied the feasibility of EDTA recovery from a complex solution containing Cu(II), Ni(II), Co(II), Cd(II), Ca(II) and Mg(II) complexes of EDTA, using Cu(II) as the precipitating agent. The subsequent removal of Cu(II) was then attained by electrolysis. The proposed method was based on replacement of the metal ions in EDTA complexes by Cu(II), and subsequent formation of the insoluble $Cu_2EDTA \cdot 4H_2O$ complex under weakly acidic conditions. The removal of Cu can be then attained by electrolysis under acidic conditions, where the protonated H_4EDTA complex is produced.

It should be noted that during electrolytic separation and reverse osmosis, colloidal particles (clays and humic matter) and microorganisms may produce clogging effects on membranes and electrodes, thus reducing the yield of the recovery process (Allen and Chen, 1993; Chang et al., 2007; Leštan et al., 2008).

2.5 Experimental Results of Chelant-Assisted Washing for Metal Removal from Sediments

In this section the experimental results of previous studies conducted by the authors of this chapter on metal removal from contaminated dredged sediments through chelant-based extraction are presented. Different dredged sediment samples were investigated for the application of the washing process, including harbor sediments from a highly polluted industrial area and sediments from a contaminated river stream.

2.5.1 Materials and Methods

Four dredged sediment samples (RS1, RS2 and RS3 – river sediment dredged at different sections of a polluted river stream in southern Italy – and MS – marine sediment dredged from a polluted harbor site in northern Italy) were used for chelant-assisted washing experiments. The sediment samples were homogenized and stored in the laboratory at 4 °C until the time of testing. Before testing, the sediment samples were oven-dried at 60 °C. The material was characterized for grain size distribution, water content, pH, total organic carbon (TOC), elemental composition, as well as metal distribution in sediment. For the analytical procedures used, the reader is referred to previous work where additional details are provided about the experiments (Ceremigna et al., 2005; Polettini et al., 2006, 2007, 2009).

Samples of oven-dried sediment were subjected to lab-scale washing experiments using ethylenediaminetetraacetic acid (EDTA) (Na_2 salt), citric acid (CIT), nitrilotriacetic acid (NTA) and the S,S-isomer of the ethylenediaminedisuccinic acid ([S,S]-EDDS) (Na_3-N,N' salt) as the washing agents. Different experiments were carried out in order to study the kinetics of metal extraction, as well as the influence of chelant type/concentration and pH on metal removal efficiency. The washing treatment was performed at a liquid-to-solid ratio of 20 L kg^{-1} on duplicate samples rotated on an end-over-end tumbler for different times ranging from 0.5 to 48 h. The chelating agents were applied at two different concentrations, 0.01 and 0.1 M for the river sediments, and 0.02 and 0.2 M for the marine sediment; such values correspond to a chelant dosage of 1 (stoichiometric) and 10 mol mol^{-1} of total metal content, respectively. Total metal content included major metal cations (Ca, Fe, K and Mg) as well as trace metals. At the end of the washing period, the slurries were centrifuged at 6000 rpm. The liquid solution obtained was then filtered through 0.45-μm membrane filters and then acidified with 1:1 HNO_3 for chemical analyses.

The influence of pH on metal removal by chelant washing was investigated by means of 2-hour pH-stat experiments on the MS sample using 1 M HNO_3 or NaOH to maintain the solution at endpoint pH values of 5, 6, 7 and 8 using an automatic titrator; control runs using deionized water as the extracting solution were also carried out at the same pH values, to serve for reference purposes.

Two- and four-stage washing experiments were also performed. Two-stage washing involved the application of EDTA in the first step and CIT in the second step at either a 0.01 or 0.1 M concentration. Four-stage experiments were conducted using a single type of chelating agent (EDTA or CIT; 0.01 or 0.1 M) by renewing the extracting solution four times. During each washing step, a 2-hour contact time was adopted, while the other operating conditions were as previously described.

2.5.2 Results and Discussion

The main physical and chemical characteristics of the investigated sediments are reported in **Table 2.1**. As highlighted in the table, for all the investigated sediments the content of Cr, Cu, Pb and Zn exceeded one or more threshold limits prescribed by the Italian regulation for soils (Decreto Legislativo 3 aprile 2006, n. 152) and sediments (Ministero dell'Ambiente, 1993); for the two harbor sediments, additional critical elements were found to include As and Cd.

The kinetic data of metal extraction from the MS sample is shown in Figure 2.2 for trace metals and in

Figure **2.3** for major elements. Significant differences in metal removal were observed depending on both the investigated metal and the type and concentration of chelating agent.

Table 2.1 Results of the physical and chemical characterization of sediments

Sample **Parameter**	**MS1**	**RS1**	**RS2**	**RS3**
Grain size distribution:				
Φ < 2 mm	98.0	89.0	79.0	100.0
Φ < 425 μm	95.2	53.0	64.5	98.0
Φ < 150 μm	87.2	17.0	35.5	62.0
Φ < 63 μm	64.2	5.4	14.4	22.6
pH	7.8	7.9	7.7	7.5
TOC (% dry wt.)	2.7	3.4	1.6	2.8
As (mg kg^{-1} dry wt.)	180.9 •,◆,★	4.6	4.4	3.3
Ca (mg kg^{-1} dry wt.)	94080	78260	117920	154840
Cd (mg kg^{-1} dry wt.)	30.7 •,◆,★	< 2.0	< 2.0	< 2.0
Cr (mg kg^{-1} dry wt.)	108.1	223.5 •	702.3 •,★	217.8 •
Cu (mg kg^{-1} dry wt.)	323.0 •	178.9 •	186.5 •	148.8 •
Fe (mg kg^{-1} dry wt.)	41110	24320	30248	27272
Mn (mg kg^{-1} dry wt.)	411.3	232.5	262.4	187.8
Pb (mg kg^{-1} dry wt.)	886.2 •,◆,★	249.4 •	324.2 •	237.8 •
Zn (mg kg^{-1} dry wt.)	3723 •,◆,★	286.4 •	386.7 •	268.2 •

• Exceeding the limit concentration for soil (residential areas) (Decreto Legislativo 3 aprile 2006)

◆ Exceeding the limit concentration for soil (industrial areas) (Decreto Legislativo 3 aprile 2006)

★ Exceeding the limit concentration for sediment (col. C Venice Lagoon Act) (Ministero dell'Ambiente, 1993)

Among the trace metals of concern, Ni (not reported graphically here) and Cr displayed the lowest removal yield. In particular, Ni in the final extracting solution was always below the analytical detection limit, while Cr was removed to an appreciable degree only by 0.2 M EDTA (~32% after 48 h) and 0.2 M CIT (~23% after 48 h). For the other metals, the measured removal efficiency was dependent on the specific affinity and selectivity of each chelating agent as well as on the applied dosage. Considering the extraction yield attained after 48 hours, the following ranking for the trace metals/metalloids of concern was derived on the basis of the experimental results:

EDTA 0.02 M:	Pb > Cd ≅ Zn ≅ Cu > As > Cr
EDTA 0.2 M:	Pb ≅ Cu > Cd ≅ Zn > Cr > As
NTA 0.02 M:	Cd > Cu > Pb ≅ Zn ≅ As > Cr
NTA 0.2 M:	As > Zn > Cd ≅ Cu ≅ Pb ≅ Cr
CIT 0.02 M:	Cd > Zn ≅ Cu > Pb ≅ As ≅ Cr
CIT 0.2 M:	Cd ≅ Zn ≅ Pb > As > Cu ≅ Cr
EDDS 0.02 M:	Pb > Cd > Cu > Zn > As > Cr
EDDS 0.2 M:	Pb > Cd > Cu ≅ Zn > As > Cr

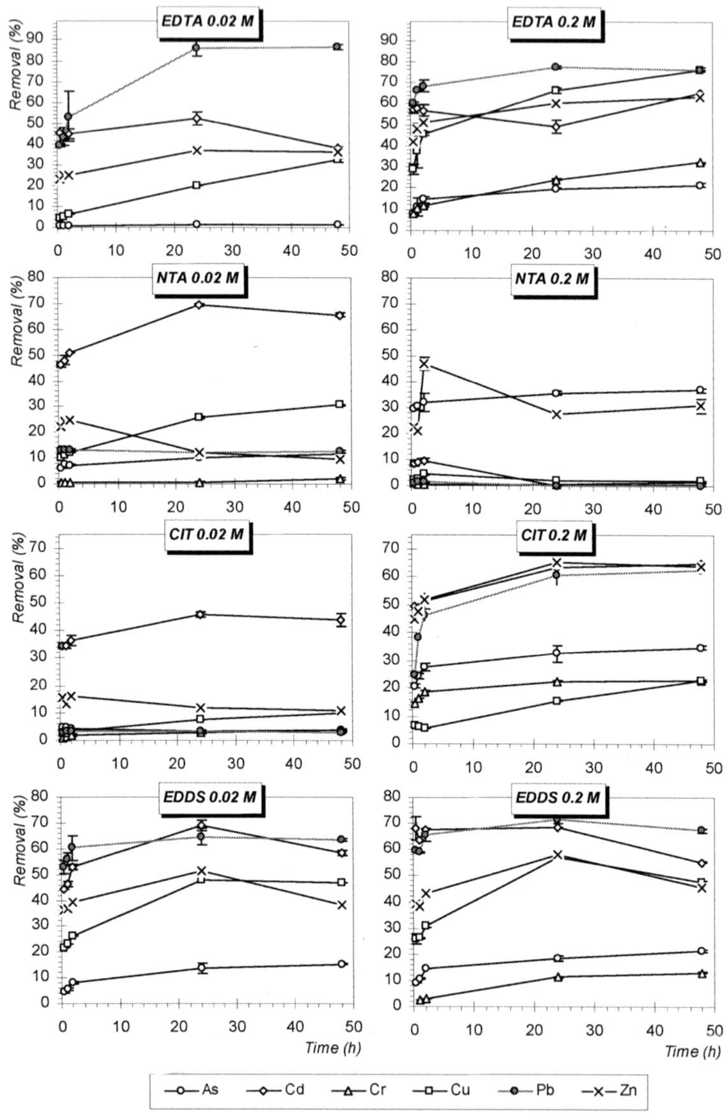

Figure 2.2 Kinetics of metal extraction by chelating agents (trace elements; sediment MS) – error bars are plotted at one standard deviation of experimental data.

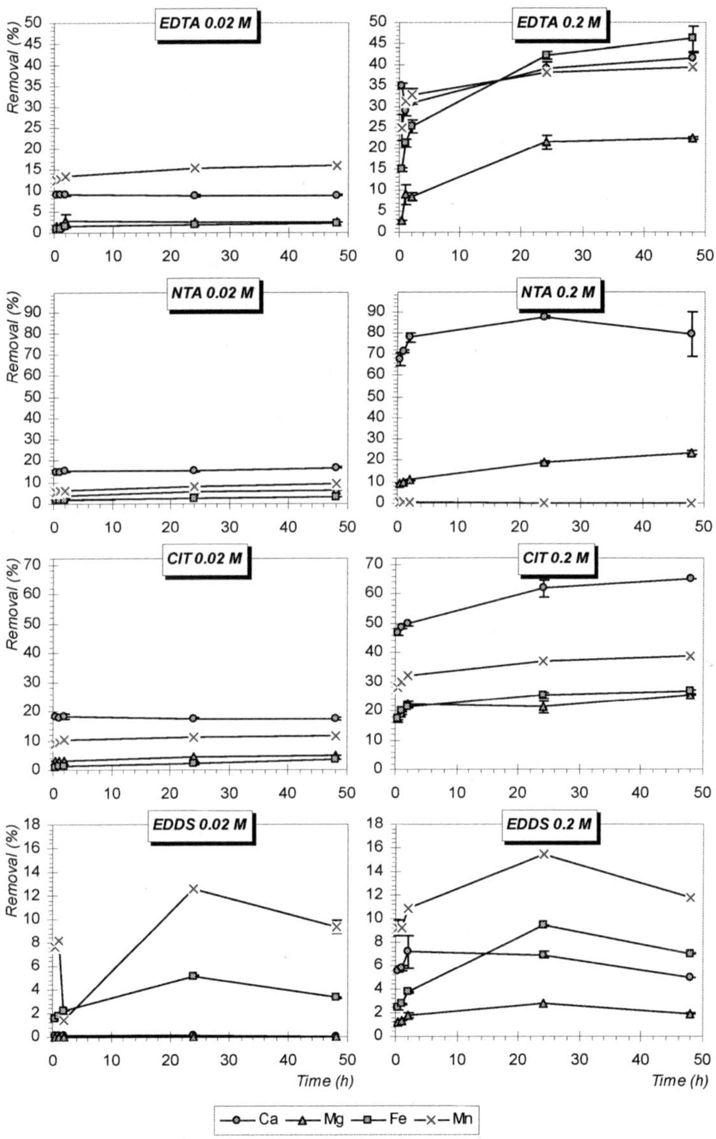

Figure 2.3 Kinetics of metal extraction by chelating agents (major elements; sediment MS) – error bars are plotted at one standard deviation of experimental data.

Concerning the individual contaminants, after 48 hours As removal was 37% using 0.2 M NTA, 35% using 0.2 M CIT, while ~21% with 0.2 M EDTA and 0.2 M EDDS. At the lower dosage, in no case was it possible to extract more than 10% of the initial As content in sediment. The poor removal of As observed in the present study is consistent with the findings of other authors (Tokunaga and Hakuta, 2002) indicating that chelating agents are not effective in extracting As. This is related to the fact that the anionic form of both As(III) and As(V) in solution does not allow for the formation of stable chelant complexes.

Among the investigated metal contaminants, the highest removal after 48-h washing (78–86% with 0.02 and 0.2 M EDTA) was displayed by Pb, followed by Cu (76% with 0.2 M EDTA and 57% with 0.2 M EDDS) and Zn (65% using 0.2 M Cit and 63% using 0.2 M EDTA).

The residual concentrations of Cd, Pb and Zn after washing with EDTA and EDDS were always below the limit values established by the Italian regulation for sediments (Ministero dell'Ambiente, 1993). Only in the case of 0.02 M EDTA, did the treated material exceed the limit concentration for Zn. Conversely, the efficiency of NTA and CIT, especially at the 0.02 M level, appeared to be in most cases inadequate to meet the regulatory limits. An important issue still to be resolved is therefore the inability of the selected chelating agents to produce appreciable As removal efficiencies. Our further studies (Polettini et al., 2009) indicated that oxalic acid applied as the extracting agent during washing was effective in removing As from contaminated sediments. The combined use of surfactants and chelating agents also improved As removal, although under the tested conditions it was still unable to reduce the final As concentration in sediment to below the regulatory limit.

Further information from kinetic experiments on sediment MS was derived on the behavior of the major cations Ca, Mg, Fe and Mn as reported in Figure **2.3**. NTA and CIT at the 0.2 M level were found to display a strong affinity towards Ca (48-h extraction yields of 80 and 65%, respectively) with a fast mobilization rate, as also documented elsewhere in the literature (Ritschel, 2003). Similarly, Mg was better extracted by 0.2 M CIT, 0.2 M NTA and 0.2 M EDTA, although to a lower extent (up to 25%). Fe and Mn were mobilized to an extent of 46 and 39% by 0.2 M EDTA, and 27 and 39% by 0.2 M CIT. However, in the case of Fe the extraction kinetics was noticeably slower if compared to the other major cations Ca and Mn. On the basis of the results from sequential extraction, it may be hypothesized that the extraction efficiency of EDTA and CIT is related to their capability of dissolving Fe and Mn oxides, although some studies report very slow kinetics of oxide dissolution by chelating agents (Elliott et al., 1989). In this study, sequential extraction on 2-h washed sediment (see Polettini et al., 2006; for further details) indicated that when the chelating agents were used at the 0.2 M level, the mobilization of Fe from the Fe+Mn oxides fraction was 94% (CIT), 89% (EDTA), 86% (EDDS) and 49% (NTA). The corresponding values observed for Mn were lower, namely 43% (EDTA), 29% (CIT), 21% (NTA) and 14% (EDDS). This indicates an appreciable dissolution, mainly of Fe oxides by, in the order, CIT, EDTA and EDDS.

The pH-dependence of metal extraction for 2-hour washing experiments on sediment MS is reported in Figure 2.4. In general (results not reported here; further details in Polettini et al., 2007), metal extraction was significantly increased when chelating agents were used instead of deionized water only. Metal concentrations in the extracting solutions were also variously affected by pH. At the 0.02 M chelant concentration, a pH decrease in the acidic range resulted in all cases in an increased removal of As, although with different shapes of the concentration-versus-pH curves. While EDDS and EDTA gave an increase in As concentration with decreasing pH over the entire range investigated, CIT and NTA showed a pH-dependent extraction with a minimum at pH 7. At the lower chelant dosage, the ranking of As removal was EDDS > CIT > NTA > EDTA at all pHs. At the 0.2 M level, pH affected As extraction particularly in the case of EDDS (removal increasing from ~5% at pH 8 to ~70% at pH 6) and NTA (removal decreasing from ~73% at pH 8 to ~5% at pH 6). At the 0.2 M chelant concentration, CIT was nevertheless observed to produce the strongest mobilization of As over the entire pH range investigated, with a relatively small influence of pH.

In the case of Cu, pH exerted a minor influence on the extraction yield, with the only exceptions of 0.02 M CIT (with a minimum at pH 7) and 0.2 M EDTA (with a maximum at pH 6). At the 0.02 M level, Cu extraction was in the order EDDS > NTA > EDTA >> CIT. At higher dosages Cu mobilization by EDTA and CIT was appreciably enhanced, which is presumably related to the speciation of the two chelating agents in the extracting solutions. At the 0.02 M level, Ca and Mg complexes with chelating agents were calculated to account for more than 72% of EDTA and CIT in solution, while the corresponding values for NTA and EDDS were appreciably lower (more specific details on such calculations can be found in Polettini et al., 2007). This indicates a more significant competition effect by Ca and Mg towards Cu extraction in the case of EDTA and CIT if compared to NTA and EDDS, demonstrating the role played by the chemical equilibria occurring in the extracting solutions during chelant-assisted washing.

At the 0.02 M chelant dosage, Pb was extracted by the four chelating agents according to the ranking EDTA > NTA > EDDS > CIT under acidic conditions and EDTA ≅ EDDS ≅ NTA >> CIT at pH 8. For 0.02 M CIT, pH was also found to largely affect Pb extraction, with Pb concentrations decreasing by approximately two orders of magnitude from pH 5 to pH 8. The pH dependence of Pb extraction by CIT is likely related to the fact that at pH values ≤ 6 the protonated complex PbH–CIT was found to be present in solution along with $Pb–CIT^{-}$ (Polettini et al., 2007). The poor influence of pH on Pb extraction by the other chelating agents can be related to the fact that protonated chelates were found to be thermodynamically unfavored over the non-protonated ones. This can be explained considering that the stability constants for the protonated free forms of EDDS, NTA and EDTA are much higher ($\log K >$ 9.8) than that of CIT ($\log K(H–CIT^{2-}) = 6.40$), reducing the conditional stability constant of protonated metal–chelant complexes. At the 0.2 M level, Pb mobilization by CIT was significantly improved if compared to the other chelating agents, and was also dependant on pH to a lower degree. On the basis of chemical speciation

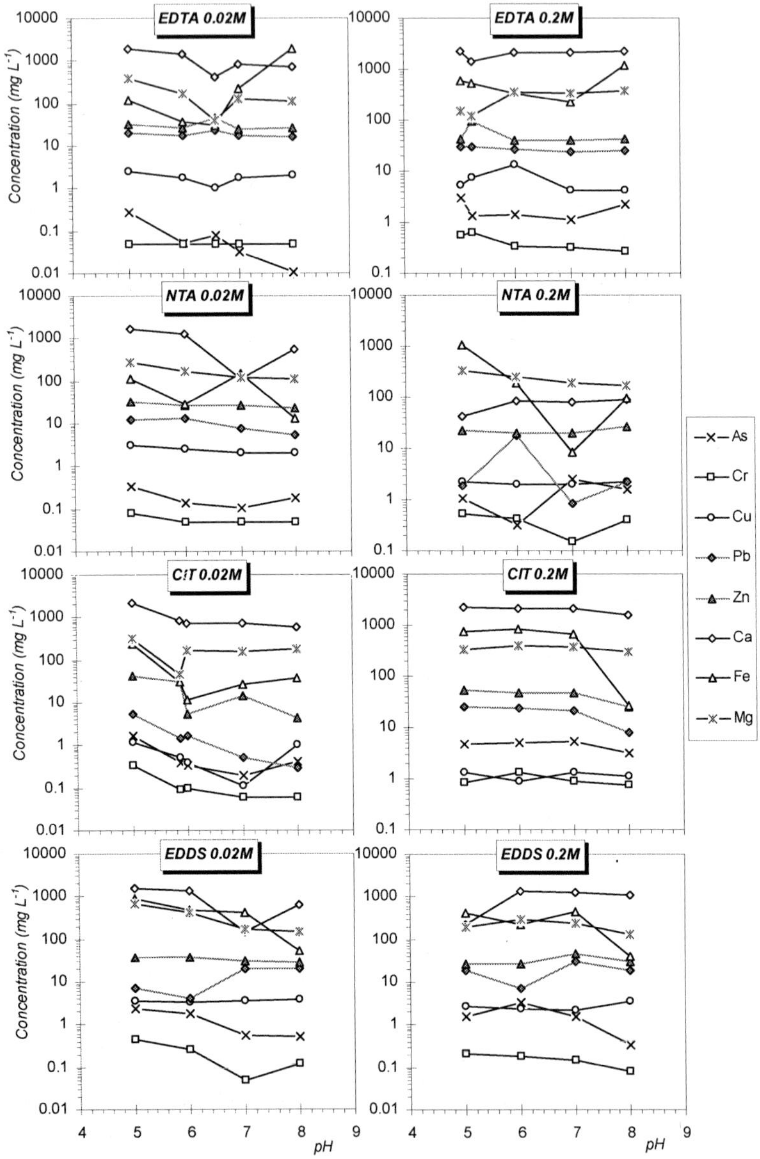

Figure 2.4 Metal extraction as a function of pH (sediment MS).

calculations (Polettini et al., 2007), the increased extraction capacity of CIT was found to be related to the formation of $Pb–(CIT)_2^{4-}$ at high pHs and $PbH–(CIT)_2^{3-}$ at low pHs, favored by the chelant excess adopted.

In the case of Zn, with the exception of 0.02 M CIT, the investigated chelating agents yielded comparable removal efficiencies, with a minor dependence on solution pH. Zn was found to be the only trace metal taking up an appreciable, although still low (up to ~3%), portion of the chelating agents (Polettini et al., 2007). This is related to the fact that, although the stability constants of Zn chelates are generally lower if compared to other metals (e.g., Cu and Pb), the Zn content in sediment was much higher than that of the other contaminants of concern, and this metal was thus more competitive for complex formation.

Chemical speciation calculations also showed that a tenfold increase in chelant dosage from 0.02 to 0.2 M remarkably increased the amount of free chelant forms in solution (CIT^{3-}; $H–EDTA^{3-}$ and $H_2–EDTA^{2-}$; $H–NTA^{2-}$; $H–EDDS^{3-}$ and $H_2–EDDS^{2-}$), but was not capable of producing a corresponding increase in the amount of metal chelates formed. This finding is obviously related to the specific characteristics of sediments in terms of interactions of metal contaminants with the solid surface, which results in the inability of chelating agents to detach metals which are tightly bound to the solid surface. Contaminants associated to the detrital fraction of the material can only be extracted provided that solid matrix dissolution is attained. As a consequence, when increasing chelant dosage, only a small portion is effectively transformed into metal chelates, while the excess remains in the free form with no gain in the remediation performance.

As for major elements, Ca, Mg and Fe were found to be strong competitors in metal chelate formation, with concentrations in the washing solution significantly higher (by more than two orders of magnitude) than those measured for trace metals. In particular, Ca extraction was found to be affected by pH to an extent that depended on chelant dosage. At the lower dosage, the investigated chelating agents produced comparable results in the acidic pH range. In the neutral to alkaline pH range, while Ca extraction was poorly dependent on pH for EDTA and CIT, both NTA and EDDS displayed a concentration minimum at pH 7. At the 0.2 M level, the pH dependence of Ca mobilization was less evident, and EDTA and CIT extracted more Ca (~50–60% of total content) than EDDS (~30%) and NTA (< 25%). The amount of Ca complexed by CIT, EDTA and NTA at the 0.02 M dosage was also calculated to exceed by far that of the other forms (including the free chelant), with the exception of NTA at pH 7 where the dominant species was $H–NTA^{2-}$. For EDDS, complexation with Ca appeared to be less relevant, and the prevalent forms of the chelating agent were found to include $H_2–EDDS^{2-}$ and/or $H–EDDS^{3-}$ (depending on pH), and $Fe-EDDS^{-}$ at pH 5. An appreciable fraction of EDDS (~20% of total chelant) was calculated to be present in the form of Ca complexes at pH 8 only (Polettini et al., 2007).

The influence of pH on Fe extraction was quite relevant at both dosages for all the investigated chelating agents, with a more than 100-fold variation in soluble Fe concentration with pH. At the 0.02 M dosage, EDDS displayed the highest Fe extraction capacity at pHs of 5–7 (with extraction efficiencies between ~44% and ~20%), whereas Fe solubilization was the highest for NTA at pH 8 (~77%). At the 0.2 M dosage, in the pH range 5–7 CIT showed the highest extraction capacity (between ~40 and ~32%), while at pH 8 Fe mobilization by EDDS was the highest, although with much lower values (~6%).

In the case of Mg, 0.02 M EDDS produced the highest extraction over the whole pH range investigated (48% at pH 5 and 10% at pH 8), while at the same dosage the other chelating agents extracted less than 25% of the total Mg content of sediment. At the 0.2 M dosage, Mg extraction by EDDS was reduced at all pH values, while that of EDTA and CIT did not vary appreciably if compared to 0.02 M level.

In Figure 2.5 the metal extraction yield attained in single-stage washing experiments on river sediments (RS1, RS2 and RS3) is reported as a function of the chelating agent dosage. As already observed for the MS sample, EDTA, NTA and

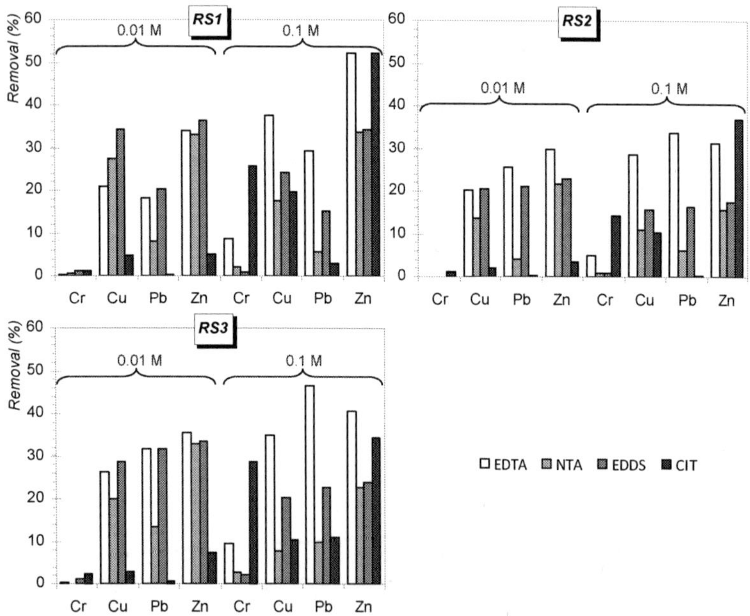

Figure 2.5 Metal removal for single-stage washing of sediments RS1, RS2 and RS3 (2-h experiments).

EDDS displayed a poor extraction capacity towards Cr for all the investigated sediments, with extraction yields always below 10%. Compared to the other chelating agents, CIT showed a higher affinity towards Cr, with removal yields of 26, 14 and 29% for the RS1, RS2 and RS3 samples, respectively.

In the case of Cu, the highest extraction capacity was shown by EDTA, followed by, in descending order, EDDS, NTA and CIT, likely due to the underlying effect of solution pH (Polettini et al., 2009). The low removal attained using CIT can be explained considering the generally low values of the stability constants for Cu-CIT complexes. For the RS3 sample, EDTA and EDDS at both dosages and 0.01 M NTA produced good remediation results (final Cu concentrations below the limit value of 120 mg kg^{-1}); the treatment on RS1 sediment met the required quality criteria for Cu using 0.01 M EDDS and 0.1 M EDTA, while it was never possible to achieve adequate remediation results for the RS2 sample.

The same extraction ranking EDTA > EDDS > NTA > CIT noted for Cu was also observed for Pb, confirming the affinity of the investigated chelating agents observed for the MS sample. The highest Pb removal yield (29%, 34% and 46% for RS1, RS2 and RS3, respectively) was displayed by 0.1 M EDTA. Although the initial Pb content in the materials was yet lower than the regulatory limit prescribed for sediments, none of the investigated chelating agents was capable of meeting the most stringent quality criterion of 100 mg kg^{-1}.

In the case of Zn, at the 0.01 M chelant concentration EDTA, NTA and EDDS displayed comparable extraction efficiencies (in the ranges 33–36% for RS1, 23–30% for RS2 and 33–35% for RS3), while CIT showed a poor mobilization capacity (5% for RS1, 3% for RS2 and 7% for RS3). With an increase in the chelant dosage, Zn was better extracted by EDTA and CIT at approximately the same level (52% in both cases for RS1, 31 and 37% for RS2, and 40 and 34% for RS3), and to a lower but comparable degree by NTA and EDDS (~34% for RS1, 16 and 18% for RS2, 23 and 24% for RS3). The observed behavior confirms the selectivity ranking as a function of chelant dosage already observed for the MS sample.

As commented above for Pb, the most stringent quality level for Zn (150 mg kg^{-1}) prescribed for soils in residential areas could not be attained using single-stage washing; the only exception was the RS1 sample washed with 0.1 M EDTA or 0.1 M CIT.

Metal removal results of multi-stage washing experiments on river sediments are depicted in Figure 2.6 (two-stage treatment) and in Figure 2.7 (four-stage treatment). In general, at the 0.01 M chelant dosage the incremental metal removal efficiency was found to decrease dramatically as the number of washing steps increased. At the higher dosage, the gain in metal removal, although still relatively low after the first washing stage, improved appreciably. As mentioned earlier, the progressive decrease in metal extraction in multi-stage washing is due to the fact that only the labile portions of contaminants are extracted by chelating agents and, as washing proceeds, removing the residual contaminants becomes progressively more difficult since these are more strongly bound to the solid matrix and are thus less amenable to chelant extraction. For these reasons, multi-stage washing experiments were capable of meeting the regulatory limits in selected cases only. In particular, the RS2 sample gave the worst results in terms of compliance with the Italian legislative quality standards. For the RS1 and RS3 sediments, Cr removal was improved by using CIT both in two- and in four-stage washing. The residual Cr concentrations in two-stage experiments using 0.1 M CIT were below the quality level of 150 mg kg^{-1} established for soils in residential areas, with an attained removal of 45% for RS1 and 42% for RS3. In four-stage washing with 0.1 M CIT (and to a certain extent with 0.1 M EDTA) appreciable Cr extraction yields were observed even in the third and fourth washing steps, so that the residual Cr content in the RS1 and RS3 samples was reduced to below the threshold limit of 50 mg kg^{-1} prescribed for sediments (85% and 98% removal for RS1 and RS3).

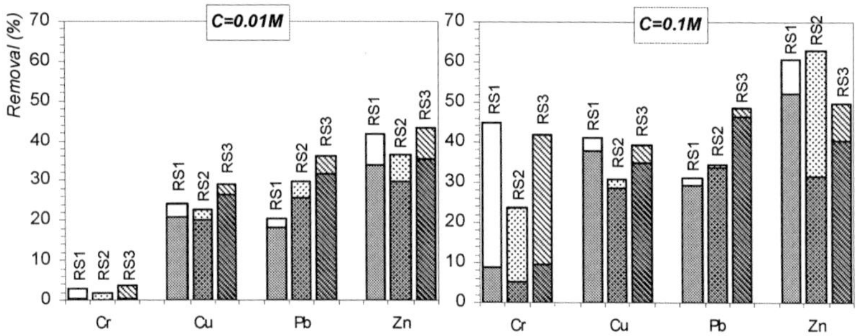

Figure 2.6 Metal removal for 2-h two-stage washing of river sediments – the lower and upper bars denote results for the first (EDTA) and second (CIT) washing steps.

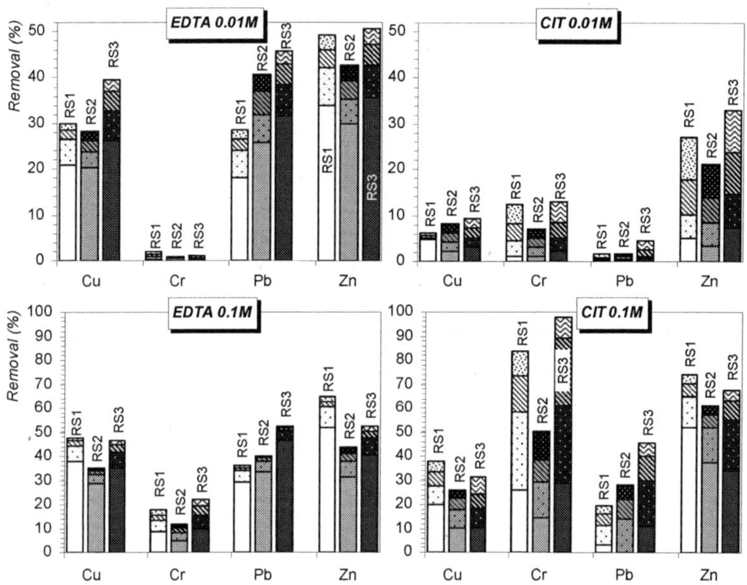

Figure 2.7 Metal removal for 2-h four-stage washing of river sediments – the four bars denote results for each washing step.

As for Cu, the most stringent criterion of 120 mg kg^{-1} was met, in addition to single-stage 0.1 M EDTA washing, through multiple washing with chelating agents at the 0.1 M level. Four-stage washing with 0.1 M EDTA was also adequate for compliance with the mentioned quality requirement for the RS2 sample, for which a total removal of 45% was attained.

Again, for samples RS1 and RS3 the extraction yield was also improved for Zn when multi-stage washing was performed. The amounts removed were 61 and 50% (0.1 M EDTA + 0.1 M CIT), 49 and 50% (four-stage 0.01 M EDTA), 65 and 52% (four-stage 0.1 M EDTA), and 74 and 68% (four-stage 0.1 M CIT). In these cases the residual Zn concentration in sediment was reduced to below the limit of 150 mg kg^{-1} for soils in residential areas.

As described above for single-stage experiments on both marine and river sediments, multi-stage washing also produced appreciable mobilization of major elements even after the first washing stage. In particular, two-stage experiments at the 0.1 M chelant concentration resulted in an overall mobilization of 60–70% for Ca and 23–36% for Fe for the three river sediments.

2.6 Conclusions

The results reported in the present study show that chelant-based washing of metal-contaminated sediments is a potentially effective process for remediation purposes. Nevertheless, important issues, including the specific characteristics of the contaminated material in terms of physical and mineralogical properties and type and history of contamination, should be evaluated if the remediation performance is to be correctly predicted. In addition, it should be considered that the typical presence of multiple contaminants in sediments appears to require a multi-stage washing sequence (in some cases using a combination of different extraction agents), possibly also coupled with treatment processes of different nature.

It should also be mentioned that in the case of strong interactions existing between metal contaminants and the sediment surface sites, as indicated by association with the less mobile fractions of the material, the applied treatment is expected to result in poor remediation performance. While this may negatively reflect on compliance with the prescribed quality criteria, it also identifies the need for careful evaluation of the real mobility of metal pollutants in contaminated sediments under different environmental conditions.

2.7 References

Ager, R., and Marshall, W. D. (2003). "Recycle of thermomechanical pulp filtrate after removal of metals: A study with EDTA." *Journal of Pulp and Paper Science*, 29, 303-307.

Allen, H. E. and Chen, P. H. (1993). "Remediation of metal contaminated soil by EDTA incorporating electrochemical recovery of metal and EDTA." *Environmental Progress*, 12, 284-293.

Amiran, M. C. and Wilde, C. L. (1994). "PAH removal using soil and sediment washing at a contaminated harbor site." *Remediation Journal*, 4, 319-330.

Andrade, M. D., Prasher, S. O., and Hendershot, W. H. (2007). "Optimizing the molarity of a EDTA washing solution for saturated-soil remediation of trace metal contaminated soils." *Environmental Pollution*, 147, 781–790.

Andreottola, G., Bonomo, L., De Gioannis, G., Ferrarese, E., Muntoni, A., Polettini, A., Pomi, R., and Saponaro S. (2010). "Lab-scale feasibility tests for sediment treatment using different physico-chemical techniques." *Journal of Soils and Sediments*, 10, 142-150.

Barceló, D., and Petrovic, M. (2007). "Preface." Sustainable management of sediment resources: Sediment quality and impact assessment of pollutants, Elsevier, Amsterdam, v-viii.

BioGenesis Enterprises and Brookhaven National Laboratory (1999). "Bio-Genesis Sediment Washing Technology Full-Scale, Sediment Decontamination Facility for the NY/NJ Harbor Region", Final Report on the Pilot Demonstration Project, BioGenesis Enterprises, Springfield, VA.

Bortone, G. (2007). "Sediment treatment – a general introduction." Sustainable management of sediment resources: Sediment and dredged material treatment, Elsevier, Amsterdam, 1-10.

Carbonaro, R. F., Gray, B. N., Whitehead, C. F., and Stone, A. T. (2008). "Carboxylate-containing chelating agent interactions with amorphous chromium hydroxide: Adsorption and dissolution." *Geochimica et Cosmochimica Acta*, 72, 3241-3257.

Ceremigna, D., Polettini, A., Pomi, R., Rolle, E., De Propris, L., Gabellini, M., and Tornato, A. (2005). Comparing sediment washing yields using traditional and innovative biodegradable chelating agents. 3rd International Conference on Remediation of Contaminated Sediments, New Orleans, LA, USA.

Chang, F. C., Lo, S. L., and Ko, C. H. (2007). "Recovery of copper and chelating agents from sludge extracting solutions." *Separation and Purification Technology*, 53, 49-56.

Chen, T.-C., Macauley, E., and Hong, A. (1995). "Selection and test of effective chelators for removal heavy metals from contaminated soils." *Canadian Journal of Civil Engineering*, 22, 1195-1197.

Colacicco, A., De Gioannis, G., Muntoni, A., Pettinao, E., Polettini, A., and Pomi, R. (2010). "Enhanced electrokinetic treatment of marine sediments contaminated by heavy metals and PAHs." *Chemosphere*, 81, 46-56.

Decreto Legislativo 3 aprile 2006, n. 152. Norme in materia ambientale (Environmental Regulation), Suppl. Ord. n. 96 Gazzetta Ufficiale Repubblica Italiana n. 88 del 14 aprile 2006 (in Italian).

De Gioannis, G., Muntoni, A., Polettini, A., and Pomi, R. (2008). "Enhanced electrokinetic treatment of different marine sediments contaminated by heavy metals." *Journal of Environmental Science and Health*, A 43, 852-865.

Dermont, G., Bergeron, M., Mercier, G., and Richer-Laflèche, M. (2008). "Soil washing for metal removal: A review of physical/chemical technologies and field applications." *Journal of Hazardous Materials*, 152, 1-31.

Detzner, H. D., Hakstege, A. L., Hamer, K., and Pallemans, I. (2007). Overview on treatment and disposal options. Sustainable management of sediment resources: Sediment and dredged material treatment, Elsevier, Amsterdam, 59-67.

Di Palma, L., Ferrantelli, P., Merli, C., and Biancifiori, F. J. (2003). "Recovery of EDTA and metal precipitation from soil flushing solution." *Journal of Hazardous Materials*, 103, 153-168.

Di Palma, L., and Mecozzi, R. (2007). "Heavy metals mobilization from harbour sediments using EDTA and citric acid as chelating agents." *Journal of Hazardous Materials*, 147, 768-775.

Elliott, H. A., Brown, G. A. (1989). "Comparative evaluation of NTA and EDTA for extractive decontamination of Pb-polluted soils." *Water, Air, and Soil Pollution,* 45, 361-369.

Elliott, H. A., Linn, J. H., and Shields, G. A. (1989). "Role of Fe in extractive decontamination of Pb-polluted soils." *Hazardous Waste and Hazardous Materials*, 6, 223-229.

Fangueiro, D., Bermond, A., Santos, E., Carapuça, H., and Duarte, A., 2002. "Heavy metal mobility assessment in sediments based on a kinetic approach of the

EDTA extraction: search for optimal experimental conditions." *Analytica Chimica Acta*, 459, 245-256.

Gyliene, O., Aikate, J., and Nivinskiene, O. (2004). "Recovery of EDTA from complex solution using Cu(II) as precipitant and Cu(II) subsequent removal by electrolysis." *Journal of Hazardous Materials*, 116, 119-124.

Hakstege, A. L. (2007). Description of the available technology for treatment and disposal of dredged material. Sustainable management of sediment resources: Sediment and dredged material treatment, Elsevier, Amsterdam, 68-118.

Hass, A., and Fine, P. (2010). "Sequential selective extraction procedures for the study of heavy metals in soils, sediments, and waste materials – a critical review." *Critical Reviews in Environmental Science and Technology*, 40, 365-399.

Hong, A., and Jiang, W. M. (2005). Factors in the selection of chelating agents for extraction of lead from contaminated soil: effectiveness, selectivity, and recoverability, Nowack, B., and Van Briesen, J.M. (eds.), Biogeochemistry of chelating agents, ACS Symposium Series, 910, 421-431.

Hong, A. K. P., Li, C., Banerji, S. K., and Regmi, T. (1999). "Extraction, recovery, and biostability of EDTA for remediation of heavy metal contaminated soil." *Journal of Soil Contamination*, 8, 81-103.

Kim, C., Lee, Y., and Ong, S. K. (2003). "Factors affecting EDTA extraction of lead from lead-contaminated soils." *Chemosphere*, 51, 845-853.

Kim, C., and Ong, S. K. (1999). "Recycling of lead-contaminated EDTA wastewater." *Journal of Hazardous Materials*, B69, 273-286.

Knepper, T. P. (2003). "Synthetic chelating agents and compounds exhibiting complexing properties in the aquatic environment." *Trends in Analytical Chemistry*, 22, 708-724.

Leštan, D., Luo, C. L., and Li, X. D. (2008). "The use of chelating agents in the remediation of metal-contaminated soils: A review." *Environmental Pollution*, 153, 3-13.

Lim, T. T., Tay, J. H., and Wang, J. Y. (2004). "Chelating-agent-enhanced heavy metal extraction from a contaminated acidic soil." *Journal of Environmental Engineering*, 130, 59-66.

McCready, S., Birch, G. F., and Taylor, S. E. (2003). "Extraction of heavy metals in Sydney Harbour sediments using 1M HCl and 0.05M EDTA and implications for sediment-quality guidelines." *Australian Journal of Earth Sciences*, 50, 249-255.

Meegoda, J. N., and Perera, R. (2001). "Ultrasound to decontaminate heavy metals in dredged sediments." *Journal of Hazardous Materials*, 85, 73-89.

Ministero dell'Ambiente (1993). Protocollo di intesa, Criteri di sicurezza ambientale per gli interventi di escavazione, trasporto e reimpiego dei fanghi estratti dai canali di Venezia (Act on environmental preservation measures in excavation, transportation and reuse of dredged sediments from Venice canals), 8 aprile 1993 (in Italian).

Mulligan, C. N., Yong, R. N., and Gibbs, B. F. (2001a). "Remediation technologies for metal-contaminated soils and groundwater: an evaluation." *Engineering Geology*, 60, 193-207.

Mulligan, C. N., Yong, R. N., and Gibbs, B. F. (2001b). "An evaluation of technologies for the heavy metal remediation of dredged sediments." *Journal of Hazardous Materials*, 85, 145-163.

Neaman, A., Walle, B., Mouélé, F., Trolard, F., and Bourrié, G. (2004). "Improved methods for selective dissolution of manganese oxides from soils and rocks." *European Journal of Soil Science*, 55, 47-54.

New Jersey Department of Environmental Protection (NJDEP) (2001). Certification Report For New Jersey Corporation For Advanced Technology (NJACT), Verification of Brice Environmental Services Corporation (Soil Washing Process).

Nowack, B., Kari, F. G., and Krüger, H. G. (2001). "The remobilization of metals from iron oxides and sediments by metal-EDTA complexes." *Water, Air, and Soil Pollution*, 125, 243-257.

Nowack, B. and Sigg L. (1996). "Adsorption of EDTA and metal-EDTA complexes onto goethite." *Journal of Colloid and Interface Science*, 177, 106-121.

Peters, R. W. (1999). "Chelant extraction of heavy metals from contaminated soils." *Journal of Hazardous Materials*, 66, 151-210.

Polettini, A., Pomi, R., and Calcagnoli, G. (2009). "Assisted washing for heavy metal and metalloid removal from contaminated dredged materials." *Water, Air, and Soil Pollution*, 196, 183-198.

Polettini, A., Pomi, R., Rolle, E., Ceremigna D., De Propris, L., Gabellini, M., and Tornato, A. (2006). "A kinetic study of chelant-assisted remediation of contaminated dredged sediment." *Journal of Hazardous Materials*, B137, 1458-1465.

Polettini, A., Pomi, R., and Rolle, E. (2007). "The effect of operating variables on chelant-assisted remediation of contaminated dredged sediment." *Chemosphere*, 66, 866-877.

Ritschel, J. (2003). "Extraction of heavy metals from soil with selected biodegradable complexing agents." Diploma Thesis, ETH Zürich.

Rulkens, W. (2005). "Introduction to the treatment of polluted sediments." *Reviews in Environmental Science and Biotechnology*, 4, 213-221.

Stumm, W., and Morgan, J. J. (1996). Aquatic Chemistry. 3rd ed., Wiley Interscience, New York.

Tandy, S., Bossart, K., Mueller, R., Ritschel, J., Hauser, L., Schulin, R., and Nowack, B. (2004). "Extraction of heavy metals from soils using biodegradable chelating agents." *Environmental Science & Technology*, 38, 937-944.

Tessier, A., Campbell, P. G. C., and Bisson, M. (1979). "Sequential extraction procedure for the speciation of particulate trace metals." *Analytical Chemistry*, 51, 844-851.

Tokunaga, S., and Hakuta T. (2002). "Acid washing and stabilization of an artificial arsenic-contaminated soil." *Chemosphere*, 46, 31-38.

U.S. EPA (United States Environmental Protection Agency) (1993a). Toronto Harbour Commissioners (THC), Soil Recycle Treatment Train. Applications Analysis Report. EPA/540/AR-93/517.

U.S. EPA (United States Environmental Protection Agency) (1993b). Selecting remediation techniques for contaminated sediments. Report EPA-823-B93-

001.

U.S. EPA (United States Environmental Protection Agency) (1994). Pilot-Scale Demonstration of Sediment Washing for the Treatment of Saginaw River Sediments, EPA 905-R94-019.

U.S. EPA (United States Environmental Protection Agency) (1995). Bergmann USA Soil Sediment Washing Technology. Applications Analysis Report. EPA/540/AR-92/075.

U.S. EPA (United States Environmental Protection Agency) (1998). EPA's contaminated sediment management strategy. EPA 823/R-98-001.

U.S. EPA (United States Environmental Protection Agency) (2004). The Incidence and Severity of Sediment Contamination in Surface Waters of the United States, National Sediment Quality Survey. 2nd ed., Report EPA-823-R-04-007.

U.S. EPA (United States Environmental Protection Agency) (2005). Contaminated sediment remediation - Guidance for hazardous waste sites, EPA-540-R-05-012.

Williams, D. R. (2005). Speciation of Chelating Agents and Principles for Global Environmental Management. Biogeochemistry of Chelating Agents, Nowack, B., and Van Briesen, J.M. (eds.), ACS Symposium Series, 910, 20-49.

Yang, Y., Ratté, D., Smets, B. F., Pignatello, J. J., and Grasso, D. (2001). "Mobilization of soil organic matter by complexing agents and implications for polycyclic aromatic hydrocarbon desorption." *Chemosphere*, 43, 1013-1021.

Yuan, S., Wu, X., Wan, J., Long, H., Lu, X., Wu, X., and Chen, J. (2010). "Enhanced washing of HCB and Zn from aged sediments by TX-100 and EDTA mixed solutions." *Geoderma*, 156, 119-125.

Zeng, Q. R., Sauvé, S., Allen, H. E., and Hendershot, W. H. (2005). "Recycling EDTA solutions used to remediate metal-polluted soils." *Environmental Pollution*, 133, 225-231.

Zhang, W., Huang, H., Tan, F., Wang, H., and Qiu, R. (2010). "Influence of EDTA washing on the species and mobility of heavy metals residual in soils." *Journal of Hazardous Materials*, 173, 369-376.

Zou, Z., Qiu, R., Zhang, W., Dong, H., Zhao, Z., Zhang, T., Wei, X., and Cai, X. (2010). "The study of operating variables in soil washing with EDTA." *Environmental Pollution*, 157, 229-236.

CHAPTER 3

Operational Conditions of Chelant-Enhanced Soil Washing for Remediation of Metal-Contaminated Soil

Weihua Zhang and Rongliang Qiu

3.1 Principle of Soil Washing

Soil washing is an ex-situ aqueous-based soil remediation technology by separating contaminants from bulk soil via one of the following two ways: dissolving or suspending them in the washing solution, or concentrating them into a smaller volume of soil through particle size separation, gravity separation, and attrition scrubbing (FRTR, 2006). The concept of the latter one is based on the findings that most contaminants tend to bind, either chemically or physically, to clay, silt, and organic soil particles (Zimdahl and Skogerboe, 1977; Raghavan, 1989).

3.2 General Procedure and Application of Soil Washing

In general, soil washing involves five steps in sequence: (1) excavation of contaminated soils; (2) remedial treatment of contaminated soils; (3) solid/liquid separation; (4) treatment or disposal of the residues; and (5) final soil deposition, as shown in Figure 3.1. To protect the facilities or polish the running, large debris or soil particles coarser than 2 mm are often separated in the pretreatment after excavation. The soil then enters a soil scrubbing unit, where fine soil particles are scrubbed from coarse and sand soil particles. If the fine silt and clay in the contaminated soil accounts for a modest percentage of the bulk soil (i.e. less than 30%), the soil washing process with particle separation is a promising technology in terms of cost-effectiveness, as a large percentage of relatively clean coarse soil is recovered, while the cost for the disposal of all contaminated soil without separation may be higher or equal to the sum of the cost related to the disposal of residues and capital investment on soil washing facilities (Willichowski, 2001).

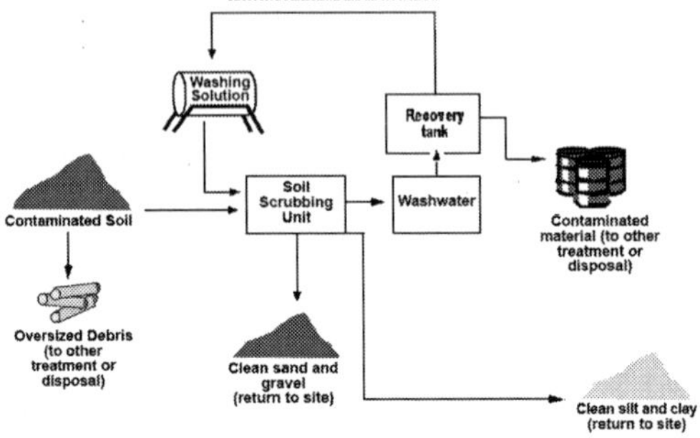

Figure 3.1 Flow chart of soil washing.

Besides particle separation, soil washing in combination with other contaminant transfer and transform techniques offers the greatest promise for application to soils contaminated with a wide variety of heavy metals, radio-nuclides, and organic contaminants (i.e., TPHs, PAHs, PCBs, and pesticides). In addition, the technology has the ability to recover target contaminants from soils. Moreover, the short to medium duration and moderate cost (\$70-187 m^{-3} for physical separation, and \$358–1717 m^{-3} for chemical extraction (FRTR, 2007), make it feasible for the remediation of industrial sites, which usually needs a fast remediation for later commercial development.

However, commercialization of the technology is not yet extensive. The first large-scale soil washing operated in the United States was completed in October, 1992 at the King of Prussia Technical Corporation site in New Jersey (Van Benschoten et al., 1997). Soil washing method has been implemented at only a few Superfund sites contaminated with metals: at King of Prussia, NJ, at Twin Cities Army Ammunition Plant, MN, as well as at Vineland Chemical Co., NJ (ASR, 2007). Several Superfund programs that were planned to apply soil washing currently deselected it due to the cost or other reasons, as listed in Table 3.1.

The popularity of soil washing may be limited by soil contamination that involves a variety of co-existing contaminants such as metals, total petroleum hydrocarbons, volatile organic compounds, and semi-volatile organic compounds. In addition, heterogeneous nature of soil textures further makes it difficult to formulate a standardized washing solution that is able to provide consistent and reliable performance in contaminant removal. Therefore, a site specific chemical formula, washing sequence, and operational conditions may be required. Besides, the applicability and performance may be limited by other factors including (1) high humic content in soils; (2) aqueous stream; (3) elevated concentration of contaminants in the treated residuals; and (4) difficulty in removing organics adsorbed onto clay-size particles.

Table 3.1 Soil washing demonstration projects applied or planned to apply on heavy metal-contaminated Superfund sites (data from FTRT, 2007)

Site location	***Washing agents***	***Target metals***	***Disposal or treatment***	***Notes***
Ewan Property, NJ	Clean water	As, Cr, Cu, Pb	Solvent extraction of organic contaminants	Soil volume was much smaller that originally projected
GE Wiring Devices, PR	KI solution	Hg	Return the clean soil	Because cost of soil washing was too high, it has been changed to incineration.
King of Prussia, NJ	Chelating agents	Ag, Cr, Cu	Sludge treatment	Completed
Zanesville Well Field, OH	Clean water	As, Cr, Hg, Pb	Return the clean soils	Due to soil volume being much smaller that originally projected; determined to be too expensive, the soil washing was deselected.
Vineland Chemical Co, NJ		As	Off-site disposal for filter cake residue	Completed
Twin Cities Army Ammunition Plant, MN	acid	Cd, Cr, Cu, Hg, Pb	Soil leaching	Completed

3.3 Chelant-Enhanced Soil Washing

Although the transfer of contaminants from soils to aqueous phase may occur when pure water-based washing is employed, the removal of contaminants is often incomplete or slow because of the poor dissolution and mobilization of contaminants, especially in the aged contaminated sites, where the soil solid phases often have a strong affinity with the contaminants (Reddy et al., 2000; Pichtel et al., 2001; Finzgar and Lestan, 2007; Zhang et al., 2008; Zou et al., 2009). Since some chemical agents such as surfactants, chelating agents, and acids, can significantly increase the dissolution and mobilization of contaminants, thus improving their release, the chemical-enhanced soil washing emerges as a more promising technique for the remediation of these soils in a more cost-effective scheme (Tobia, 1993; Paff, 1994;

Roy et al., 1997; Davis and Singh, 1995; Griffiths, 1995; Robin and Krishna, 1996; Abumaizar and Smith, 1999; Saponaro et al., 2002; Chu, 2003; Tandy et al., 2004).

The chemicals used to enhance the separation between soil particles and contaminants vary depending on the contaminant type and binding characteristics. Potential chemical agents include acids, bases, surfactants/co-solvents, chelating agents and reducing agents. The surfactants and solvents often effectively facilitate the removal of organic contaminants in soils by increasing the apparent solubility of contaminants. Both strong acids and chelating agents are able to enhance the mobility of heavy metals in the soils and thus improve their removal. Unlike the strong acids that always destroy the soil structure, chelating agents can form various complexes with heavy metals, which is soluble and stable in solution. This chelating effect substantially increases the nominal solubility of heavy metal in the washing solution and is, therefore, conducive to removal of heavy metals from soils (Pichtels, 1997).

There are five major criteria in the selection of chelating agents for soil washing to remove heavy metals: (1) the selected chelating agent should be able to form highly stable and appreciably soluble complexes over a wide pH range; (2) the biodegradability of the chelating agent and metal complexes should be slow, if the chelating agent is to be recovered and reused; (3) the formed metal complex should be non-sorbable on soil surfaces; (4) the chelating agent should have a low toxicity and minimal impact on environment; and (5) the chelating agent should be cost-effective (Peters and Shem, 1992).

3.4 Factors Influencing Chelant-Enhanced Soil Washing

3.4.1 Soil Properties

Soil properties significant influence the applicability of soil washing technology. As mentioned above, the soil washing typically is applicable for sandy soils, since heavy metals often concentrate in the silty and clayey fractions, which have a strong affinity with heavy metals. After particle size separation, a great amount of clean sands can be obtained, and the volume of soil required for further treatment significantly decreases. If the percentage of silts and clays (< 63 μm) exceeds approximately 30% in mass, the cost for the disposal of all contaminated soil without separation may be less or equal to the sum of the cost of the residual disposal and capital investment of soil washing. Consequently, the soil washing is economically acceptable in treating soil with less than 30% of silt and clay (Willichowski, 2001).

Besides silt and clay, the organic matter content and pH buffer strength of aimed soils will also limit the application of soil washing, because the organic matter often has a strong complexation with heavy metals, and the high pH buffer strength will potentially compromise the proton-attack process, which is one of the main mechanisms for contaminant removal in soil washing.

3.4.2 Species and Properties of Contaminants

The properties and species of target heavy metals also play a key role in the performance of soil washing. In general, heavy metals are retained in soils by different mechanisms including electrostatic ion-exchange, inner-sphere complexation on the oxides of soil minerals or organic matters, surface precipitation or co-precipitation. Metals vary in the extent to which they are adsorbed by these mechanisms, and soil constitutes also show differences in selectivity for metals.

In general, the metal ions with smaller hydrated radius and/or greater charge tend to exchange with other ions or preferentially adsorb on a surface because of stronger electrostatic attraction and higher proximity to the particle surface (Impellitteri et al., 2001; Santamarina et al., 2001). Metals with the greater electro-negativity and/or Lewis acid softness often form the stronger covalent bonds with oxygen atoms on any particular mineral surface or with soft functional groups of organic matter. Pb, for example, was less labile in the soil than Zn. In addition, the aging time and heavy metal loading also affect the fate of heavy metals in soils (McBride, 1994), and less labile fraction is expected with the increased loading and aging.

Sequential extraction procedures have been widely applied to estimate the metal species and fractions since Tessier et al. (1970) initially developed them. In general, the metal retained by electrostatic ion exchange and bound to carbonates, which is readily extracted by acetic acid, is considered as the labile fraction that is of great concern in terms of ecologic risk.

3.4.3 Operational Conditions for Chelant-Enhanced Soil Washing

Heavy metal removal by chelating agents is known to be highly influenced by the sorption behavior of heavy metal on soil, which is dependant on the pH, CEC, the constituents of soils, the original speciation of heavy metals in solution, the degree of aging and heavy metal loading, as discussed in Section 3.4.1. In general, weakly-bound heavy metals are relatively easy to be removed by chelating solution. Due to the different chelating capability, metal type also affects its removal efficiency. The chelating capability can be preliminarily evaluated by examining the stability constants of heavy metal-chelant complexes. In engineering practice, the operating conditions such as contact time, pH, chelatant concentration as well as the electrolyte added in washing solution are adjustable. Therefore, studies on the effects of different operational conditions are essential to improve the performance of chelant-enhanced soil washing.

Chelating Agent

Chelating agents are chemicals that form soluble, complex molecules with certain metal ions, inactivating the ions, so that the metals cannot normally react with other elements or ions to produce precipitates or scale. In general, common chelating agents include natural organic compounds, and synthetic chelating agents as well as some inorganic salts.

Natural organic acids often contain a single or several carboxyl groups, which is prone to supply electrons to form metal complex. The commonly used organic acids in soil washing include citric acid, tartaric acid, oxalic acid and acetic acid. These organic acids have distinct priority for different metals. For example, citric acid has selectivity for metal extraction in the following order: Cd > Cu > Zn > Hg >> Pb ~ As > Cr ~ Fe, while the priority of oxalic acid and salts is as follows, Hg >> Cu > Cd > Zn, As > Cr, and Fe > Pb. Both of them have priority over the Cd complex, as evidenced by the higher removal efficiency for Cd using these two acids than EDTA. Zou et al. (2008) found that 0.1 mol L^{-1} oxalate achieved the highest removal of As (54%) and Cd (49%) among all the studied washing agents, including strong inorganic acids and EDTA, because oxalate is able to promote greater Al/Fe/Mn dissolution than that obtained by an equivalent EDTA through ligand-promoted oxide dissolution and reductive dissolution mechanisms.

As an alternative, many synthetic chelating agents are developed to overcome the limited chelating capacity of natural organic acids. These synthetic chelants include ethylene diaminetriacetic acid (EDTA), nitrilo-triacetic acid (NAT), trans-1,2, cyclohexanediaminetetraacetic acid (CDTA), ethylenediaminedisuccinic acid (EDDS), methylglycinediacetic acid (MGDA), glycoletherdiamine tetraacetic acid (EGTA), N-2 acetamido-iminodiacetic acid (ADA), pyridine-2, 6-dicarboxylic acid (PDA), ethylene diamine N,N'-bis-2-hydroxyphenyl acetic acid (EDDHA), and so on. All of these chelating agents have multiple carboxyl groups, which can supply electrons to complex ions.

Among them, EDTA and its salts have been studied. It was reported that EDTA can appreciably increase the dissolution and mobility of heavy metals (Peters, 1999; Bassi et al., 2000; Tandy et al., 2004; Xia et al., 2009), rendering the target heavy metals a low aquatic toxicity and bioaccumulation risk to organisms throughout the food chain (ECB, 2004). Under a wide pH range, EDTA can form a soluble and mobile complex with most metal cations found in soils, thus increasing their mobility. However, at the insufficient dosage, EDTA shows the selectivity for the metals in the following order: Cu > Pb ~ Zn > Cd ~ Hg >> Cr ~ Fe ~ As (Peters, 1999).

In general, other synthetic chelating agents, such as PDA, DTPA, EGTA, and NTA, have lower removal efficiencies than EDTA, and in turn often require a longer contact time or higher concentration to obtain an equivalent metal removal (Huang et al., 1997; Maculey and Hong, 1995). However, some agents may have some priority over contain target metals, and result in a better removal performance than EDTA. For example, the extraction capacity is DcyTA> EGTA> EDTA >NTA in the Cd-contaminated soil (Hong and Pintauro, 1996). On the other hand, EDDS has become another focus in the soil washing recently due to its bio-degradability and comparable extraction efficiency to EDTA (Yip et al., 2010).

Besides the organic compound containing carboxyl groups, other inorganic ligands, such as ammonia, or iodide and chloride can also form complexes with some target metal cations. Reddy et al. (2003) and Zou et al. (2008) both found that KI can result in a substantial removal of Hg from contaminated soils. Cline and Reed (1995) found that 36% of Pb can be mobilized by 0.1 mol L^{-1} $CaCl_2$ through cation exchange and complexation between Pb^{2+} and Cl^-. Ammonia citrate and acetate can increase the mobility of Cu and Pb in soils through complexation with ammonia and carboxyl ions, and they have the selectivity for metals in the following order Cd >> Zn >> Hg ~ Cu ~ Pb > Cr ~ As > Fe. The Cd removal efficiency was even higher than that using EDTA in some cases (Peters, 1999).

Dosage of Chelating Agents

In general, chelants are ligands that contain one or several electron-donor groups which can forms 1:1 molar ratio complexes with several metal ions. Many studies used the molar ratio of chelant:metal (i.e., [EDTA]/[Me]) to represent the 1:1 molar ratio stoichiometric dose of chelating agent. In many cases this ratio was calculated from the concentration and volume of chelating agent and the mass of soils (Tandy et al., 2004). However, many chelants are nonspecific chelating agents, as major compositional elements in soils are also extractable and they may compete against the target metals over chelating agents. The dosage of chelating agents must therefore exceed that for the target heavy metal presented in soils. Not surprisingly, the ratio in the reported literature was always much larger than 1, since this calculated ratio excluded other metals and major soil cations like Ca, Fe, Mg, Mn, etc. Elliott and Brown (1989) found that EDTA could virtually extract all of the non-detrital Pb when at least a stoichiometric amount of EDTA was employed. Increasing EDTA above the stoichiometric requirement could cause more Pb release, and the maximum removal of Pb occurred at the molar ratio of EDTA to Pb between 1.5 and 2.5; however, when the complete release of Pb was approached, the release of Pb diminished with each incremental EDTA concentration. Papassiopi et al. (1999) found the highest metal removal by 0.25 mol L^{-1} Na_2-EDTA with two-step washing occur at an EDTA/Me ratio of 20. The excessive EDTA was inferred to have promoted the dissolution of aluminum and iron oxides in soils by formation of soluble Al- or Fe-EDTA complexes (Scheckel and Sparks, 2001).

If the liquid/soil ratio is fixed, the dose is directly correlated to the concentration of chelating agent. Qiu et al. (2010) found that the removal of five cation metals increased with increasing EDTA concentration from 0.005 to 0.1 mol L^{-1} (liquid/solid 20) but the increase was not linear; and EDTA concentration had little additional effect on the removal efficiency of heavy metals, when the concentration exceeded 0.02 M. This was consistent with the findings of Elliott and Brown (1989) and Steele and Pichtel (1998). Nevertheless, Papassiopi et al. (1999) found that a quite different behavior that the improvement of metal removal was almost linear with increasing EDTA concentration.

On the other hand, if the concentration of EDTA is fixed, the dose is directly correlated to the liquid/soil ratio. Qiu et al. (2010) found that increasing the liquid/soil ratio progressively promoted the metal extraction. The removal of Cd, Cu and Pb increased as the ratio increased from 5 to 40 (0.01 mol L^{-1}), then no substantial increase was observed at the larger ratio after which almost all the mobile metals had been extracted.

Contact Time

In general, chelating agents can enhance the metal mobilization by two different mechanisms: a fast thermodynamically favorable complexation between target metals and chelating agents, and a slow chelant-promoted dissolution. The former can directly break down some weak bonds between metals and soils, while the latter can indirectly mobilize metals that are bound to oxides and organic matter through partially destructing the soil structure. The short-term mobilization of metals is dominated by the most labile species, while the long-term removal is determined by the replenishment of the labile pool from more recalcitrant species (Bermond and Ghestem, 2001; Kirpichtchikova et al., 2006). Some studies on the artificially contaminated soils, where metals were weakly associated with soil fractions due to limited chemical weathering, found that metal extraction by chelating agent was rapid, reaching equilibrium in a relatively short time (Lo and Yang, 1999). The removal can reach a relatively high level at a given time, and then no additional appreciable removal was identified with further increasing the contact time (Zhang and Lo, 2006). The rate of extraction in the rapid phase was reported to be highly pH dependent, as found in the study of Yu and Klarup (1994), and a lower pH led to a faster extraction rate.

However, in real contaminated soil, a significant portion of metals are strongly bound with soils, and their desorption mainly is via ligand-promoted dissolution, which starts by a fast adsorption of free or complexed chelating agent onto specific surface sites, and then is followed by a rate-limiting metal detachment from the oxide structure. Therefore, heavy metal extraction process by chelating agents consists of the rapid desorption in the first minutes to hours followed by a slow release of the remainder (Qiu et al., 2010). Therefore, the extraction time plays a very important role in soil washing. A longer contact time can facilitate the detachment of the chelant-stabilized metals from the soil surface, and reduce their residual concentration in the washed soil (Zhang et al., 2010). However, a longer contact time usually requires an increase in the volume of reaction containers or a decrease in their treatment capacities.

Washing Mode

The washing mode also influences the performance of soil washing. For a certain dose (moles) of chelating agent, a higher concentration is often accompanied with lower liquid/soil ratio, or a lower concentration with a higher liquid/soil ratio, or a lower concentration with more washing cycles. Using a low concentration with a higher liquid/soil ratio was found to more effectively mobilize the target metals than a high concentration with a low liquid/soil ratio (Zou et al., 2009). A higher liquid/soil ratio often results in a low metal-chelant complex concentrations in the washing solution, and thus accelerate the detachment of metal-chelant complex from the soil surface and also limit their re-adsorption(Strawn and Sparks, 2000).

Similarly, consecutive extractions using low concentrations are more effective than using a single extraction with concentrated EDTA under the same EDTA dosage (Moutsatsou et al., 2006; Finžgar and Leštan, 2007; Qiu et al., 2010). Besides the above mentioned reasons, the renewal solution can also remove some residual metal-chelants which may be trapped in the pore solution of the soil (Sun et al., 2001) or re-adsorbed onto the soil particles during the first washing of the sequential extraction procedure. Moreover, the initial step(s) may remove the major portion of the competitive cations, and the interference in metal–chelant complex formation is thus mitigated in the following steps (Polettini et al., 2007). As a result, consecutive extraction with renewal solution often leads to an increase in metal desorption.

However, as the rounds of successive extractions increased, only a small percentage of the total metals were extracted. After the previous extractions, where the mobile forms were extracted, the metal release from the less mobile forms (i.e., bound to organic matter, bound to Fe/Mn oxides or residual fraction) occurred at a much slower rate. This is presumably related to the dissolution of mineral constituents of soil (including oxides and silicates) initially retaining the contaminants (Polettini et al., 2007).

Using a lower concentration with a high liquid/soil ratio has some advantages. Andrade et al. (2007) considered that reducing the liquid/soil ratio and increasing the concentration of the washing solution could increase the extraction efficiency of trace metals and produce smaller amounts of leachate. Furthermore, the toxicity of chelating agent and its complexes for soil microorganisms and plants was lower in soil treated with a lower concentration of chelating agent (Grcman et al., 2001).

pH and Electrolytes of Washing Solution

Solution pH is another important factor determining the efficiency of chelant-enhanced soil washing, because it can influence metal retention by the soils as well as the capability of chelating agent to extract the contaminants from soil through different mechanisms (Peters, 1999). The mechanisms include: (1) the solubility of chelating agent; (2) chelating capacity of chelating agent with metal ions in the solutions; (3) trace metal sorption/desorption and ion exchange (Kim and Ong, 1999).

When the pH in 0.01 M Na_2EDTA solution is less than 3, Na_2EDTA becomes less soluble. Naturally, improving the solubility and mobility of Na_2EDTA at a proper pH during the extraction will promote the formation of stable and soluble complexes with heavy metals (Pichtel and Pichtel, 1997; Qiu et al., 2010).

Due to the competition of OH^- and dissociation of chelating agent, the thermodynamic capacity of chelating agent to complex free metal ions varies under different pH. It can be evaluated using the conditional stability constants of metal-chelant complexes. Figure 3.2 provides a comparison of the conditional stability constants for various metal-EDTA complexes as a function of pH. In general, at high pH (10 or higher), the hydrolysis of heavy metals is favored over the complexation with chelating agent, so the conditional stability constants of metal-EDTA complex decrease, suggesting a decrease in the ability of chelating agent to enhance metal solubility (Peters, 1999).

Besides, the different pH sensitivity of metal-chelant complexes also changes their priority. Based on the conditional stability constants illustrated in Figure 3.2, the tendency for metal-EDTA is $Fe^{3+} > Cu^{2+} > Pb^{2+} > Al^{3+} > Zn^{2+}$ under low pH conditions (less than 3). EDTA will prefer to complex with Fe^{3+} over the other cations at that pH. Under high pH conditions (e.g. pH 7), however, Pb(II)-EDTA and Zn(II)-EDTA would be dominant over the Fe(III)–EDTA complex.

In addition, metals bound to soil hydrous oxides can often be retrieved simply by lowering pH because protons can promote oxide dissolution. Hydrogen ions are rather weak competitive cations which can replace the adsorbed heavy metals via a cation exchange mechanism. As the H^+ ion concentration increases, the particle surface becomes increasingly protonated and acquires a positive charge, which will promote desorption of metals (Peters, 1999).

The combined effects of these mechanisms generally lead to a higher extraction efficiency under a lower solution pH. Van Benschoten et al. (1997) found EDTA addition significantly improved soil-washing performance at pH 3. Elliott et al. (1989) also observed the highest removal was generally achieved at pH < 5. A substantial increase in Pb release was observed below pH 5 even in the absence of EDTA. Zou et al. (2009) found that the removal of five metals decreased with the increased pH from 2.0 to 10.0, and the highest efficiency was obtained at pH 1.

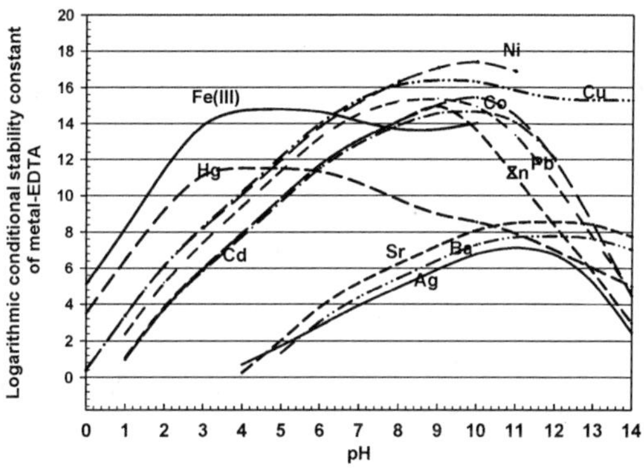

Figure 3.2 Conditional stability constants of metal-EDTA complexes (data from Lo and Zhang, 2005).

However, in practice the pH is normally controlled within the range of 5-9 in order to avoid the damage of soils under too acid or alkaline conditions. More importantly, an acid environment favored the release of Fe from ferric oxides in soils via ligand-promoted dissolution. As a result, the enhanced mobility of target heavy metals would be depressed (Nowack and Sigg, 1997), since chelating agents, such as

EDTA, thermodynamically prefer to complex with Fe^{3+} over other cations under such pH conditions.

Besides pH, Brown and Elliott (1992) found that the presence of electrolytes such as sodium, lithium or ammonium perchlorate in the EDTA washing solution increased the release of Pb at pH 5-9. Divalent electrolyte, calcium or magnesium perchlorate resulted in a similar improvement in Pb release at an acidic pH, but suppressed Pb mobilization in a more alkaline media (Theodoratos et al., 2000).

3.5 Risk Assessment and Control of Chelant-Enhanced Soil Washing

3.5.1 Metal Mobility in the Washed Soils

As reported in the previous studies (Nowack and Sigg, 1997; Sposito, 2004; Tsang et al., 2007), chelating agents can enhance the metal mobilization by two mechanisms: a fast, thermodynamically favorable complexation between chelating agents and the labile metal fractions, and a slow ligand-promoted dissolution for less labile metal fractions, which are often bound to oxides and organic matter. In aged contaminated sites, the majority of the metal species were bound to oxides or organic matter, ligand-promoted dissolution can play a substantial role in the overall metal removal under the chelating agent adequacy (Yip et al., 2009). Ligand-promoted dissolution itself also includes two steps: a fast adsorption of free or complexed chelating agent onto specific surface sites via surface complexation, which can destabilize the metal-oxygen bonds within the mineral structure, and a rate-limiting metal detachment from the oxide structure. Therefore, the kinetics of metal detachment may potentially increase the mobility of chelant-destabilized metals in the washed soil.

Many studies investigated this risk of soil washing, and found that the metals in the washed soils became labile or more weakly-adsorbed (Barona et al., 2001; Lei et al., 2008). Some research showed that the mobility of some residual heavy metals, i.e. Pb, increased after seven days abiotic aging (Udovic and Lestan, 2009). Zhang et al. (2010a) found that the soil washing combination of 0.0005 M EDTA and half an hour-washing increased the instant mobility of Ni, Zn and Pb of a contaminated soil possibly owing to the slow detachment of EDTA-destabilized metals. Metal fractionation also exhibits the corresponding increase in their labile exchangeable fractions. Therefore, a more concentrated EDTA solution for a longer duration can completely destabilize the strongly bound metals into the aqueous solution, resulting in the decreases of their mobility in washed soils. Therefore, we should prudently select the chelating agent concentration and washing duration to control the mobility and availability of the remaining heavy metals in the system design.

3.5.2 Soil Dissolution during Washing

Due to non-specificity or non-selectiveness of the added chelating agent, some soil componential metals, such as Ca, Mg, Mn, Al and Fe, are often released together with the target metals. In fact, the soil dissolution is also an important mechanism involved in the target metal removal, especially at a low molarity ratio of the target metals and chelating agent.

Zhang et al. (2010a) investigated the soil dissolution under different EDTA concentrations and different contact times in a laboratory test, and found that the most significant soil compositional element released during EDTA-enhanced soil washing was Ca, while other elements (i.e., Fe, Mg, Al, and Mn) were less substantial. A similar phenomenon was observed by other researchers (Papassiopi et al., 1999; Theodoratos et al., 2000; Polettini et al., 2007), who also identified the tremendous amount of Ca dissolution from calcareous soils. The Ca dissolution was likely ascribed to the proton-promoted dissolution of Ca (pH 4.7~5.2 for 0.005-0.0005 M EDTA solution) and the cation exchange between Ca^{2+} and Na^{+} in the solution (Palma, 2009). Significant amount of dissolved Ca can then quickly complex with the free EDTA to form Ca-EDTA. The slow kinetic process of metal exchange reaction between the Ca-EDTA and the sorbed target metals on the soils significantly diminished the EDTA extraction capability for target metals (Nowack and Sigg, 1997; Papassiopi et al., 1999; Theodoratos et al., 2000; Tsang et al., 2007). The low levels of Al, and Mn and Fe in the solution revealed that the EDTA-promoted dissolution was limited, since EDTA-promoted dissolution could considerably increase the Al, Mn and Fe release from the soil (Tsang et al., 2007). Therefore, the contribution related to EDTA-promoted dissolution on the overall metal removal was also limited.

An increase in pH during soil washing has been reported in many studies (Tsang et al., 2007; Zou et al., 2009; Zhang et al., 2010a), and such an increase indicates the limited kinetic reaction between EDTA solution and some soil compositional metals, such as Al and Fe, since the dissolution of these oxides always consumes the bulk of H^{+} in the solution. Moreover, Zhang et al. (2010a) found a higher concentration of EDTA solution was consistently accompanied by more soil componential elements in the solution, suggesting that the extent of soil dissolution was dependent on EDTA concentration.

3.5.3 Recovery of Washing Solution

One of the main drawbacks of soil washing is the vast volume of wastewater produced, which contains the mobilized metallic species that have been complexed with chelating agents and must subsequently be treated before it can be safely discharged. The high cost of chelating agents is another factor hindering its wide application. Some chelating agents, such as EDDA, are biodegradable, and thus the extracted metals will be dissociated and recoverable, and is ready to be recovered. On the other hand, some chelating agents are biologically persistent and not ready to be biodegraded when it is chelated with heavy metals (Davis and Green, 1999), such as EDTA and NTA. However, based on the risk assessment report on EDTA approved by the Technical Meetings of European Union member state representatives(ECB, 2004), EDTA was reported to have a low aquatic toxicity and no bioaccumulation in living organisms through the food chain. The concerns about EDTA application mainly arise from the enhanced mobility of heavy metals, eutrophication and nutrient deficiency of plants. The use of EDTA should be discouraged in those sites that use large volumes of EDTA in the absence of effective treatment of wastewater. These problems necessitate the recovery and treatment of these chelating agents after soil washing.

Recently, different strategies have been proposed to recover chelating agents (such as EDTA), including electrochemical procedures, metal precipitation and chelating agent regeneration via addition of the chemical agents.

Electrochemical Procedures

Allen and Chen (1993) presented the concept of electrolytic separation of metals and chelating agents in the soil washing solution. A cell was used allowing the anode and cathode compartments to be separated by a cation exchange membrane, to prevent electrochemical oxidation of EDTA during electrolysis. Tests with Pb-EDTA demonstrated that over 95 % recovery of both EDTA and lead could be achieved by electrolysis. Juang and Wang (2000) and Huang et al. (2000) also applied electrolysis in conjunction with a cation-exchange membrane for the EDTA recovery from metal–EDTA complexes. Their results showed that the recovery of metal and EDTA was approximate 99% and 91%, respectively. However, the pH in the cathode compartment was found to increase with time as a result of hydroxyl ion production, which caused the degradation of the membrane (Allen and Chen, 1993).

Precipitation

Chelating agents can also be recovered by acidifying the metal-chelant complex solution or precipitating metals with hydroxide, sulfide, or oxalate. As mentioned in Section 3.4.3, the solubility of chelating agents is pH dependent, and an acidic environment can reduce the aqueous solubility. Therefore, the concentrated chelating agent solution precipitates due to over-saturation after being acidified. Palma et al. (2003) precipitated more than 93% of the used EDTA in the washing effluent of Pb or Cu "artificially" contaminated soils, after an initial evaporation process that concentrated the diluted metal-EDTA by reducing about 75% of solution volume. This method is practically feasible for a concentrated EDTA solution rather than a dilute EDTA solution, as the latter requires a costly concentrating process with a high energy consuming operation; otherwise, a pre-concentration treatment is required.

Moreover, metal-EDTA complexes can be dissociated by precipitating metals as a hydroxide, sulfide, or oxalate based on the stabilities of metal-EDTA complexes and produced precipitates. For example, the EDTA in Fe-EDTA complex can be completely dissociated and recovered by $Fe(OH)_3$ precipitation under a moderately alkaline condition, such as pH 10.0 (Lo and Zhang, 2005). Rudd et al. (1995) found sulfide precipitation could reduce the lead concentration to an acceptable level in the EDTA washing effluent. However, this method is only effective for few metal complexes due to the high stability of metal-EDTA complexes. The precipitation on the Zn-EDTA solution using hydroxides or sulfides is thermodynamically ineffective (Lo and Zhang, 2005). Therefore, trans-complexation processes are often conducted before metal precipitation. The trans-complexation typically involves three steps: metal substitution by Fe^{3+}, Mg^{2+} or Ca^{2+} in chelating agent complexes, target metal precipitation, and precipitation of substitutive metals. Fe^{3+} is the most commonly used substitutive metal for EDTA recovery, because the conditional stability constant of Fe(III)-EDTA is larger than most other heavy metal-EDTA complexes (i.e. such as Cu, Zn, Ni, Co, Cd and Pb) at pH 2-4, thus Fe can successfully replace most heavy metals in EDTA complex, and can also be easily dissociated as $Fe(OH)_3$ precipitates in an alkaline environment (Kim and Ong, 1999; Lo and Zhang, 2005; Lim et al., 2005). Lim et al. (2005) reported that 85%, 89%, and 90% of the extracted Pb, Cd, and Ni were recovered from EDTA solution, respectively. The reused EDTA was slightly less efficient than fresh EDTA in soil washing for metal removal (Lo and Zhang, 2005). Ager and Marshall (2003) applied zero-valent Mg and Pd to substitute metals in EDTA complexes. Jiang et al. (2008) developed a simple $Ca(OH)_2$-based replacement-precipitation process, and applied it to the removal of EDTA-chelated copper in wastewater. Their experimental results showed that the $Ca(OH)_2$-based replacement-precipitation process could efficiently remove the chelated copper in wastewater at a pH of 12~13 and a molar ratio of Ca^{2+} to Cu(II) more than 2, the removal efficiency could reach 99%, close to the theoretically predicted results. Although trans-complexation allows partial EDTA recycling, it involves consumption of other chemicals and does not provide a cost-effective solution for the final treatment of EDTA extracts.

Fu et al. (2007) developed a new dithiocarbamate-type heavy metal precipitant, sodium 1,3,5-hexahydrotriazinedithiocarbamate (HTDC), and found that it can effectively precipitate copper to less than 0.5 mg L^{-1} from both synthetic and real industrial wastewater containing Cu-EDTA in the pH range of 3~9 by forming a kind of coordination supramolecular compound, $[Cu_3(HTDC)_2]^n$. The toxicity characteristic leaching procedure and semi-dynamic leaching test indicated that the supramolecular precipitate was non-hazardous and stable in weak acid and alkaline conditions.

Pre-concentration Technology

In order to improve the recovery efficiency, washing solutions containing metal complexes are often concentrated by membrane filtration, ion exchange with resin, or evaporation before the recovery treatment. Beside evaporation used by Palma et al. (2003a), Tejowulan and Hendershot (1998) proposed a simple procedure to transfer negatively charged metal-EDTA complexes from the washing effluent using an anion exchange resin. However, the high price of anion exchange resins (more than $100 per 500 g) becomes a constraint that has to be overcome by effective resin recycling. Palma et al. (2003b) proposed reverse osmosis for the reduction of the extract volume. But colloidal particles (clays and humic materials) and bacteria could clog the membranes and thus comprise the performance and shorten the life-time of the membranes.

Jiang (2007) used reverse osmosis or ultra-filter to concentrate the metal-EDTA solution before the replacement-precipitation treatment. His research showed that ultra-filtration is ineffective for Cu-EDTA concentration, but it can function as the pretreatment to remove the particles or colloids. The reverse osmosis was found to not only lower the electrical conductivity by effectively removing the soluble inorganic salts, but also to effectively intercept the EDTA complex. When the operating pressure is 1.20 MPa and volume concentration ratio is 9.92, the residual Cu^{2+} concentration in effluents could be reduced from 2 mg L^{-1} to 0.012 mg L^{-1}

3.5.4 Treatment of Chelating Agents in the Effluent

To treat decontaminated wastewater from the nuclear industry and other aqueous effluents contaminated with EDTA, the chemical destruction of EDTA and its complexes using advanced oxidation processes (AOP) with H_2O_2, Fenton's reagent or Fenton-like reagent has been proposed (Pirkanniemi et al., 2007; Fu et al., 2009).

For example, Jiang (2009) added the H_2O_2 solution to treat the concentrated Cu-EDTA solution separated by reverse osmosis, and found that 55.6%COD in the solution can be destroyed in 30 min. When coated $FeSO_4$ was applied to catalyze the oxidation reaction of H_2O_2, the removal rate of COD could reach 68.5%and almost all the chelated copper ions were converted into free copper ions, which could be easily removed by hydroxide precipitation.

Finžgar and Leštan (2006) developed a novel EDTA-based soil leaching method that involved the treatment and reuse of extractants in a close-loop process for soil washing: an AOP in combination of ozone and UV was used to generate hydroxyl radicals for the oxidative decomposition of EDTA-metal complexes. The dissociated metals were then removed from the washing solution by absorption on a zeolite-based commercial metal absorbent. Finally, the extracts contained fairly low concentrations of Pb, Zn and EDTA and could be directly discharged. The method was successfully applied to a washing solution for soils contaminated with Pb, Zn, Cd and Cu (Leštan and Finžgar, 2007), resulting in substantial removal of metals from the soil and a pronounced reduction of the mobility and bio-accessibility (toxicity) of metals left.

3.5.5 Disposal or Further Treatment of the Washed Soil Residue

After washing, a proportion of heavy metals still remains in the washed soil residue, especially if the soil is rich in organic matter or clay minerals, which have a strong affinity for the target heavy metals (Pichtel et al., 2001; Finzgar and Lestan, 2007; Zou et al., 2009). If these remaining heavy metals are present as stable mineral forms or bound to non-labile soil fractions, they are less mobile and bio-available, and thus less toxic. In fact, heavy metal mobility and bioavailability are increasingly used to assess the success or failure of soil remediation instead of the total metal content (Kumpiene et al., 2007). However, the remaining metals may become even more weakly associated with soil components or more readily mobile, if improper operating conditions or washing solution are applied, leading to the kinetic ligand-attacked metal detachment, soil dissolution, or the cation exchange between the chelating agent complexes and the soil particles. How to dispose of or further treat these washed soil residue has become a great challenge for the soil washing.

The most commonly used disposal method is landfilling as hazardous wastes, which occupies land space and is expensive. When other disposal methods such as solidification or stabilization are applied, the influence of chelating agents on their mobilization effectiveness should be considered. Zhang et al. (2010b) investigated the chemical immobilization on the EDTA-washed soil residue, and found that soil washing increased the mobility of Zn and Pb in the soils subsequently immobilized by $Ca(OH)_2$, likely owing to the continued mobilization of the destabilized metals during the ageing period. Based on Zhang et al. (2010b), the influence of soil washing on the subsequent immobilization performance is dominated by three competitive processes: the removal of metal labile fractions, the destabilization of less labile fractions, and chemical immobilization. If chemical immobilization is highly effective, soil washing may hardly influence the sequential chemical immobilization. Soil washing may facilitate the immobilization if the metal initial labile fractions are not negligible and the immobilizing agents are not so effective. Therefore, the sequential combination of EDTA soil washing and chemical immobilization may have advantages for treating contaminated soils containing substantial amounts of labile metals, like in freshly contaminated sites.

3.6 A Case Study of EDTA-Enhanced Soil Washing for Remediation of a Demolished Electroplating Site

The site is a demolished electroplating plant located in the north of Guangzhou city, China, with a total area about 600 m^2 (20 m×30 m). The site was closed about ten years ago, and then had been left idle until 2006. The site investigation started in early 2006, and found that the soil in some locations was severely contaminated by heave metals including Cr, Cu, Ni, Pb, or Zn at concentrations exceeding Class B Standard of Soil Quality Assessment for Exhibition Sites (HJ 350-2007). In addition, cyanide concentrations in some samples were also found at concentrations exceeding the referred standard. Inadequate ventilation may have contributed the persistence of cyanide in the subsurface environment at the site.

3.6.1 Treatability Study

Soil collected from different sampling locations, where metal concentrations exceeded the referred standard, were mixed and homogenized for the treatability study. The characteristics of the soil, including pH, density, organic content, metal contents and particle size distribution, were summarized in Table 3.2.

As listed in Table 3.2, five heavy metals and cyanide were identified at appreciably different concentrations, especially Cr and Ni concentrations were 5 and 33 times higher than the referred standards, respectively. Soil particles more than 200 mesh (< 75 μm) accounted for about 81% of total mass. This kind of soil was generally considered feasible for soil washing in terms of particle size distribution. Moreover, the short time span required for the site cleanup (e.g. half a year) and the potential commercial value of the site made soil washing an acceptable option.

Table 3.2 Characteristics of the mixed soils for treatability study

Soil Properties	Value	Soil Properties	Value
pH	6.76	Soil Particle Distribution (%, by mass)	
CEC (cmol kg^{-1})	6.83	Coarse	20.04
Density (g cm^{-3})	2.41	Sand	61.03
Organic Content	1.06 %	Silt	15.96
		Clay	7.04
Metal (mg kg^{-1})	Contents in Soil	Regulatory Permitted Levels*	
Pb	1740	500	
Cu	1037	400	
Zn	1474	500	
Cr	2783	500	
Ni	6726	200	
CN-	17.35	8.00	

* Class B Standard of Soil Quality Assessment for Exhibition Sites (HJ 350-2007)

To estimate the soil volume for washing treatment after particle size separation and to further determine the threshold sieving size, the heavy meal distribution on soil particles of different size was investigated. Results (data not shown herein) indicated that metal concentrations on particles smaller than 2 mm were higher than the referred standard. In addition, at least one metal concentration in the H_2SO_4/HNO_4 acid extraction from these soil particles exceeded the levels to be identified as hazardous waste, described by CEPM (2007). Therefore, all particles less than 2 mm need the chelant-enhanced soil washing, and the single particle size separation can not directly remedy them. So, in the following study, the mixed soil was employed without further particle size separation.

The cyanide ion (CN^-) had different properties from other cationic heavy metals, so it should be oxidized before adding chelating agents. The different oxidizing agents, such as ClO_2, H_2O_2 solution (containing about 30% H_2O_2) and leaching powder (containing $Ca(ClO)_2$) were investigated under different operating conditions (various dosages and contact time). Results are presented in Figure 3.3. It was found that 1:200 ratio of ClO_2 solution (mL) to soil (g) can cost-effectively oxidize cyanide in the soil in 0.5 h to reach the permitted level (less than 8.0 mg kg^{-1}).

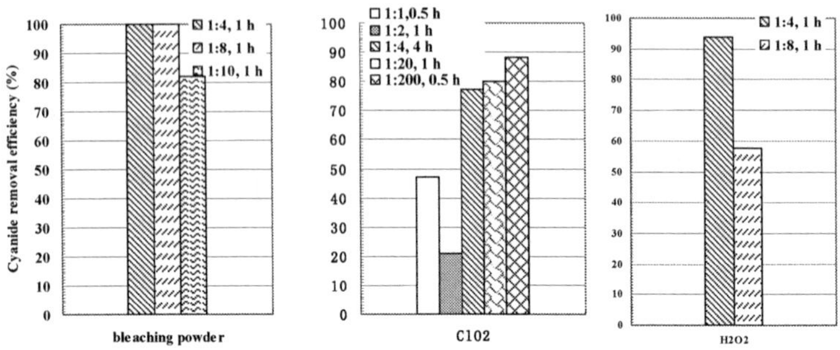

Figure 3.3 Cyanide removal by the different oxidizing agents under various operating conditions.

To compare the effectiveness of washing agents to remediate the studied soil, a wide range of potential washing reagents including H_2SO_4, HCl, acetic acid, oxalic acid, and EDTA, were investigated under various operating conditions. The results are shown in Figure 3.4. It was found that the inorganic acid, especially HCl, resulted in a substantial removal of all five target metals with the efficiencies of Cu, Cr, Ni, and Zn removal even higher than EDTA, although EDTA has been reported as an excellent washing agent for these metals. However, considering the residual cyanide in the soil, a neutral or slightly acid condition was required, since an extreme acid environment would facilitate the formation of HCN, which is volatile and highly toxic, posing a great health and safety risk to workers in the field. Therefore, EDTA was selected as the washing combination for the demonstration project, as the pH of di-sodium EDTA is slightly acidic (pH 5.8 in the 0.01 M EDTA solution).

The influence of EDTA concentration on metal removal was sequentially investigated in the laboratory, and the results are illustrated in Figure 3.5. The metal removal efficiencies generally increased with respect to EDTA concentrations at a fixed soil/solution ratio of 20 mL per 1 g soil. However, when EDTA increased from 0.01 mol L^{-1} to 0.02 mol L^{-1}, the increment became minimal. Therefore, taking into account of the solubility of EDTA and the metal removal efficiency, 0.01 M EDTA was chosen in the field application.

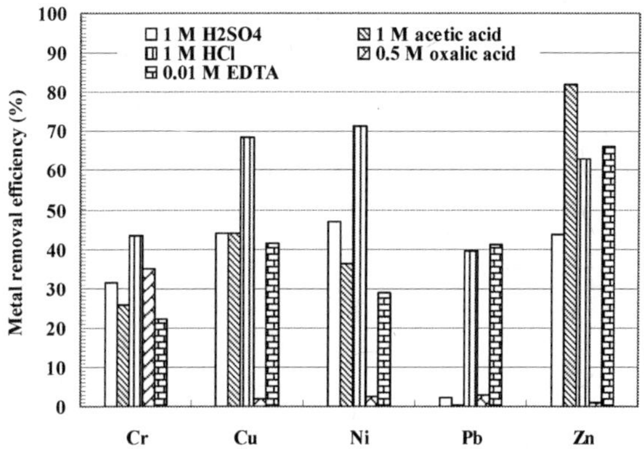

Figure 3.4 Metal removal by different washing agents.

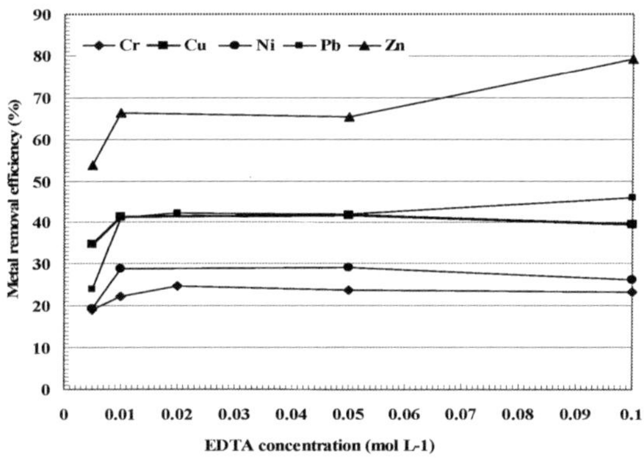

Figure 3.5 Metal removal by different EDTA concentrations.

3.6.2 Demonstrative Soil Washing System in the Field

Overview of Demonstrative Soil Washing System

The overall process of soil washing in the field is shown in Figure 3.6, and the main facilities used are listed in Table 3.3. The soil washing system comprised of a series of sequential batch treatments, where about 250 kg of contaminated soils were treated in each batch. The effluent treatment system was operated in a continuous scheme with a treatment capacity of about 1 m^3 effluent per hour.

First, the contaminated soils were manually excavated, and sieved using a 10-mesh sieve to remove the coarse fractions to facilitate the sequential running and to protect the facilities. Then, the sieved soil was transferred to the sieving tank, which was equipped with a 500-mm cylindrical trammel screen with the sieve size of 2 mm, and then a flushing tank of 1700 mm×1200 mm×1000 mm (L×W×H), a spraying pipe (1000 mm of length) on top of the trammel screen. In the sieving tank, the ClO_2 solution was sprayed first to oxidize the cyanide in the soil and solution. Then the EDTA solution was added. The particles of larger than 2 mm size were retained by the trammel screen, while the finer particles entered into the sieving tanker and were pumped into the washing tank. In the washing tank with a mechanical shaker, the finer soil particles were completely mixed with EDTA solution, and metal was dissociated there. After a 4-hour soil washing, the mixture was settled for 1 hour via gravity separation. The supernatant was discharged into the effluent treatment system, while the sludge entered a sieve filter for sand and silt-clay separation. The separated sand was returned to the sites, while the silt/clay fraction was dewatered using a pressure filter. The wastewater produced in the pressed filter was discharged into the effluent treatment system for the recovery. After recovery, the washing effluent was reused for another batch of soil washing. In order to minimize the land area occupied by the whole system, the effluent treatment system, sludge storage tank, and treated solution storage tank were built into a single tank, with effluent treatment system on the top, and the sludge and solution storage tank on the bottom.

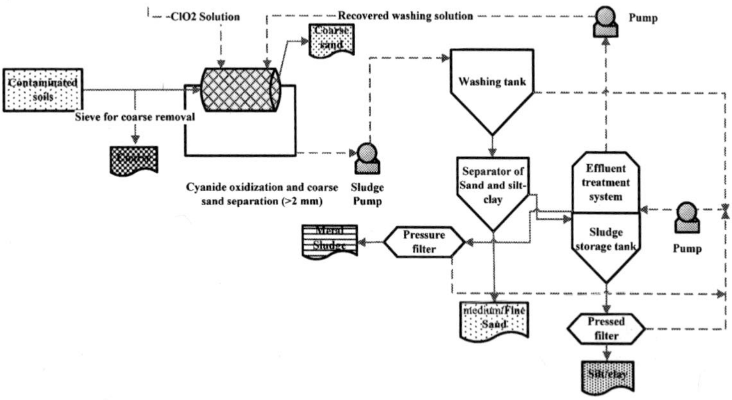

Figure 3.6 The flow chart for soil washing.

Table 3.3 Main facilities used in the demonstrative soil washing system

Facilities	Parameters	Materials	Number
trammel screen	500×1000, pore size 2 mm	Iron with anti-rust treatment	1
Sieving tanker	1700 mm × 1200 mm × 1000 mm	Iron with anti-rust treatment	1
Washing tanker	1500×2500 with adjustable shaker of 40~60 rpm	Iron with anti-rust treatment	2
Chemical storage tanker	1000×1000	PVC	4
Pressure filter	MAYJ5. 32/630, 0.6 Mpa, 20 m^2	Reinforced polypropylene	1
Separator of sand and silt/clay	500×500×1000 with 4 sieve filters	Iron with anti-rust treatment	1

The effluent from washing tanks were treated by $Ca(OH)_2$-based replacement and precipitation, where Fe^{3+} was also added as a flocculation agent to facilitate the precipitation process. Every four or five cycles, Na_2CO_3 solution was added to precipitate Ca^{2+} as $CaCO_3$, and free the EDTA from the Ca-EDTA to improve the chelating capacity of the recovered EDTA solution.

Performance of Soil Washing

Heavy metal removal efficiencies are presented in Figure 3.8. Compared with the laboratory treatability test illustrated in Figure 3.7, it was found that the Zn removal efficiencies by EDTA in the demonstrative system were higher, and the Cu removal efficiency was similar. However, the removal efficiency of Ni, Cr, and Pb seemed less effective. After 4 hours soil washing, the metal removal efficiencies did not further increase significantly over time. Therefore, a contact time of 4 hours was applied in the field demonstration.

Figure 3.8 presents the metal concentrations in the effluent of wastewater treatment system. It was found that Cr and Ni concentrations were a bit high, and increased with the number of cycles, while the other metals, i.e., Pb, Zn, and Cu, were at trace levels, and did not increase significantly with the number of cycles. Table 3.4 summarizes the average metal concentrations in the influents and effluents of the wastewater treatment system in 23 cycles. Based on Figure 3.9 and Table 3.4, it was found that the wastewater treatment system was effective to remove the Cu, Pb and Zn, with removal efficiencies over 85%, but the system seemed ineffective for Cr and Ni. The recovered washing solution maintained a certain chelating capacity to remove metals from the soils, after the labile portion of metal was removed.

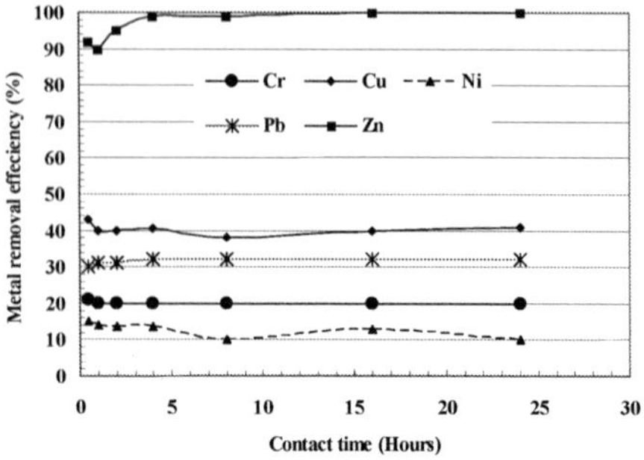

Figure 3.7 Metal removal efficiency with fresh 0.01 mol L^{-1} EDTA in the demonstrative soil washing system (data from Zou, 2009).

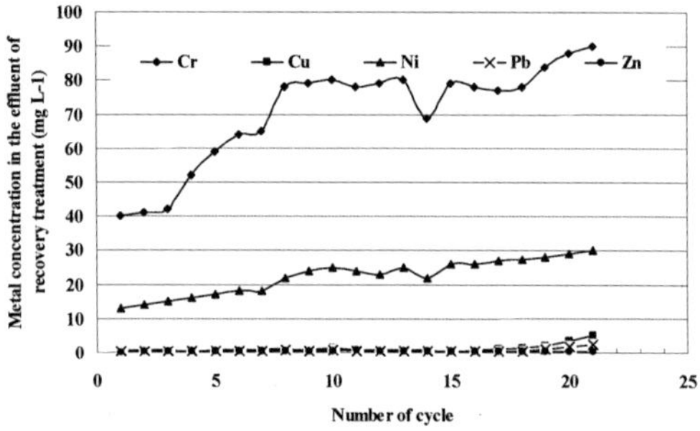

Figure 3.8 Metal concentration in the effluent of wastewater treatment system (data from Zou, 2009).

Table 3.4 Metal concentration in the influent and effluent of wastewater treatment system (data from Zou, 2009)

Metal average concentration (mg L^{-1})	Cr	Cu	Ni	Pb	Zn
In influents	91.53	11.37	32.69	4.32	11.20
In effluents	71.02	1.34	24.03	0.91	0.21

In order to evaluate the performance of chelant-enhanced soil washing, the separated coarse sands from the trammel sieve, medium/fine sands from the sand and silt/clay separator, and the silt/clay in sludge cake from the pressure filter, were collected for the H_2SO_4/HNO_3 acid extraction test as described by CEMP (2007), the results on coarse sand and silt/clay are listed in Table 3.5, and the results on the middle/fine sand are illustrated in Figure 3.9.

As listed in Table 3.5, heavy metal concentrations in the coarse sands were less than the permitted levels. Therefore, it could be disposed of as general waste or reused as construction material. In the silt/clay fractions, the released Cu, Ni and Pb were all below the permitted levels, while Cr and Ni were beyond it. So, this portion of soil was a hazardous waste, and further treatment was required. However, the volume only accounted for about 18% of the total volume of the original soil.

Table 3.5 H_2SO_4/HNO_3 acid extraction of the washed soils (data from Zou, 2009)

Soil fraction	*No of cycle*	*Total Cr (mg L^{-1})*	*Cu (mg L^{-1})*	*Ni (mg L^{-1})*	*Pb (mg L^{-1})*	*Zn (mg L^{-1})*
Silt/Clay	1st	34.01	1.15	14.44	0.05	Undetectable
	23rd	24.47	6.62	9.77	1.08	5.06
Coarse sand	1st	6.96	0.02	0.06	Undetectable	Undetectable
	11th	7.83	0.07	0.90	Undetectable	0.13
	23rd	4.34	0.06	1.13	0.01	2.15
CEPA Standard		15	100	5	5	100

As illustrated in Figure 3.9, only the released Cr in the 11th and 17th cycles and the released Ni in the 8th cycle from separated medium/find sands exceeded the permitted levels. Moreover, the other metals released, i.e., Cu, Pb, Zn were all below the permitted levels. Therefore, most of the medium/find sands after washing were not considered hazardous waste any more, and could be disposed of as general waste.

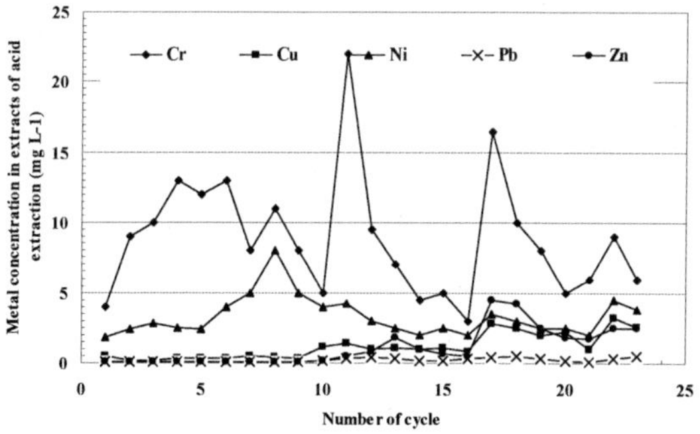

Figure 3.9 Metal concentration in the extracts of H_2SO_4/HNO_3 acid extraction on the medium /find sands (data from Zou, 2009).

3.7 Conclusions

Over recent years, considerable effort has been devoted to the development of chelant-enhanced soil washing. The soil properties, metal nature and species in soils, and soil washing operating conditions, such as chelating agent type and dosage, contact time and washing mode, solution pH and electrolyte, were all found to influence the performance of soil washing to different extents. In addition, the mobility of residual metals, soil dissolution during washing, the recovery and treatment of chelating agents in washing solution, as well as the disposal of washed soils also influence the application of soil washing. Especially, the strategies to optimize efficiency in the extraction of toxic metals and in the recovery of chelating agents need to be improved, because the current methods proposed to mobilize metals from the soil particles, and to recover chelating agents from the spent washing solution work well only within a narrow range of metal contamination and soil types.

An EDTA-enhanced soil washing system was designed and applied as a demonstrative project on a demolished electroplating site, where aged multiple metal contamination occurred. After particle separation and EDTA extraction, the soil sand fraction became general waste, and only the silt and clay fractions (18% of the total) needed further disposal as a hazardous waste. The less effectiveness can be in part ascribed to the ineffective removal of Cr and Ni during the washing effluent treatment.

3.8 Acknowledgments

This work was supported in part by grants from the Natural Science Foundation of China (No. 40802088), and the Fundamental Research Funds for the Central Universities.

3.9 References

Abumaizar, R. J., and Smith, E. H. (1999). "Heavy metal contaminant removal by soil washing." *Journal of Hazardous Materials,* 70, 71-86.

Ager, R., and Marshall, W.D. (2003). "Recycle of thermomechanical pulp filtrate after removal of metals: A study with EDTA." *Journal of Pulp and Paper Science,* 29, 303-307.

Allen, H. E., and Chen, P. H. (1993). "Remediation of metal contaminated soil by EDTA incorporating electrochemical recovery of metal and EDTA."

Environmental Progress, 12, 284-292.

Andrade, M. D., Prasher, S. O., and Hendershot, W. H. (2007). "Optimizing the molarity of an EDTA washing solution for saturated-soil remediation of trace metal contaminated soils." *Environment Pollution,* 147, 781-790.

Barona, A., Aranguiz, I., and Elias, A. (2001). "Metal associations in soils before and after EDTA extractive decontamination: implications for the effectiveness of further clean- up procedures." *Environment Pollution,* 113, 79-85.

Bassi, R., Prasher, S. O., and Simpson, B. K. (2000). "Extraction of metals from contaminated sandy soil using citric acid." *Environmental Progress,* 19, 275-282.

Bermond, A., and Ghestem, J. P. (2001). "Kinetic study of trace metal EDTA-desorption from contaminated soils." In: Selim, H.M., Sparks, D.L. (Eds.), Heavy Metals Release in Soils. Lewis Publishers, Boca Raton, FL, pp. 131–147.

Blesa, M. A., Morando, P. J., and Regazzoni, A. E. (1994). Chemical Dissolution of metal oxides. CRC, Boca Raton, FL, pp 281.

Brown, G. A., and Elliott, H. A. (1992). "Influence of electrolytes on EDTA extraction of Pb-polluted soil." *Water, Air, & Soil Pollution*, 62, 157-165.

Chinese Ministry of Environment Protection (CMEP), (2007). Solid Waste-extraction Procedure for Leaching Toxicity: Sulfuric Acid and Nitric Acid Methods (HJ/T299-2007), China Environmental Science Press, Beijing.

Chu, W. (2003). "Remediation of contaminated soils by surfactant-aided soil washing." *Practice Periodical of Hazardous, Toxic, and Radioactive Waste Management,* 7, 19-24.

Cline, S. R., and Reed, B. R. (1995). "Lead removal from soils via bench-scale soil washing techniques." *Journal of Environmental Engineering,* 121, 700-705.

Davis, A. P., and Green, D. L. (1999). "Photo-catalytic oxidation of cadmium-EDTA with titanium dioxide." *Environmental Science & Technology,* 33, 609-617.

Davis, A. P., and Singh, I. (1995). "Washing of Zinc (II) from contaminated soil column." *Journal of Environmental Engineering,* 121, 174-185.

Elliott, H. A., and Brown, G. A. (1989). "Comparative evaluation of NTA and EDTA fro extractive decontamination of Pb-polluted soils." *Water, Air, & Soil Pollution*, 45, 361-369.

Elliott, H. A., and Shastri, N. L. (1999). "Extractive decontamination of metal-polluted soils using oxalate." *Water, Air, & Soil Pollution*, 110, 335-346.

European Chemical Bureau, (ECB) (2004). European Union Risk Assessment Report on Na_4EDTA, Office for Official Publications of the European Communities, Italy.

Federal Remediation Technologies Roundtable (FRTR) (2006). "Compiled from Remediation Technology Screening Matrix and Reference Guide."4th version, available at http://www.frtr.gov/matrix2/section1/toc.html, assessed on April 29.

Federal Remediation Technologies Roundtable (FRTR) (2007). Remediation Technologies Screening Matrix and Reference Guides, Version 4.0, http://www.frtr.gov/matrix2/top page.html, March.

Finžgar, N., and Leštan, D. (2006). "Heap leaching of Pb and Zn contaminated soil using ozone/ UV treatment of EDTA extractants." *Chemosphere,* 63, 1736-1743.

Finžgar, N., and Leštan, D. (2007) "Multi-step leaching of Pb and Zn contaminated soils with EDTA." *Chemosphere,* 66, 824-832.

Fu, F. L., Zeng, H. Y., Cai, Q. H., Qiu, R. L., Yu, J., and Xiong, Y, (2007). "Effective removal of coordinated copper from wastewater using a new dithiocarbamate-type supramolecular heavy metal precipitant." *Chemosphere,* 69, 1783-1789.

Fu, L. F., Wang, Q., and Tang B. (2009). "Fenton and Fenton-like reaction followed by hydroxide precipitation in the removal of Ni(II) from Ni-EDTA wastewater: a comparative study." *Chemical Engineering Journal,* 155(3), 769-774.

Grcˇman, H., Velikonja-Bolta, Sˇ., Vodnik, D., Kos, B., and Leštan, D. (2001). "EDTA enhanced heavy metal phytoextraction: metal accumulation, leaching and toxicity." *Plant & Soil,* 235, 105-114.

Griffiths, R. A. (1995). "Soil-washing technology and practice." *Journal of Hazardous Materials,* 40(2), 175-189.

Hong, J., and Pintauro, P. N. (1996). "Selective removal of heavy metals from contaminated kaolin by chelating agents." *Water, Air, & Soil Pollution*, 87, 73-91.

Huang, C. P., Hsu, M. C., and Miller, P. (2000). "Recovery of EDTA from power plant boiler chemical cleaning wastewater." *Journal of Environmental Engineering,* 126, 919-924.

Huang, J. W., Chen, J., Berti, W. R., and Cunningham, S. D. (1997). "Phytoremediation of lead-contaminated soils: role of synthetic chelates in lead phytoextraction." *Environmental Science & Technology*, 31, 800-805.

Impellitteri, C. A., Allen, H. E., Yin, Y. J. You, S. J, and Saxe, J. K. (2001). "Soil properties controlling metal partitioning." *Heavy Metals Release in soils,* Selim, H. M. and Sparks, D. L. ed., Lewis Publishers.

Jiang, S. X. (2009). New Method for Advance Treatment of Wastewaters containing Chelated Copper. PhD Dissertation, Sun Yat-sen University, Guangzhou, China.

Jiang, S. X., Fu, F. L., Qu, J. X., and Xiong, Y. (2008). "A simple method for removing chelated copper from wastewaters: $Ca(OH)_2$-based replacement-precipitation." *Chemosphere,* 73, 785-790.

Juang, T. S., and Wang, S. W. (2000). "Metal recovery and EDTA recycling from simulated washing effluents of metal contaminated soil." *Water Research,* 34(15), 3796-3808.

Kim, C., and Ong, S. K., 1999. "Recycling of lead-contaminated EDTA wastewater." *Journal of Hazardous Materials,* 69, 273-286.

Kirpichtchikova, T. A., Manceau, A., Spadini, L., Panfili, F., Marcus, M. A., and Jacquet, T. (2006). "Speciation and solubility of heavy metals in contaminated soil using X-ray microfluorescence, EXAFS spectroscopy, chemical extraction, and thermodynamic modeling." *Geochimica et Cosmochimica Acta*, 70, 2163-2190.

Kumpiene, J., Lagerkvist, A., and Maurice, C. (2007). "Stabilization of Pb- and Cu-contaminated soil using coal fly ash and peat." *Environmental Pollution,* 145, 365-373.

Lei, M., Liao, B. H., Zeng, Q. R., Qin, P. F., and Khan, S. (2008). "Fraction distributions of lead, cadmium, copper, and zinc in metal-contaminated soil before and after extraction with disodium ethylenediaminetetraacetic acid." *Soil Science Plant Analysis,* 39 1963-1978.

Lim, T. T., Chui, P. C., and Goh, K. H. (2005). "Process evaluation for optimization of EDTA use and recovery for heavy metal removal from a contaminated soil." *Chemosphere,* 58, 1031-1040.

Lo, I. M. C., and Zhang, W. H. (2005). "Study on the Optimal Conditions for the Recovery of EDTA from Soil Washing Effluents." *Journal of Environmental Engineering,* 131, 1507-1513.

Lo, I. M. C., and Yang, X. Y. (1999). "EDTA extraction of heavy metals from different soil fractions and synthetic soils." *Water, Air, & Soil Pollution,* 109(1-4), 219-236.

Macauley, E., and Hong, A. (1995). "Chelation extraction of lead from soil using pyridine-2,6-dicarboxylic acid." *Journal of Hazardous Materials,* 40, 257-270.

MCbride, M. B. (1994). Environmental Chemistry of Soils, Oxford University Press Inc, New Yorks, NY, USA.

Moutsatsou, A., Gregou, M., Matsas, D., and Protonotarios, V. (2006). "Washing as a remediation technology applicable in soils heavily polluted by mining–metallurgical activities." *Chemosphere, 63*, 1632–1640.

Nowack, B. and Sigg, L. (1997). "Dissolution of Fe (III) (hydr)oxides by metal-EDTA complexes." *Geochimica et Cosmochimica Acta,* 61, 951-963.

Paff, S. W., Bosilovich, B. E., and Kardos, N. J. (1994). "Site emerging techniques: acid extraction treatment system for treatment of metal contaminated soils." *EPA/540/SR-94/513, Superfund Innovative Technology Evaluation,* Washington DC, USA.

Palma, L. D. (2009). "Influence of indigenous and added iron on copper extraction from soil." *Journal of Hazardous Materials,* 170, 96-102.

Palma, L. D., Ferrantelli, P., Merli, C., and Biancifiori, F.J., (2003a). "Recovery of EDTA and metal precipitation from soil flushing solutions." *Journal of Hazardous Materials,* 103, 153-168.

Palma, L. D., Ferrantelli, P., Merli, C., Petrucci, E., (2003b). "Treatment of the solution extracted from metal contaminated soils by reverse osmosis and chemical precipitation." *Annali di Chimica,* 93, 1005-1011.

Papassiopi, N., Tambouris, S., and Kontopoulos, A. (1999). "Removal of heavy metals from calcareous contaminated soils by EDTA leaching." *Water, Air, & Soil Pollution*, 109, 1-15.

Peters, R. W. (1999). "Chelating extraction of heavy metals from contaminated soils." *Journal of Hazardous Materials,* 66, 151-210.

Peters, R. W., and Shem, L. D. (1992). "Use of chelating agents for remediation of heavy metal contaminated soil." *Environmental Remediation-Removing Organic and Metal Ion Pollutants, ACS Symposium Series 509*, Vandegrift, G. F., Reed, D. T., and Tasker, I. R. ed.

Pichtel, J., and Pichtel, T. M. (1997). "Comparison of solvents for ex-situ removal of chromium and lead from contaminated soil." *Environmental Engineer Science.* 14, 97-104.

Pichtel, J., Vine, B., Kuula-Vaisanen, P., and Niskanen, P. (2001). "Lead extraction from soils as affected by lead chemical and mineral forms." *Environmental Engineer Science*, 1891-1898.

Pirkanniemi, K., Metsarinne,S., and Sillanpaa, M. (2007). "Degradation of EDTA and novel complexing agents in pulp and paper mill process and waste waters by Fenton's reagent." *Journal of Hazardous Materials,* 147, 556-561.

Polettini, A., Pomi, R., and Rolle, E. (2007). "The effect of operating variables on chelant-assisted remediation of contaminated dredged sediment." *Chemosphere,* 33, 866-877.

Qiu, R. L., Zou, Z. L., Zhao, Z. H., Zhang, W. H., Zhang, T., Dong, H. Y., and Wei, X. G. (2010). "Removal of trace and major metals by soil washing with Na_2EDTA and oxalate." *Journal of Soil & Sediment,* 10, 1, 45-53.

Raghavan, R. (1989). "Cleaning excavation solid using extraction agents: a state–of-the–art review." *EPA/600/2-89/034*, Risk reduction engineering laboratory, Office of research and development, Cincinnati, OH, USA.

Reddy, K. R., and Chinthamreddy, S. (2000). "Comparison of different extractants for removing heavy metals from contaminated clayey soils." *Soil Sediment Contamination*, 9, 449-462.

Reddy, K. R., Chaparro, C., and Saichek, R. E., (2003). "Removal of mercury from clayey soils using electro-kinetics." *Journal of Environmental Science & Health.*, 38, A (2), 307-338.

Robin, S., and Krishna, R. R. (1996). "Evaluation of soil washing process to remove mixed contaminated from a sandy loam." *Journal of Hazardous Materials,* 45, 45-57.

Roy, D., Kommalapati, R. R., Mandava, S. S., Valsaraj, K. T., and Constant, W. D. (1997). "Soil washing potential of a natural surfactant." *Environmental Science & Technology*, 31, 670-675.

Rudd, B., Hasnson, A., Heil, D., and Samani, Z. (1995). "Reuse of EDTA used in soil washing." *Proc. 5th Annual WERC Technology Development Conference*, Las Cruces, NM.

Santamarina, J. C., Klein, K. A., and Fam, M.A. (2001). "Particle-fluid interactions." *Soils and Waves*, John Wiley & Sons, New York, 59-102.

Saponaro, S., Bonomo, L., Barbafieri, M., and Petruzzelli, G. (2002). "Soil washing feasibility at a manufacturing gas plant site." *Soil Sediment Contamination,* 11(5), 751-767.

Scheckel, K. G., and Sparks, D. L. (2001). "Dissolution kinetics of nickel surface precipitates on clay mineral and oxide surfaces." *Soil Science Society American Journal,* 65, 719-728

Song, Q. J. and Greenway, G. M. (2004). "A study of the elemental leachability and retention capability of compost." *Journal of Environmental Monitor,* 6, 31-37

Sposito, G. (2004). The Surface Chemistry of Natural Particles, Oxford University Press, Oxford, UK, 2004.

Steele, M. C., and Pichtel, J. (1998). "Ex-situ remediation of a metal-contaminated superfund soil using selective extractants." *Journal of Environmental Engineering,* 124, 639-645.

Strawn, D. G., and Sparks, D. L. (2000). "Effects of soil organic matter on the kinetics and mechanisms of Pb(II) sorption and desorption in soil." *Soil Science Society of America Journal 64*, 144-156.

Sun, B., Zhao, F. J., Lombi, E., and McGrath, S. P. (2001). "Leaching of heavy metals from contaminated soils using EDTA." *Environmental Pollution,* 113, 111-120.

Tandy, S., Bossart, K., Mueller, R., Ritschel, J., Hauser, L., Schulin, R., and Nowack, B. (2004). "Extraction of heavy metals from soils using biodegradable chelating agents." *Environmental Science & Technology*, 38, 937-944.

Tejowulan, R. S., and Hendershot, W. H. (1998). "Removal of trace metals from contaminated soils using EDTA incorporating resin trapping techniques." *Environmental Pollution,* 103, 135-142.

Tessiar, A., Campbell, P. G. C., and Bisson, M. (1979). "Sequential extraction procedure for the speciation of particulate trace metals." *Analytical Chemistry,* 51, 844-851.

Theodoratos, P., Papassiopi, N., Georgoudis, T., and Kontopoulos, A. (2000). "Selective removal of lead from calcareous polluted soils using the Ca-EDTA salt." *Water, Air, & Soil Pollution*, 122, 351-368.

Tobia, R. J. (1993). "Final report: pilot-scale soil washing study." *EPA/68-03-3482*, Pensacola, Florida, USA.

Tsang, D. C. W., Zhang, W. H., and Lo, I. M. C. (2007). "Copper extraction effectiveness and soil dissolution issues of EDTA-flushing of contaminated soils." *Chemosphere,* 68, 234-243.

Udovic, M., and Lestan, D. (2009). "Pb, Zn and Cd mobility, availability and fractionation in aged soil remediated by EDTA leaching." *Chemosphere,* 74, 1367-1373.

USEPA (1996). "Engineering Bulletin: Technology Alternatives for the Remediation of Soils Contaminated with Arsenic, Cadmium, Chromium, Mercury, and Lead." Office of Emergency and Remedial Response, Cincinnati, OH.

USEPA (2000) "Innovative remediation technologies: field-scale demonstration projects, 2nd Edition." *EPA 542-B-00-004,* Office of Solid Waste and Emergency Response, Washington DC, USA.

Van Benschoten, J. E., Matsumoto, M. R., and Young, W. H. (1997). "Evaluation and analysis of soil washing for seven lead-contaminated soils." *Journal of Environmental Engineering,* 123(3), 217-224.

Willichowski, M. (2001). "Remediation of soils by washing process and historical overview." *Treatment of Contaminated Soil Fundamentals, Analysis and Application,* Berlin, Springer, 417-434.

Xia, W. B., Gao, H., Wang, X. H., Zhou, C. H., Liu, Y. G., Fan, T., and Wang, X. (2009). "Application of EDTA decontamination on soils affected by mining activities and impact of treatment on the geochemical partition of metal contaminants." *Journal of Hazardous Materials,* 164, 936-940.

Yip, T. C. M., Tsang, D. C. W., Ng, K. T. W., and Lo, I. M. C. (2009). "Kinetic interactions of EDDS with Soils: 1. Metal re-sorption and competition under EDDS deficiency." *Environmental Science & Technology,* 43, 831-836.

Yu, J. Z., and Klarup, D. (1994). "Extraction kinetics of Cu, Zn, Fe and Mg from contaminated sediment using disodium EDTA." *Water, Air, & Soil Pollution,* 75(3-4), 205-225.

Zhang, W. H., and Lo, I. M. C. (2006). "EDTA-enhanced soil washing for the remediation of Pb- and /or Zn- contaminated soil." *Journal of Environmental Engineering,* 132, 1282-1288.

Zhang, W. H., Huang, H., Tan, F. F., Wang, H., and Qiu, R. L. (2010a). "Influence of EDTA-washing on the species and mobility of heavy metals residual in soils." *Journal of Hazardous Materials,* 173, 369-376.

Zhang, W. H., Tsang, D. C. H., and Lo, I. M. C. (2008). "Removal of Pb by EDTA-washing in the presence of hydrophobic organic contaminants or anionic surfactant." *Journal of Hazardous Materials,* 155, 433-439.

Zhang, W.H., Tong, L.Z., Yuan, Y., Liu, Z.Y., Huang, H., Tan, F. F., Qiu, R. L. (2010b). "Influence of soil washing with a chelating agent on subsequent chemical immobilization of heavy metals in a contaminated soil" *Journal of Hazardous Materials,* 178, 578-587.

Zimdahl, R. L., and Skogerboe, R. K. (1977). "Behavior of lead in soils." *Environmental Science & Technology,* 11, 1202-1207.

Zou, Z. L. (2009). Soil Washing in Abandoned Industrial Heavy Metal Contaminated Area and Demonstration Engineering. PhD Dissertation, Sun Yat-sen University, Guangzhou, China.

Zou, Z. L. Qiu, R. L., Zhang, W. H., Dong, H. Y., Zhao, Z. H., Zhang, T., Wei, X. G., and Cai, X. D. (2009). "The study of operating variables in soil washing with EDTA." *Environmental Pollution,* 157, 229-236.

Zou, Z. L., Zhang, W. H., Yang, X. H., Wei, X. G., and Qiu, R. L. (2008). "Soil Washing for Remediation of a Multiple Heavy Metal-Contaminated Site." *Proceeding of 9th International Symposium on Environmental Geotechology,* June 1-4, Hong Kong.

CHAPTER 4

Electrochemical Treatment and Recovery of Chelating Agents

Domen Lestan

4.1 Remediation of Metal Contaminated Soils

In soils, toxic metals are present in various chemical forms and generally exhibit different physical and chemical behavior in terms of chemical interactions, mobility, biological availability and potential toxicity (Bohn et al., 1979, Tandy et al., 2004). Unlike organic compounds, toxic metals are not degradable in the environment, and persist in soils for centuries. It is therefore highly desirable to apply suitable remedial approaches to polluted soil, which can reduce the risk of metal contamination. The excavation and disposal of soil is no longer considered to be a permanent solution. The demand for soil treatment techniques is consequently growing and the development of efficient remediation technologies has become one of the key research activities in environmental science and technology.

Since metals cannot be degraded, any cleanup requires their immobilization and toxicity reduction or removal. Various approaches have been suggested: toxic metals can be contained to prevent their movement, i.e. by leaching through the soil or by soil erosion. This can be achieved by capping the site with asphalt or other impermeable materials to prevent the infiltration of water, by planting permanent plant cover (e.g., phyto-stabilization) or by covering the site with unpolluted soil (Guo et al., 2006). Stabilization involves fixing up the contaminants in stable sites by mixing or injecting inorganic or organic soil amending agents (e.g., liming agents, organic materials, aluminosilicates, phosphates, iron and manganese oxides, coal fly ashes, etc.). Due to the effects of a change in pH, such agents are effective at decreasing the bioavailability of metals by introducing additional binding sites for toxic metals. Stabilized metals then become less available for plants, and their bioconcentration through the food chain is reduced (Guo et al., 2006). Another immobilization method is vitrification by heating the contaminated soil to up to 2000 ^{o}C. Vitrification usually involves imposing an electrical current between electrodes inserted into the contaminated soil. Due to its low electrical conductivity, the soil begins to heat and produces a melt that hardens into a blocks of glasslike material.

Vitrification is expensive but applicable to soils with mixed organic and metallic contamination, for which few technologies are available (Buelt and Farnsworth, 1991).

However, if after remediation toxic metals remain in the soil they can be harmful when soil dust is ingested or inhaled. Remediation approaches which remove harmful metals from soil are therefore preferred. For example, smaller, but usually more polluted, soil particles can be removed from the rest of the soil by various separation techniques developed and used in the mining industry. These include the use of hydrocyclones, which separate larger particles from smaller ones using centrifugal force; and solid-liquid separation techniques, such as gravimetric settling and flotation, which are based on the different surface characteristics of particles (Mulligan et al., 2001; Vanthuyne and Maes, 2002). Electrokinetic extraction has been proposed as an in situ method for the remediation of blocks of contaminated soil. Electrokinetic extraction involves the electrokinetic movement of charged particles suspended in a soil solution, initiated by an electric gradient. The target metals can be removed by precipitation at the electrodes (Hicks and Tondorf, 1994). Phytoextraction is a publicly appealing (green) remediation technology. It can be effectively applied for soils contaminated with specific metals and metalloids, e.g. Ni, Zn and As, which are readily bioavailable for hyper-accumulating plants. Common crop plants with a high biomass can be triggered to accumulate large amounts of low bioavailability metals (e.g. Pb, Cr, U, Hg) when the mobility of these metals in the soil is enhanced by the addition of mobilizing agents (Huang et al., 1997; Luo et al., 2005). Different chelating agents desorb toxic metals from soil solid phases by forming strong water-soluble complexes, which can be removed from the soil by plants through enhanced phytoextraction or by using soil washing techniques.

4.1.1 Soil Washing Using Chelating Agents

Soil washing is one of the most promising and studied metal removal remediation techniques. It involves the separation of toxic metals from soil solid phases by solubilizing the metals in a washing solution. Acids and chelating agents are the most prevalent removal agents used in soil washing (Peters, 1999). Acids dissolve carbonates and other metal-bearing soil material and exchange trace metals from soil surfaces where H^+ ions are attracted more strongly than the cations of toxic metals. Since acidic solutions can cause deterioration in the physico-chemical properties of the soil, using chelating agents is considered to be less environmentally disruptive than using acids (Xu and Zhao, 2005).

Chelating agents desorb trace metals from soil solid phases by forming strong and water-soluble metal-chelant coordination compounds (complexes). These complexes are very stable, prevent the precipitation and sorption of metals, and do not release their metal ions unless there is a significant drop in soil pH. The efficacy of a chelant in the extraction of metals is usually rated with the stability constants K_s of the chelant-metal complexes. According to Elliott et al. (1989), the order of magnitude of the values of K_s can be used to rank different chelants according to their general efficacy, but not to rank the efficacies of a specific chelant toward different metals because the latter is also influenced by the metal speciation in a given soil matrix. Many different chelants (mostly aminopolycarboxylic acids) have been tested for soil washing. In the literature, Di-sodium salt of ethylenediamine tetraacetate (EDTA) is the most frequently cited chelating agent for extracting toxic trace metals from soils, because of its efficiency, availability and relatively low cost.

Several soil washing technologies has been proposed. Soil flushing is an *in situ* soil washing technology applicable to specific soil conditions, in which the contaminated zone is underlain by non-permeable materials, allowing the washing solution to be pumped and treated. The method is suitable for sandy soil or sediment with high hydraulic conductivity. The washing solution is forced through the in-place soil matrix via injection wells or is infiltrated into the soil using surface sprinklers or similar devices. The washing solution is pumped from the soil using a set of recovery wells installed down gradient of the contaminated area. The washing solution must be treated to remove toxic metals and the process water reused in the flushing process. The disadvantage of in situ soil flushing is the low degree of control over the movement of contaminants into undesirable areas. The hydrology of the site must therefore be precisely understood. The extraction of soil slurry refers to the batch treatment of soil slurry in a reactor. Following an initial screening of the excavated soil to remove the surface debris, the soil is vigorously mixed with the chelating agent solution, separated by a second screening step (filtration), and then returned to the ground. In soil leaching, the washing solution is gravitationally percolated through a soil heap or column *ex situ*. Soil is excavated, screened and placed in a mound on a pad. Metals are removed by passing washing solution through the soil using some type of liquid distribution system. The extractant is collected in a pregnant solution pit and processed to remove metals. Soil leaching is operationally simple and holds the potential for the economical treatment of large amounts of soil. The leaching efficiency is higher for soils with higher hydraulic conductivity.

In soil washing technologies the strategies for developing chelanting agent washing solutions to achieve optimal efficiency in the extraction of toxic metals still need to be improved and effective treatment methods for waste washing solutions are urgently needed.

4.2 Chemical Treatments of Waste EDTA Soil Washing Solution

In practice the use of EDTA in full-scale soil washing is prohibited by large volumes of waste washing solution generated, which must be treated before disposal. EDTA is not overly expensive (in Europe, it costs less than 5 € kg^{-1} for the technical-grade chemical from a major European manufacturer). However, toxic wastewaters containing complexed EDTA cannot be treated using conventional methods such as filtration, flocculation and participation (Jiraroj et al., 2006).

To treat the washing solution, Finzgar and Lestan (2006a, 2006b) used a combination of ozone and UV, an advanced oxidation process (AOP). AOP generated hydroxyl radicals (˙OH) for the oxidative decomposition of EDTA–metal complexes (˙OH are one of the most powerful oxidants in aqueous solutions). The released metals were then removed from the washing solution by adsorption. The method produced a discharge solution with a low concentration of EDTA and toxic metals (Pb, Zn, Cd and Cu). However, coloration and particles in the washing solution absorb and scatter UV light (Shu and Chang, 2005). Ozone-UV based AOP was therefore efficient only for fairly colorless and nonturbid solutions, however, soil washing solutions typically have intensive yellow-brown coloration (due to the formation of Fe–EDTA complexes). Another practical problem was removal of released metals, which consumed a significant quantity of expensive adsorbent. In another study Di Palma et al. (2003) proposed reverse osmosis for the separation of EDTA complexes from the spent soil washing solution. However, the soil colloidal particles tend to clog the membranes. In yet another study Tejowulan and Hendershot (1998) separated EDTA using an anion exchange resin. The practical means to recycle and reuse expensive resin were not proposed.

Technologies aiming to recycle and reuse EDTA would further improve the economy of the soil washing remediation processes. Ager and Marshall (2001) used zero-valent bimetallic mixtures (Mg^0-Pd^0, Mg^0-Ag^0) to precipitate Pb from the solution while liberating EDTA in alkaline pH. Metals liberated from the EDTA complex were cemented to the surfaces of the excess magnesium or removed from the solution as insoluble hydroxides. The method is efficient but could be economically prohibitive. Hong et al. (1999) separated Pb from EDTA with Na_2S and $Ca(OH)_2$ at alkaline (pH 10) conditions, resulting in almost complete recovery of metals through precipitation in the form of insoluble metal sulphides. While the $Ca(OH)_2$ provided Ca^{2+} ions to compete for the EDTA ligand (by replacing the chelated contaminant metal), Na_2S was used as anionic precipitant to provide HS^- and S^{2-} to compete with EDTA for the contaminating metals. This method has found limited application due to the hazardous nature of the reagents and the sludge produced, cost and operational difficulties. Kim and Ong (1999) recycled chelant from Pb-EDTA solution by substituting Pb with Fe^{3+} in acidic conditions, followed by precipitation of the released Pb with phosphate (Na_2HPO_4) near neutral pH. Fe^{3+} ions were then precipitated as hydroxides at high pH using NaOH, thus liberating EDTA. Alkaline precipitation is the simplest way to separate metals from chelant and the process does not use expensive or hazardous reagents. However, the cost of the process is affected by a high reagent consumption. For Pb precipitation from EDTA soil washing solution, Kim and Ong (1999) found than a phosphate / Pb molar ratio of about 30 was necessary. Di Palma et al. (2003) reduced reagent consumption by evaporating the soil washing solution volume by 75% and then recycled the EDTA by substituting Cu in the EDTA complex with Fe^{3+} in acidic conditions, followed by alkaline precipitation. Although EDTA recycling procedures described above have been demonstrated on a laboratory scale, there is currently no practical and commercially available chemical method.

4.3 Electrochemical Oxidation

Using electricity to treat water was first proposed in the UK in 1889 (Chen, 2004). Today electrochemical technologies are known as simple and efficient methods for the treatment of many wastewaters, characterized by a compact size of the equipment, simplicity of operation, and low capital and operating costs (Chen, 2004). Electrochemical oxidation of contaminants through anodically generated chlorine and hypochlorite is well known. This technique is efficient at high chloride concentration (>3 g/l). Electrochemical oxidation of contaminants in waste waters oxidation can also occur directly on anodes by generating active oxygen absorbed into the oxide lattice on the anode. This process is usually called direct anodic oxidation. Active oxygen can cause the complete combustion of organic compounds and formation of selective oxidation products. The anodic oxidation does not require addition of chemicals to waste water, which is an advantage over other electrooxidation processes. In an early study Johnson et al. (1972) reported that using a Pt anode in a conventional electrolytic cell oxidized EDTA into CO_2, formaldehyde and ethylendiamine, and could thus potentially be used for treating waste soil washing solutions.

The important parameter of an anodic oxidation process is the anode material. During anodic oxidation process mainly molecular oxygen is produced during water electrolysis if the oxygen overvoltage is not sufficiently high. In electrochemical AOP (EAOP) however, anode material has sufficient oxygen overvoltage before H_2 (cathode) and O_2 (anode) form. This electrochemical window allows production of hydroxyl radicals at the anode according to the equation below directly from the electrolyzed water at a high current efficiency (Kraft et al., 2003; Oliveira et al., 2007).

$$H_2O \rightarrow {}^{\cdot}OH + e^- + H^+$$

In general, •OH is more efective for oxidation of contaminants than active oxygen in the anode oxide lattuce. Different anode materials have been therefore studied for effective •OH production: graphite, Pt, and different noble metal oxides (PbO_2, IrO_2, TiO_2, SnO_2) on titanium substrate. However, EAOP became really feasible with the recent development of a large area, boron-doped diamond anode (BDDA) (Troster et al., 2002). BDDA has an extreme oxygen overvoltage of >3 V. BDDA is also extraordinary chemically inert and therefore suitable for treating various wastewaters. Yamaguchi et al. (2006) reported that EDTA was oxidized through sequential removal of the acetate groups until an unidentified small size hydrocarbon product was formed.

The feasibility of a EAOP with BDDA was evaluated using spent EDTA (20 mol kg^{-1}) washing solution after leaching the soil contaminated with Pb (1374 mg kg^{-1}), Zn (1007 mg kg^{-1}), and Cd (9.1 mg kg^{-1}) (Finzgar and Lestan, 2008). EDTA removed 44% Pb, 14% Zn and 35% Cd from the soil. Oxidative decomposition of metal-EDTA complexes was measured at constant current density of 15 mA cm^{-2}.

The released metals were removed from the solution by filtration as insoluble participate and by electro-deposition on the cathode. After treatment the discharge solution was clear and almost colorless, with pH 7.73 and 0.47 mg L^{-1} Pb, 1.03 mg L^{-1} Zn, below the limits of quantification of Cd and 0.023 mM EDTA. In yet another feasibility study the spent washing solution from EDTA leaching of Cu (365 mg kg^{-1}) contaminated vineyard soil was used (Pociecha nad Lestan, 2009). During remediation 20 mmol kg^{-1} EDTA removed 26% of Cu from the soil, mostly from carbonate and oxide soil fractions (58% and 40% Cu reduction). The soil Cu oral availability (in vitro Physiologically Based Extraction Test) was reduced after remediation by 42% and 51% in the simulated stomach and intestinal phases. Spent washing solution was treated using EAOP with BDDA at a constant current density of 40 mA cm^{-2}. Again the released Cu was removed from the solution mostly as an electro-deposit on the cathode. The discharge solution was clear, almost colorless, with pH 8.4 and 0.5mg L^{-1} Cu and 0.07mM EDTA.

In both studies remediation method comprised of two separate phases (Figure 4.1). In the first phase we used EDTA for metal leaching, while in the second phase we used an EAOP for the treatment and reuse of the washing solution for soil rinsing (removal of soil-retained, chelant-mobilized metal complexes) in a closed loop.

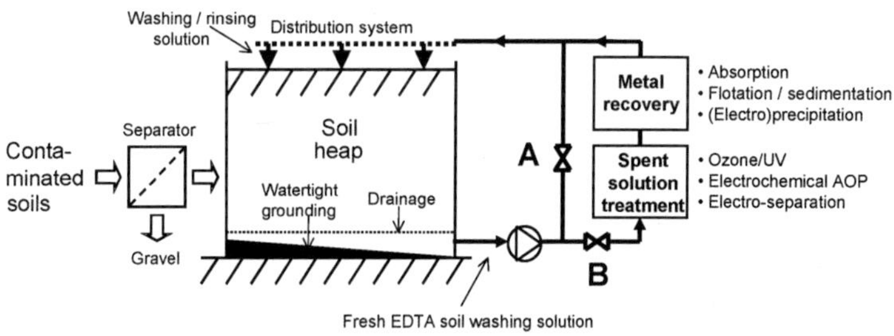

Figure 4.1 Conceptual presentation of the two-phase method for remediation of metal contaminated soils: (A) leaching of heavy metals from soil with chelants and (B) treatment and removal of heavy metals/chelants from the washing solution and soil rinsing.

In the last decade a special type of electrochemical reactions, the electro-Fenton system, has attracted considerable research interest (Pratap and Lemley, 1998; Tezcan Un et al., 2006). Traditionally the Fenton system is a mixture of ferrous salt and hydrogen peroxide. In electro-Fenton ferrous ion is produced from a sacrificial iron anode via the oxidation reaction:

$$Fe \rightarrow Fe^{2+} + 2e^{-}$$

Hydrogen peroxide is either added into the electrolytic cell or electrogenerated from two-electron reduction of sparged oxygen on the cathode. In this case, the cathode is made of porous carbonpolytetrefluorethylene. Ferrous ion reacts with hydrogen peroxide added externally to produce hydroxyl radicals, which then participate in other oxidation reactions:

$$H_2O_2 + Fe^{2+} \rightarrow Fe^{3+} + {}^{\cdot}OH + OH^-$$

However, our recent research has demonstrated that electro-Fenton treatment of the soil washing solution containing 570 mg L^{-1} Pb and 5.6 g L^{-1} EDTA at current density of 47.5 mA cm^{-2} and with added 3.0 g L^{-1} H_2O_2 was significantly less efficient than EAOP treatment with BDDA (Figure 4.2).

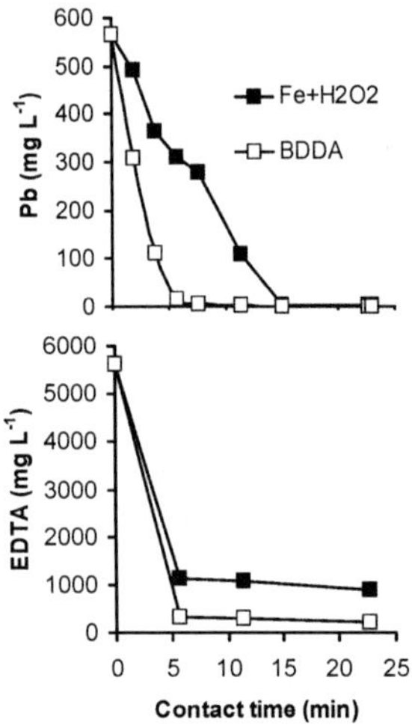

Figure 4.2 Removal of EDTA and Pb from the soil washing solution using electro-Fenton and BDDA EAOP. The released metal was removed from the solution by filtration as insoluble participate and by electro-deposition on the stainless steel cathode. Error bars represent standard deviation from the mean value (n=3).

Previous use of BDDA EAOP includes destruction of persistent organic substances (i.e. benzenes, phenols, various pesticides and pharmaceutical drugs) in waste aqueous solutions (Canizares et al., 2005; Polcaro et al., 2005; Oliveira et al., 2007). Wastewaters containing EDTA other than spent soil washing solution have also been successfully treated with BDDA EAOP (Kraft et al., 2003).

4.4 Electro-coagulation

Electro-coagulation is an efficient method for the purification of wastewaters containing organic or metallic pollutants. Electro-coagulation with aluminum and iron electrodes was patented in the US in 1909 and first applied on a large scale in the US in 1946 (Chen, 2004). When a potential difference is applied between the Al or Fe anode and a cathode, Al or Fe ions are generated from the anode and hydroxyl ions from the cathode. The reactions for the electrochemical system with an Al anode are as follows:

At the anode: $Al \rightarrow Al^{3+} + 3^{e-}$

At the cathode: $2H_2O + 2e^- \rightarrow H_2 + 2OH^-$

Al^{3+} and OH^- ions react further to form various monomeric Al hydroxides such as $Al(OH)^{2+}$, $Al(OH)_2^+$ or $Al_2(OH)_2^{4+}$ and polymeric Al hydroxides such as $Al_6(OH)_{15}^{3+}$, $Al_7(OH)_{17}^{4+}$, $Al_8(OH)_{20}^{4+}$, $Al_{13}O_4(OH)_{24}^{7+}$ or $Al_{13}(OH)_{34}^{5+}$ (Zaied and Bellakhal, 2009). Jiang et al. (2006) reported that Al_{13} polymers comprised 43% of all Al hydroxide species, with a long lasting positive charge. Finally, they all transform into amorphous $Al(OH)_3$ and combine to form flocks with a large surface area and considerable adsorption capacity for pollutants and for particle aggregation (Maximova and Dahl, 2006). Metals are removed from the solution by adsorption and co-precipitation on Al hydroxide flocks:

$Al(OH)_3 + M^{2+} \rightarrow Al(OH)O_2M + 2H^+$

Pociecha and Lestan (2010) evaluated electrocoagulation with an Al sacrificial anode using spent EDTA washing solution that was obtained after leaching of Pb (3200 mg kg^{-1}), Zn (1100 mg kg^{-1}), and Cd (21mg kg^{-1}) contaminated soil. A constant current density of 16-128 mA cm^{-2} was applied between the Al anode and the stainless-steel cathode. In the process negatively charged metal-EDTA complexes adsorb on flocks of various monomeric and polymeric positively charged Al hydroxides, which form during electro-coagulation, and were afterwards removed from the soil washing solution by sedimentation and centrifugation. In the experiment we removed up to 95% Pb, 68% Zn and 66% Cd. The discharge solution was clear and colorless, with pH 7.5. After treatment chelant balance showed that some EDTA was absorbed on the soil solid phases during soil extraction, up to 17% of the EDTA was adsorbed on Al coagulant and precipitated, the rest (more than half) of the initial EDTA remained in the washing solution, presumably after exchange of contaminating metal in the EDTA complex with Al. This trans-complexation phenomenon in an electric field was used to develop electrochemical EDTA recycling using electrolyis with sacrificial Al anode, as described below.

Electro-coagulation has a long history as a water treatment technology, but it has never been accepted as a mainstream technology (Holt et al., 2005). Nevertheless, electro-coagulation has been successfully used to treat special wastewaters, such as oil wastes (Adhoum and Monser, 2004), black liquor from the paper industry (Zaied and Bellakhal, 2009), dye solutions and textile wastewaters (Kaboya et al., 2003) and nutrient enriched wastewaters (Vasudevan et al., 2009). Although electro-coagulation is primarily used to remove organic and suspended materials from various types of effluents, its use for metal removal, for example Cu, Zn and Cr from electroplating wastewaters (Adhoum et al., 2004) and Pb from acidic soil leachate (Meunier et al., 2004) has also been recorded.

4.5 Electrochemical EDTA Recycling

As already mentioned soil washing of metal-contaminated soil by chelation with the recovery and reuse of the chelating agent represents an attractive objective towards commercialization of remediation technology. Currently, chemical hydroxide precipitation remains the more widely practiced approach on the basis of performance, ease of operation, and cost. To allow electrochemical EDTA recycling from the spent soil washing solution, treatment in a two chamber cell separated by a cation-selective membrane was proposed (Allen and Chen, 1993). During the treatments metals are liberated from the complex with EDTA, possibly on the outside of the electric double layer of the cathode:

$$M\text{-}EDTA^{2-} \rightarrow M^{+} + EDTA^{4-} \quad (M = Pb, Zn, Cd, Cu)$$

The anionic EDTA species cannot pass the cation-selective membrane into the anode chamber, which prevents them from oxidation (Figure 4.3). Metals (M) are reduced and deposited onto the cathode and the EDTA simultaneously recycled.

$$M^{2+} + 2e^{-} \rightarrow M_{(s)}$$

When the cell voltage is applied the anolyte provides Na^{+} to carry the current through the membrane into the cathode chamber, where protons are generated. A reasonably high feed of the spent washing solution is therefore required to prevent EDTA precipitation in acidic media and deposition on the membrane surface (Juang and Wang, 2000). This method is however prone also to operational problems such as membrane fouling and degradation (Di Palma et al., 2003).

The feasibility studies of the method are presented in several papers. Juang and Wang (2000) for example attempted electrolytic recovery of Cu and Pb from equimolar EDTA solutions in a two-chamber cell separated with a cation exchange membrane Neosepta CM-1. The iridium oxide coated on titanium (Ti / IrO_2) and stainless steel electrodes were used as the anode and cathode. The recovery of Cu was faster and more efficient than that of Pb. Under the optimal current density (139 A m^2, 50 min reaction time) and catholyte pH of 2.08 the recovery rate of Cu was higher than 96% when the initial Cu concentration was beyond 18 mM.

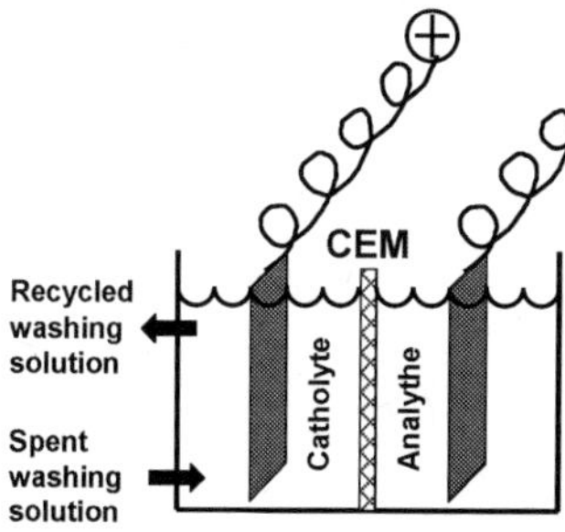

Figure 4.3 Electrochemical EDTA recycling from spent soil washing solution in a two chamber electrolytic cell. The catholyte (spent washing solution) and anolyte ($NaNO_3$) solutions are separated by a cation exchange membrane (CEM).

To overcome problems with the use of CEM membrane Pociecha and Lestan (2010) and Voglar and Lestan (2010) proposed treatment of the spent washing solution obtained after EDTA soil extraction in conventional single chamber electrolytic cell (without membrane) in alkaline conditions using a sacrificial Al anode. During the treatment Al ions are generated and form networks of Al-hydroxide flocks, as described above (Equation 1). To investigate the effect of pH on the electrochemical treatment process, the pH of the Pb and Cu washing solution was adjusted to 6 and 10. After treatment at pH 10, the EDTA in both solutions remained almost entirely preserved (Figures 4.4 and 4.5). On the other hand, approximately one half of the initial EDTA was removed from the solution treated at pH 6. This happened presumably by electro-coagulation of negatively charged EDTA complexes (i.e. Pb- and Cu-$EDTA^{2-}$) by Al-hydroxides. $Al(OH)_3$ is, however, a typical amphoteric metal hydroxide and in alkaline conditions forms negatively charged Al hydroxide:

$$Al(OH)_3 + OH^- \rightarrow Al(OH)_4^-$$

This negative charge of Al hydroxide flocks explains why negatively charged EDTA complexes were not removed but remained in the washing solution treated at pH 10 (Figures 4.4 and 4.5).

Due to the high Al reactivity (electro-positivity) Al is oxidized at the anode (Eq. 1) preferentially to EDTA oxidation. Thus while the EDTA remained almost completely preserved in the washing solution, electrochemical treatment at pH 10 efficiently removed both Pb and Cu (Figures 4.4 and 4.5). This was explained by the replacement of Pb and Cu from the complex with EDTA, removal of liberated Pb and Cu from the solution and formation of Al-EDTA complex. Although Al-EDTA has a lower log *Ks* than Pb-EDTA and Cu-EDTA (16.3, 18.0 and 18.8, respectively (Martell and Smith, 2003)), Al ions formed in abundant concentrations during electro-corrosion of the Al anode. Furthermore, the stability of Al-EDTA complex has been reported to be higher in a solution with pH 9 than in solutions with pH 7 and 4 (Treacy et al., 2000) while the stability of Pb- and Cu-EDTA complex slightly decreases in solutions with pH > 9 (Chang et al., 2007). Pb and Cu are released from the EDTA complex after the reduction reaction at the cathode, as explained in the equation above. After trans-complexation, metals liberated from the EDTA complex are removed from the solution by direct electro-deposition on the cathode, precipitation as insoluble hydroxides, or absorption and co-precipitation on Al hydroxide flocks as explained (in equations) above. Pociecha and Lestan (2010) and Voglar and Lestan (2010) reported that the majority of Pb (70%) and Cu (94%) was removed from the spent washing solution by electro-deposition on the cathode. Some Pb (17%) and a small amount of Cu (5%) were precipitated. Pb and Cu removal from the solution treated at pH 10 was faster and more efficient than from solutions with pH 6, resulting in a final Pb and Cu concentration of 150 and 0.3 mg L^{-1} (Figures 4.4 and 4.5).

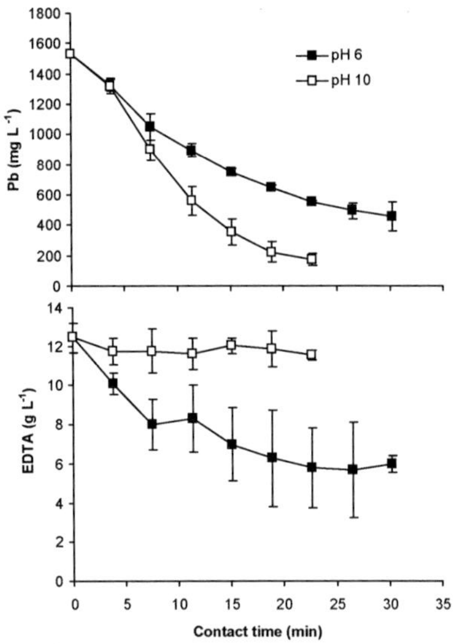

Figure 4.4 Concentrations of Pb and EDTA in the washing solution during electrochemical treatment at pH 6 and 10. Error bars represent standard deviation from the mean value (n=3).

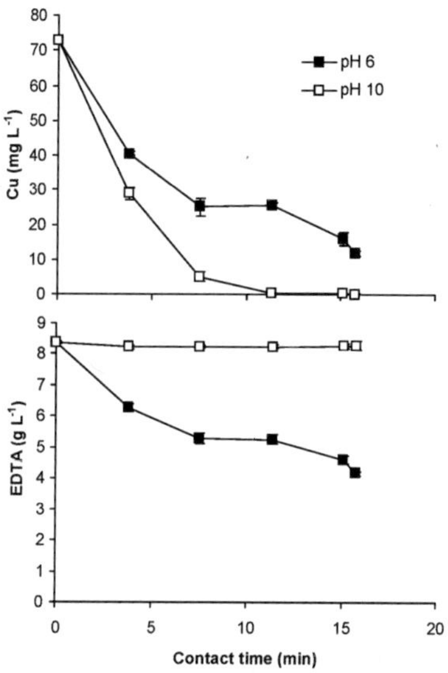

Figure 4.5 Concentrations of Cu and EDTA in the washing solution during electrochemical treatment at pH 6 and 10. Error bars represent standard deviation from the mean value (n=3).

The efficiency of EDTA recycled from a washing solution (electrochemically treated at pH 10) to extract Pb and Cu from the soil is shown in Figures 4.6 and 4.7. After adjustment to pH 4.3 (pH of the fresh EDTA washing solution), the treated washing solution retained almost 90% of the Pb and 80% of Cu extraction potential (from original soil) compared to freshly prepared EDTA solution of the same molarity and pH (Pociecha and Lestan, 2010; Voglar and Lestan, 2010).

The efficiency decrease of the recycled washing solution could partly be explained by the adsorption of EDTA into the soil during extraction and EDTA lost during the solution treatment phase (i.e. some initial anodic oxidation during electrode activation). Furthermore the Pb and Cu extraction efficiency of Al-EDTA (formed during electrochemical treatment) could be somewhat lower compared to the Na_2-EDTA in the fresh washing solution.

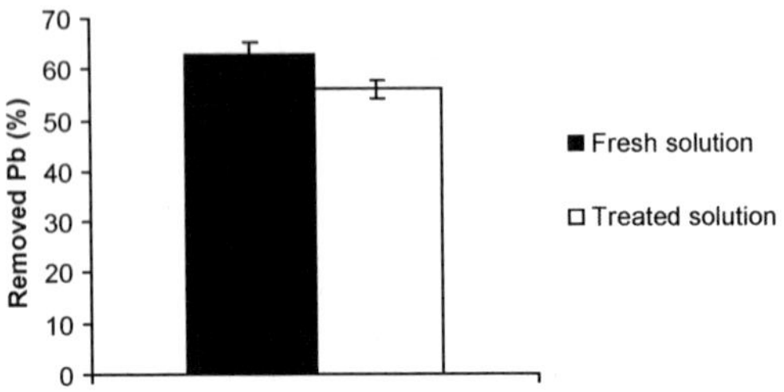

Figure 4.6 Removal of Pb from the soil using fresh and recycled EDTA solution (pH 4.3). Error bars represent standard deviation from the mean value (n=3).

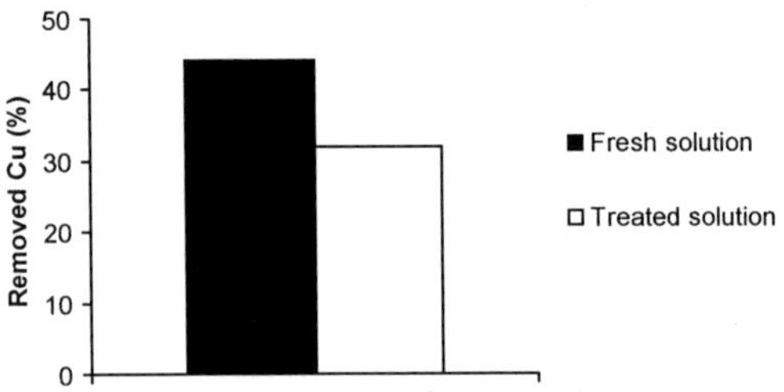

Figure 4.7 Removal of Cu from the soil using fresh and recycled EDTA solution (pH 4.3).

4.6 Conclusions

The contamination of soils with toxic metals has become a major environmental concern. For some soils washing off metals with solutions of chelating agents could represent a permanent remedial solution. In practice however, the use of chelating agents is prohibited by the large volumes of waste washing solution generated, which must be treated before safe disposal or reuse. The methods currently being proposed to treat spent soil washing solution before it can be safely discharged in the environment and recycle chelating agents are still encountering operational difficulties and work well only within a narrow range of contamination and soil types. Electrochemical methods: electro-oxidation, electro-coagulation and electrochemical recycling of chelating agent through trans-complexation have potential for the cost-effective treatment of the spent soil washing solutions as a part of soil remediation technologies.

4.7 References

Adhoum N., and Monser, L. (2004). "Decoloration and removal of phenolic compounds from olive mill wastewaters by electrocoagulation." *Chemical Engineering and Processing.*, 43, 1281–1287.

Adhoum, N., Monser, L., Bellakhal,N., and Belgaied, J. E. (2004). "Treatment of electroplating wastewater containing Cu^{2+}, Zn^{2+} and Cr(VI) by electrocoagulation." *Journal of Hazardous Materials*, 112, 207–213.

Ager, P., and Marshall, W. D. (2001). "The removal of metals and release of EDTA from pulp wash water." *Wood Science and Technology*, 21, 413-425.

Allen, H. E. and Chen, P. H. (1993). "Remediation of metal contaminated soil by EDTA incorporating electrochemical recovery of metal and EDTA." *Environmental Progress*, 12, 284-293.

Bohn, H. L., McNeal, B. L., and O'Connor, G. A. (1979). Soil Chemistry. Wiley, New York.

Buelt, J. L., and Farnsworth, R. K. (1991). "*In situ* vitrification of soils containing various metals." *Nuclear Technology*, 96, 178-184.

Canizares, P., Lobato, J., Paz, R., Rodrigo, M. A., and Saez, C. (2005). "Electrochemical oxidation of phenolic wastes with boron-doped diamond anodes." *Water Research*, 39, 2687–2703.

Chang, F.- C., Lo, S.- L., and Ko, C.- H. (2007). "Recovery of copper and chelating agents from sludge extracting solutions." *Separation and Purification Technology*, 53, 49-56.

Chen, G. (2004). "Electrochemical technologies in wastewater treatment." *Separation and Purification Technology*, 38, 11-41.

Di Palma, L., Ferrantelli, P., Merli, C., and Bianifiori, F. (2003). "Recovery of EDTA and metal precipitation from soil washing solutions." *Journal of Hazardous Materials*, 103, 153–168.

Di Palma, L., Ferrantelli, P., Merli, C., and Petrucci, E. (2003). "Treatment of the solution extracted from metal contaminated soils by reverse osmosis and chemical precipitation." *Annali di Chimica*, 93, 1005–1011.

Elliott, H. A., and Brown, G. (1989). "Comparative evaluation of NTA and EDTA for extractive decontamination of Pb-polluted soils." *Water, Air, and Soil Pollution*, 45, 361-369.

Finzgar, N., and Lestan, D. (2006). "Advanced oxidation for treatment of aqueous extracts from EDTA extraction of Pb and Zn contaminated soil." *Journal of Environmental Engineering*, 132, 1376–1380.

Finzgar, N., and Lestan, D. (2006). "Heap leaching of Pb and Zn contaminated soil using ozone/UV treatment of EDTA extractants." *Chemosphere*, 63, 1736–1743.

Finzgar, N., and Lestan, D. (2008). "The two-phase leaching of Pb, Zn and Cd contaminated soil using EDTA and electrochemical treatment of the washing solution." *Chemosphere*, 73, 1484-1491.

Guo, G., Zhou, Q., and Ma, L. Q. (2006). "Availability and assessment of fixing additives for the in situ remediation of heavy metal contaminated soils: A review." *Environmental Monitoring and Assessment*, 116, 513-528.

Hick, R. E., and Tondorf, S. (1994). "Electrorestoration of metal contaminated soils." *Environmental Science & Technology*, 28, 2203-2210.

Holt, P. K., Barton, G. W., and Mitchell, C. A. (2005). "The future for electrocoagulation as a localised water treatment technology." *Chemosphere*, 59, 355–367.

Hong, A. P. K., Li, C., Banerji, S. K., and Regmi, T. (1999). "Extraction, recovery, and biostability of EDTA for remediation of heavy metal-contaminated soil." *Journal of Soil Contamination*, 8, 81-103.

Huang, J. W., Chen, J. J., Berti, W. R., and Cunningham, S. D. (1997). "Phytoremediation of lead- contaminated soils: role of synthetic chelates in lead phytoextraction." *Environmental Science & Technology*, 31, 800-805.

Jiang, J. Q., Xu, Y., Quill, K., Simon, J., and Shettle, K. (2006). "Mechanisms of boron removal with electrocoagulation." *Environmental Chemistry*, 3, 350–354.

Jiraroj, D., Unob, F., and Hagege, A. (2006). "Degradation of Pb–EDTA complex by H_2O_2/UV process." *Water Research*, 40, 107–112.

Johnson, J. W., Jiang, H. W., Hanna, S. B., and James, W. J. (1972). "Anodic oxidation of ethylenediaminetetraacetic acid on Pt in acid sulphate solution." *Journal of the Electrochemical Society*, 119, 574-580.

Juang, R.-S., and Wang, S.-W. (2000). "Electrolytic recovery of binary metals and EDTA from strong complexed solutions." *Water Research*, 34, 3179-3185.

Juang, R-S., and Wang, S-W. (2000). "Electrolytic recovery of binary metals and EDTA from strong complexed solution." *Water Research*, 34, 3179-3185.

Kim, C. S. and Ong, S. K. (1999). "Recycling of lead-contaminated EDTA wastewater." *Journal of Hazardous Materials*, 69, 273-286.

Kobya, M., Can, O. T., and Bayramoglu, M. (2003). "Treatment of textile wastewaters by electrocoagulation using iron and aluminium electrodes," *Journal of Hazardous Materials*, 100, 163–178.

Kraft, A., Stadelmann, M., and Blaschke, M. (2003). "Anodic oxidation with doped diamond electrodes: a new advanced oxidation process." *Journal of Hazardous Materials*, 103, 247–261.

Luo, C. L., Shen, Z. G, and Li, X. D. (2005). "Enhanced phytoextraction of Cu, Pb, Zn and Cd with EDTA and EDDS." *Chemosphere*, 59, 1-11.

Martell, A. E., and Smith, R. M. (2003). NIST Critically Selected Stability Constants of Metal Complexes. Version 7.0. NIST, Gaithersburg.

Maximova, N., and Dahl, O. (2006). "Environmental implication of aggregation phenomena: current understanding." *Current Opinion in Colloid and Interface Science*, 11, 246–266.

Mulligan, C. N., Yong, R. N., and Gibbs, B. F. (2001). "Remediation technologies for metal-contaminated soils and groundwater: an evaluation." *Engineering Geology*, 60, 193-207.

Oliveira, R. T. S., Salazar-Banda, G. R., Santos, M. C., Calegaro, M. L., Miwa, D. W., Machado, S. A. S., and Avaca, L. A. (2007). "Electrochemical oxidation of benzene on boron-doped diamond electrodes." *Chemosphere*, 66, 2152–2158.

Peters, R. W. (1999). "Chelant extraction of heavy metals from contaminated soils." *Journal of Hazardous Materials*, 66, 151-210.

Pociecha, M., and Lestan, D. (2009). "EDTA leaching of Cu contaminated soil using electrochemical treatment of the washing solution." *Journal of Hazardous Materials*, 165,533-539.

Pociecha, M., and Lestan, D. (2010). "Electrochemical EDTA recycling with sacrificial Al anode for remediation of Pb." *Environmental Pollution*, doi:10.1016/j.envpol.2010.04.014.

Pociecha, M., and Lestan, D. (2010). "Using electrocoagulation for metal and chelant separation from washing solution after EDTA leaching of Pb, Zn and Cd contaminated soil." *Journal of Hazardous Materials*, 174, 670-678.

Polcaro, A. M., Vacca, A., Mascia, M., and Palmas, S. (2005). "Oxidation at boron doped diamond electrodes: an effective method to mineralise triazines." *Electrochimica Acta*, 50, 1841–1847.

Pratap, K., and Lemley, A. Z. (1998). "Fenton electrochemical treatment of aqueous atrazine and metolachlor." *Journal of Agricultural and Food Chemistry*, 46, 3285-3291.

Shu, H. Y., and Chang, M. C. (2005). "Pilot scale annular flow photoreactor by UC/H_2O_2 for the colorization of azo dye wastewater." *Journal of Hazardous Materials*, 125, 244–251.

Tandy, S., Bossart, K., Mueller, R., Ritschel, J., Hauser, L., Schulin, R., and Nowack, B. (2004). "Extraction of heavy metals from soils using biodegradable chelating agents." *Environmental Science & Technology*, 38, 937-944.

Tejowulan, R. S., and Hendershot, W. H. (1998). "Removal of trace metals from contaminated soils using EDTA incorporating resin trapping techniques." *Environmental Pollution*, 103, 135–142.

Tezcan Un, U., Ugur, S., Koparal, A. S., and Bakur Ogutveren, U. (2006). "Electrocoagulation of olive mill wastewaters." *Separation and Purification Technology*, 52, 136–141.

Treacy, G. M., Rudd, A. L., and Breslin, C. B. (2000). "Electrochemical behavior of aluminium in the presence of EDTA-containing chloride solutions." *Journal of Applied Electrochemistry*, 30, 675-683.

Troster, I., Schafer, I., and Fryda, M. (2002). "Recent development in production and application of DiaChem electrodes for wastewater treatment." *New Diamond and Frontier Carbon Technology*, 12, 89–97.

Vanthuyne, M., and Maes, A. (2002). "The removal of heavy metals from contaminated soil by a combination of sulfidisation and flotation." *Science of the Total Environment*, 290, 69-80.

Vasudevan, S., Lakshmi, J., Jayaraj, J., and Sozhan, G. (2009). "Remediation of phosphate contaminated water by electrocoagulation with aluminium alloy and mild steel anodes." *Journal of Hazardous Materials*, 164, 1480–1486.

Voglar, D., and Lestan, D. (2010). "Electrochemical separation and reuse of EDTA after extraction of Cu contaminated soil." *Journal of Hazardous Materials*, 180, 152-157.

Xu, Y., and Zhao, D. (2005). "Removal of copper from contaminated soil by use of poly (amidoamine) dendrimers." *Environmental Science & Technology*, 39, 2369-2375.

Yamaguchi, Y., Yamanaka, Y., Miyamoto, M., Fujishima, A., and Honda, K. (2006). "Hybrid electrochemical treatment for persistent metal complexes at conductive diamond electrodes and clarification of its reaction route." *Journal of the Electrochemical Society*, 153, 1123–1132.

Zaied, M., and Bellakhal, N. (2009). "Electrocoagulation treatment of black liquor from paper industry." *Journal of Hazardous Materials*, 163, 995–1000.

CHAPTER 5

Extraction of Metals from Spent Catalyst Using Fresh and Recovered Chelating Agents

Shina Goel and Alok Gautam

5.1 Introduction

A large quantity of catalysts is used in the fertilizer industry (i.e., ammonia plants), petroleum refineries, chemicals sector, various conversion processes, and automotive catalytic converters for pollution control. These catalysts include metals such as Ni, Mo, Co, Rh, Pt, and Pd in supported form (e.g., Ni-Mo sulfide/Al_2O_3, NiO/Al_2O_3, Pt/alumina, and/or Pt-Rh/alumina monolith). Catalysts contain 2-35% metals according to the requirement of the process. This amount is quite reasonable to recycle after discarding the catalyst as spent catalyst. The recovery of limited and expensive metals like Pt from spent catalyst could be a significant source of metals. Palladium is unique in that it absorbs 900 times its volume of hydrogen at room temperature. Furthermore, its chloride salt can absorb large amounts of carbon monoxide gas, therefore, it is commonly used in CO detectors. Palladium also serves as a catalyst for hydrogenation and de-hydrogenation processes (Marafi and Stanislaus, 2003a; Marafi and Stanislaus, 2007).

Similarly, spent catalysts are generated in the fertilizer industry as well, where Ni catalysts deactivate over a lifespan of about 5-7 years because of the harsh conditions in the primary and secondary reformer. Another example is the case of the low temperature water-gas shift reaction in ammonia plants, where the Cu-containing catalyst has a lifetime considerably lower than that of the Ni catalyst and needs to be recovered and reused. It is estimated that, because of the high volume of fertilizer required, more than 3000 t/year of spent catalyst is generated by China and India and 150000-170000 t/year of spent catalyst is generated worldwide. In Kuwait alone 7000 t/year of spent catalyst is produced from the hydroprocessing unit from petroleum industry (Eichler et al., 1996; Ho and Chi, 2004). In the case of refineries, the generation of spent catalysts has increased significantly because of a steady increase in the processing of heavy feedstock, containing high contents of sulfur, nitrogen, and metallic heteroatoms.

Metals present in the catalysts are generally in the form of oxides which form different complexes with oxygen, sulphur, hydrogen and carbon present in the process. Metals no longer stay in the oxide form. The catalysts in this condition do not work properly and are deactivated.

5.1.1 Deactivation of Catalysts

Catalysts are evaluated for their activity and stability. The performance of the catalyst is determined by parameters like lower pressure drop, lower tube wall temperature and longer operation close to equilibrium methane conversion. These parameters can be achieved by optimizing the properties of catalyst such as better coke resistance, easy reducibility, higher crushing strength, higher metal dispersion, higher surface area, higher pore volume, higher geometric surface area, resistance to thermal shocks, better heat transfer properties. The three most common causes of catalyst decay are fouling, poisoning or thermal degradation.

Fouling involves the deposition of material on a catalyst surface to block active sites. Coke deposition is the most common process but the deposition of rust and scale from elsewhere in the system is not uncommon. Traditional tubular nickel steam reforming catalysts deactivate over time due to either coke formation, the presence of fouling agents such as sulphur, chlorine or iron can be the result of physical breakdown due to thermal cycling and physical integrity (Furimsky, 1996; Bartholomew, 1984). The activity of a catalyst declines below an acceptable level due to degradation.

Catalyst poisoning involves strong interactions between a component of the feed or products and the active sites of the catalyst. Sulphur poisoning of metals is the most widely quoted example (McCulloch, 1983; Dowden, 1986) but depending on the catalyst, deactivation may be caused by a wide range of chemicals.

Catalysis involves interfaces, and heterogeneous catalysts are prepared with high surface areas, this is a thermodynamically unstable condition. If a suitable condition arises, such as high temperatures in the absence or presence of a suitable chemical environment, catalyst will rearrange to form the more favourable lower surface area agglomerates, which is known as sintering (Al-Dalama and Stanislaus, 2006; Furimsky and Massoth, 1999; Trimm, 1990; Marafi et al., 2007). Component interaction can also occurs due to overheating. The formation of nickel aluminate and nickel spinel from the reaction between nickel and alumina is a good example, with the catalytic activity of Ni-aluminate being much lower than that of the metal (Wagner et al., 2003). Alloy formation or phase separation can also occur which could lower overall catalytic activity. Thermal degradation, and particularly sintering is to reverse. Prevention is better than much after treatment (Chinchen, 1985). Measures to stabilize supports include the addition of lanthana and baria to alumina, stabilizers that fill vacancies in the lattice to reduce sintering. When online/in situ regeneration is not possible, further regeneration might not be economically feasible.

In such cases, spent catalysts tend to be discarded as solid wastes. Disposal of spent catalyst is a problem as it falls under the category of hazardous industrial waste.

5.1.2 Disposal of Catalyst

Although regeneration and re-use may be possible, eventually all catalysts will have to be replaced when product quality is compromised. Disposal of spent catalyst then depends both on economics and environmental laws. If the economic driving force is sufficiently large, recovery of catalyst component is justified, if not, the catalyst must be disposed off in a manner consistent with local legislation. Requirements vary from place to place but, since catalysts are often disposed of at landfill, the most relevant environmental legislation is based on leachability. Acceptable limits may vary, but values for drinking water provide a good guide. These include less than 0.7 ppb Co, 7 ppb Mo, 13.4 ppb Ni, 7 ppb V and 0.3 mg/l Fe. Encapsulation is another method to dispose the catalyst, if these limits exceed in the disposed environment. Encapsulation involves surrounding the waste with an impervious layer of sealant, such as bitumen, polyethylene or concrete. The encapulant should be stable over a long period of time, both with mechanical shear or weathering. The encapsulant adds the mass to waste and increases the volume for disposal (Marafi and Stanislaus, 2003; Noyes, 1991).

The problem of disposing of such a great amount as described above is not possible. If it is disposed each year by landfill or encapsulation, available land on the earth will no longer remain for human purposes within 5-8 years. It is now becoming necessity to recycle the catalyst by regeneration or by removing the metals from the catalyst by eco-friendly processes.

5.2 Methods to Extract Metals from Spent Catalyst

There are different methods available for the extraction of metals from spent catalyst discarded from various industries. Table 5.1 shows some examples of acid leaching, bioleaching, pyrometallurgy, and chelating agent extraction of metals from spent catalyst these processes.

Table 5.1 Selected examples of metal recovery from different spent catalyst by various processes

Spent catalyst	Metals present	Reagent and conditions	Recovery of metals	References
Primary reforming unit, fertilizer industry	Ni (7.12%)	50% H_2SO_4 conc. (100^0C, 5 h, 1:12 S:L, 800 rpm)	Ni (99%)	(Al-Mansi and Abdel Monem, 2002)
NiO catalyst	Ni (12%)	50% H_2SO_4 conc. (85^0 C, 150 min, 1:20S:L)	Ni (94%)	(Abdel-Aal and Rashad, 2004)
Ammonia plant	Ni	80% H_2SO_4 conc.(70^0 C, 50 min, 0.09 mm particle size)	Ni (99%)	(Invascanu and Roman, 1975)
Urea production	Ni (20%)	8% H_2SO_4 conc. (10% pulp density, 2 h, 90^0 C, 152 μm particle size)	Ni (98%)	(Sahu et al., 2005)
Low grade catalyst	Ni (17.7%)	28.8% HCl conc. (without external heating, 1 h)	Ni (73%)	(Chaudhary et al., 1993)
Sulphuric acid unit	Ni (0.63%) V (3.5%)	• H_2SO_4(0.3-1M) • NaOH(4M) followed by H_2SO_4(0.5M)	• Ni (96%) V (59%) • Ni (88%) V (78%)	(Ognyanova et al., 2009)
Naptha reforming from petroleum refinery	Pt (0.29%)	Aqua regia (3:1 37% HCl: 65% HNO_3), (100^0 C, 120 min, 800 rpm, 100μm particle size)	Pt (99%)	(Baghalla et al., 2009)
Petroleum industry	Ni (2%) V (9%) Mo (1.4%)	• H_2SO_4(1M), 1h • H_2SO_4 and NH_4CO_3 • LIX-841(10%)	• Ni (95%) V (95%) Mo (95%) • Mo (100%) • Mo (98%)	(Mishra et al., 2010)
Sulphuric acid unit	V (5.18%) Mo (17.84%)	0.5M Aliquat-336 in Kerosene, 25^0 C	V (93%)	(El-Nadi et al., 2009)

Spent catalyst	Metals present	Reagent and conditions	Recovery of metals	References
Refinery processing	Ni (6.09%) Mo (13.72%) Al (33.3%)	Oxalic acid secretion by Aspergillus Niger, 30 days	Ni (62.8%) Mo (78.9%) Al (58%)	(Santhiya and Ting, 2005)
Fischer-Tropsch process	Al (65%), Co (19.6%), Pt (0.05%)	• NaOH (pressure 9 bar, 190^0 C) • NaOH (12.5 M, 15 bar pressure, 190^0 C) • HCl (4.1M), HNO_3 (5M)(90^0 C)	• Al (89%) • Al (97%) • Al (99.97%) Co (99.7%) Pt (37.5%)	(Matjie et al., 2005)
Hydrodesulphurization unit	Mo (12.5%) CO (1.78%) Ni (0.56%)	Cyanex 272, H_2SO_4 (2M), pH 3.2	Mo (98%) Co (93%) Ni (90%)	(Park et al., 2007)
Palladium based catalyst	Pd (0.5%)	Cyanex 302, supercritical CO_2 (8 MPa, 40-80^0 C, 10 min)	Pd (100%)	(Iwao et al., 2007)
Automobile catalytic converter unit	Pt Pd Rh	Supercritical CO_2 with Tributyl Phosphate (60^0 C, 60 min, 20 MPa pressure)	Pd (96%)	(Faisal et al., 2008)
Hydrorefining unit	Ni (5%) Mo (20%)	NaCl (900^0 C, 60 min)	Mo (80%)	(Kar et al., 2005)
Primary reforming unit	Ni (22%)	Fresh and recovered EDTA (100^0 C, 700 rpm, 1:50 S:L, 10 h)	Ni (95%)	(Goel et al., 2009)

5.2.1 Basic Mechanism of Extraction from Existing Processes

The alkaline leaching step is followed by an acidic leaching treatment Eq. (1-4), (Ognyanova et al., 2009).

$$V_2O_3 + H_2SO_4 + O_2 \longrightarrow (VO_2)_2SO_4 + H_2O \quad (1)$$
$$NiO + H_2SO_4 \longrightarrow NiSO_4 + H_2O \quad (2)$$
$$2FeSO_4 + H_2SO_4 + 1/2\ O_2 \longrightarrow Fe_2(SO_4)_3 + H_2O \quad (3)$$
$$V_2O_3 + 2NaOH + O_2 \longrightarrow 2NaVO_3 + H_2O \quad (4)$$

A two stage leaching process for the maximum solubility of Mo and Co/Ni oxides at different pH regions. A first stage leaching with an alkali results in preferential dissolution of Mo into the liquid phase, with other metals remaining in the solid. A second acidic media leaching step will subsequently dissolve Co and Ni. In both reactions subsequent amount of Al_2O_3 is dissolved (Park et al., 2007).

(A) Basic media

$$MoO_3 + Na_2CO_3 \longrightarrow Na_2MoO_4 + CO_2 \quad (5)$$
$$Al_2O_3 + Na_2CO_3 \longrightarrow 2NaAlO_2 + CO_2 \quad (6)$$
$$CoO + Na_2CO_3 \longrightarrow Na_2CoO_2 + CO_2 \quad (7)$$
$$NiO + Na_2CO_3 \longrightarrow Na_2NiO_2 + CO_2 \quad (8)$$

(B) Acidic media

$$Al_2O_3 + 3H_2SO_4 \longrightarrow Al_2\ (SO_4)_3 + 3H_2O \quad (9)$$
$$CoO + H_2SO_4 \longrightarrow CoSO4 + H2O \quad (10)$$
$$NiO + H_2SO_4 \longrightarrow NiSO4 + H2O \quad (11)$$

The reactions mechanism of Pt extraction from aqua regia undergo through reactions (Baghalla et al., 2009).

$$HNO_3 + 3HCl \longrightarrow NOCl + Cl_2 + H_2O \quad (12)$$
$$NOCl + H_2O \longrightarrow HNO_2 + HCl \quad (13)$$

Dissolution of Pt from the catalyst is a redox reaction that undergoes according to reaction (14)

$$8H^+ + 8Cl^- + 2NO_3^- + Pt \longrightarrow PtCl_6^{2-} + 4H_2O + 2NOCl \quad (14)$$

The roasting of molybdenum spent catalyst with sodium chloride leads to the formation of soluble sodium molybdite (Kar et al., 2005). The reaction involves are:

$$MoO_3 + 2NaCl + \tfrac{1}{2}O_2 \longrightarrow Na_2MoO_4 + 2HCl \quad (15)$$
$$Na_2MoO_4 + NH_3 \longrightarrow (NH_3)_6[MoO_4] \quad (16)$$

The role of EDTA is to facilitate the transport of metal ions, exploiting the high mobility of anionic EDTA complexes. In some cases, dissolved complexes prevent adsorption. These complexes could solubilise metals that are trapped within the catalyst under certain conditions. Metals could be more adsorbed in the presence of free EDTA than the coordinated EDTA. The expected reactions are given as below. S is the solid matrix of catalyst, M is the metal content and L is the ligand or chelating agent (Goel and Gautam, 2010).

$$S\text{-}OM^{+} + H^{+} \longleftrightarrow S\text{-}OH + M^{2+} \quad (17)$$

$$Y^{4-} + M^{2+} \longleftrightarrow MY^{2-} \quad (18)$$

$$S\text{-}OH + ML^{2-} + H^{+} \longleftrightarrow S\text{-}LM^{-} + H_2O \quad (19)$$

5.2.2 Pros and Cons of Available Methods

Sulphuric acid and the addition of organic extractant in kerosene diluents were used to extract Mo, Ni and V. This process has enhanced Mo kinetics, sulphur recovery and processing of leach liquor to recover metal values. Processing of leach liquor has a long process of acid and alkali addition and stripping. The stripping step is also a repetitive step until get the final metal values. This process is time consuming and only organic extractant can be separated at the end of the process easily while extracted metals present in aqueous solution is a tedious work to separate (Mishra et al., 2010).

The selective recovery of Ni from an appreciable amount of Al from weakly acidic sulphate solution was achieved by means of ion exchange with Dowex XFS 4195 resin and complexane type of chemically modified chitosan, Figure 5.1. Modified chitosan were prepared by making EDTA and DTPA anhydride and reacted with chitosan. From batchwise adsorption of Ni, Cu, Co Zn and Al, all the metals can be separated (Nagib et al., 1999).

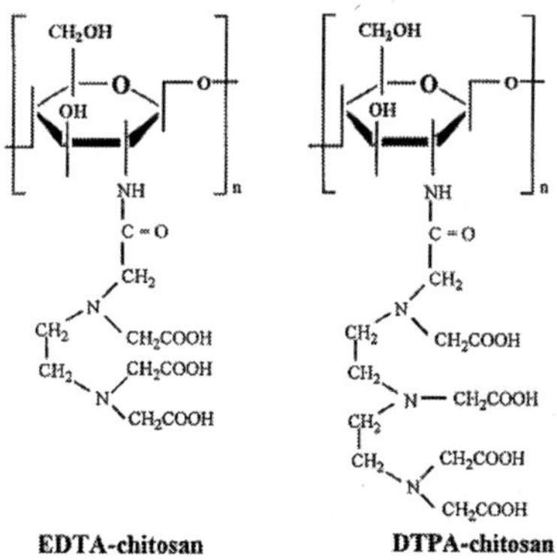

Figure 5.1 Chemical structure of modified chitosan (adapted from Nagib et al., 1999).

Extraction of Pd with supercritical CO_2 was obtained up to 100% with the addition of Cyanex 302. The extraction of Pd was also done with tributyl phosphate and HNO_3 but this process requires more pressure than the Cyanex 302 and the extraction is not up to 100% (Iwao et al., 2007). In another study, a catalyst containing Pd and other metals had less extraction efficiency with tributyl phosphate at longer stirring time. It shows that other metals present in the catalyst affected the extraction of Pd with supercritical fluids, while other metals were not extracted at all with this fluid (Faisal et al., 2008). Supercritical fluid with chelating agent can extract total amount of Pd present in catalysts but this is not applicable for the other metals present in those catalysts. Other metals present were not extracted and these metals also influence the extraction of Pd. This fluid is suitable only for a group of metal ions including Pd and Au in different media (Pourmortazavi et al., 2004). The only advantage of using the Cyanex compounds is the lower extraction time than any other processes like acid leaching or pyrometallurgy.

Liquid-liquid extraction of spent autocatalysts was done by using cyanex 923 from aqueous hydrochloric acid media, and 98% of the extraction was achieved for the Ir and Rh which were present at 0.1% each in the catalyst (Chavan and Dhadke, 2002). The amount of Ir and Rh was present in very small amount compared to other metals like Fe, Mo, Ni, Co. Cyanex compounds which have 'O' and 'S' as donor atoms and hence are useful for extraction of class *b* metal ions, can not work solely. It needs some carrier compound to react with metal and make neutral coordinated compounds.

Acid leaching by different acids like sulphuric acid, hydrochloric acid, aqua-regia can extract up to 98-99% of the metal values. These acids are not expensive but handling of these hazardous materials at large scale is difficult. The acids are not recovered in any of the process and more than one dissolved in the acid solution is difficult to separate. The process available to separate metals by liquid-liquid extraction is a long process. The metal extraction by chelating agents is comparatively easier, eco-friendly if the chelating undergoes for recycling, less energy consuming, less expensive, and easy to recover chelating agent and metals by change of pH. This chapter will describe the chelating agents advantage, recycling, kinetics and basic process to extract the metals from spent catalyst.

5.3 Advantages of Chelating Agents

Chelating agents are the most effective extractants that can be introduced to spent catalyst to enhance heavy metal extraction. The advantages of chelating agents in soil cleanup and metal removal from catalyst include high efficiency of metal extraction, high thermodynamic stabilities and good solubilities of metal complexes, and low adsorption of the chelating agents on soils. Chelating agents cause only minor impact on the physical and chemical properties of the catalyst matrix compared to acids (Fisher et al., 1998).

Ethylenediaminetetraacetic acid (EDTA) is the most widely used synthetic chelating agent in soil washing. It is an effective, recoverable and reusable chelating agent that has great potential for full scale application. Many studies have reported that EDTA could extract very high percentages of Pb and Cd from contaminated soils (Steele and Pichtel, 1998; Papassiopi et al., 1999; Garrabrants and Kasson, 2000; Kim and Ong, 1999; Wassay et al., 2001). A disadvantage arises associated with EDTA usage when it has to be destroyed before discharge. The compound is generally regarded as non-biodegradable and can be found in sewage effluents, and accumulates in surface waters and groundwater (Groot et al., 1995; Kari and Giger, 1995). To overcome this disadvantage, a cost-effective technique is required to recover the EDTA from the waste stream to reuse. Removal of metals from spent catalyst by chelation can be a valid remediation method. Important properties of the chelating agent used are: strength of the chelation bonding, reusability, biostability during the operation.

Chelating agents have been used for many years by industry and analytical chemists because of their sequestering or masking properties: that is, the ability to suppress the activity of a dissolved metal ion without its physical removal from the solution. The three most commonly used chelating agents are the polyphosphate, hydroxycarboxylic acid and amino poly carboxylic acid. Amino poly carboxylic acid chelating agents (e.g. EDTA, NTA, EGTA, and DTPA) are used most frequently, because they bind metal ions more strongly than polyphosphates and maintain their sequestering ability over a wider pH range than hydrolytic acid types (Bell, 1977). These are used more often than DTPA (diethyleneaminetriaminepentaacetic acid) and EGTA (ethylene glycol bis (2-aminoethylether) tetra acetic acid) because of their ability to form stable, water soluble complexes with a wide variety of metal ions. Anthropogenic organic chelating agents of the aminopolycarboxylates, such as EDTA and DTPA, are affected by the presence of metals, which are coordinated. These complexes can substantially affect chelating agent adsorption, precipitation, ligand assisted dissolution, metal mobilization, chemical degradation, photodegradation and biodegradation. It was emphasised that the reactions of both anthropogenic and natural chelating agents depend on the metals already coordinated with the chelator in the solution (Nowack, 2002; and Lim et al., 2005).

5.4 Mechanism of Complex Formation of Metal-Chelating Agent

A chelant is a ligand that contains two or more donor groups so that more than one bond is formed between the metal ion and the ligand. The unusual property of EDTA is its ability to chelate or complex metal ions in 1:1 metal-EDTA complexes (Figure 5.2).

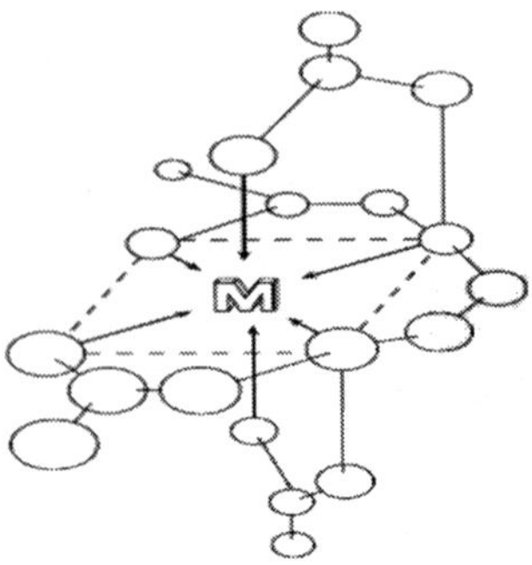

Figure 5.2 Metal and EDTA 1:1 complex.

The fully deprotonated form (all acidic hydrogens removed) of EDTA binds to the metal ion. The equilibrium or formation constants for most metals, especially the transition metals, are very large, hence the reactions favour complex formation. Many of the reactions are pH dependent, especially the weaker forming complexes with Ca^{2+} or Mg^{2+}.

$$Mn^{+} + L^{4-} \longrightarrow ML^{n-4} \quad K_f = (ML^{n-4})/(Mn^{+})(L^{4-}) \qquad (20)$$

Chelating agents show acid base properties and equilibrium because of having amide and acid groups together. Effective chelating agents typically have multiple coordination sites (ligand atoms) available for complexation with a metal centre. They have multi-protic acids (H_nL) capable of undergoing acid-base equilibrium reactions in the aqueous phase (Peters, 1999):

$$H_nL = H^{+} + H_{n-1}L^{-} \qquad (21)$$
$$H_{n-m}L^{m-} = H^{+} + H_{n-(m+1)}L^{-(m+1)} \qquad (22)$$

The important issues concerning the selection of chelants and the development of washing solutions are summarized as follows:

* Extraction strength: The chelant should be able to form strong and stable complexes with toxic metals over a wide pH range.
* Extraction selectivity towards desired toxic metals should be higher.
* Potential for recovering the spent chelant, If the chelant is to be recycled and reused in the process several times, it should have low biodegradability in soil.
* The metal-chelant complexes should have low adsorption affinity towards solid surfaces.
* The chelant should have low toxicity and low potential to harm the environment.
* The chelant should be cost effective (Fisher et al., 1998; Peters, 1999).

5.5 Extraction of Metals from Spent Catalyst with Fresh Chelating Agents

Existing literature of metal removal from spent catalyst by chelating agents is limited. Chelating agents with supercritical fluid was studied with some researchers as described in available methods but only chelating agents were used by very few researchers. Vuyyuru et al. (2010) studied the extraction of nickel metal from primary reforming spent catalyst by pressure assisted autoclave with chelating agent. The extraction of Ni was higher when the chelation step was carried out at higher temperatures in an autoclave. The fact that there was autogenous pressure build up in the autoclave with increase in temperature had little bearing on the extraction efficiency. It was primarily the effect of enhanced kinetics due to higher operating temperatures. However, Hong et al., (2008) reported break up of soil particles in the presence of high pressure and enhancement of extraction and up to 100% of the metal extraction from soil was acheived. A temperature of 150 °C was considered optimal by Vuyyuru et al. (2010) as higher temperatures (170 °C) did not significantly enhance the metal recovery and can adversely affect process economics, in terms of consumption of steam utilities. Furthermore, it was assumed that, at temperatures higher than 170 °C, chemical reaction kinetics might not be rate-limiting and external or internal mass-transfer effects or equilibrium limitations might begin to play a role. Because of an exothermic reaction, the equilibrium uptake of Ni is expected to decrease with increasing temperature.

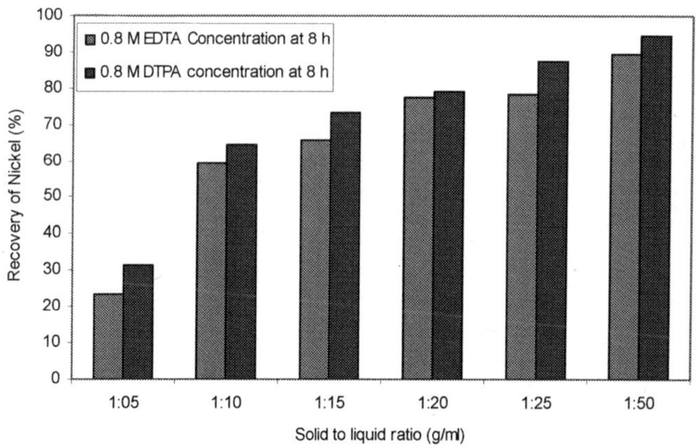

Figure 5.3 Extraction of nickel from spent catalyst using two different chelating agents at atmospheric pressure and 100^0 C with change in catalyst to chelating agent (EDTA and DTPA) solution ratio at 8 h of contact time (Goel and Gautam, 2010). Copyright Elsevier 2010, reproduced with permission.

The extraction of nickel from primary reforming catalyst was done at atmospheric pressure at 100^0 C by Goel and Gautam (2010). The removal of nickel is shown with two different chelating agents in (Figure 5.3). The extraction of metal is higher with the DTPA than the EDTA at 1:50 solid to liquid ratio by approximately 5%. It is evident from Figure 5.3 that solid to liquid ratio is a quiet an important factor for the extraction of metal; increase in ratio from 1:05 to 1:50 enhanced the extraction from 20-90% and 31-95% with EDTA and DTPA, respectively. Ni was extracted up to 87-95 % in one cycle and the rest of the Ni was still present in the catalyst. To make this 5-8 % Ni free, second cycle of experiment was done with 0.008 M concentration of chelating agent keeping other conditions the same. Concentration for the second cycle was calculated according to the Ni remaining in the catalyst. The first cycle extracted 77 % of total amount present and rest of the Ni was extracted in the second stage. Chelating agent concentration was chosen to be 0.008 M according to metal to chelating agent ratio and it was kept at 1:2 for the second cycle. This second stage of extraction released 99-99.5% of Ni present in the catalyst.

5.6 Recycle or Regeneration Methods of Chelating Agents

Chelating agents are organic compounds that could be subject to biodegradation when exposed to the environment. The biodegradation process of EDTA is very slow or almost non biodegradable. To restore the environment it is necessary to recycle and reuse the EDTA and other chelating agents. The cost of the chelating agent can be an important issue in metal extraction from spent catalyst. Methods that recycle process solution may make the use of chelants more economically feasible. There are several methods available for regeneration of chelating agents. Regeneration of chelating agents is highly dependent on the pH of metal chelating agent complex solution, the concentration of the metal and EDTA and the presence of electrolytes (Chang, 1995; Brown and Elliot (1992); Wanninen and Ringbom, 1979). A comparison of the conditional stability constants for various metal-EDTA complexes as a function of pH is given by Brown and Elliot, 1992. Chemicals are added to lower the pH conditions (<3), the tendency for metal-EDTA complexes to form may be assumed to follow the following sequence: $Fe^{3+} > Cu^{2+} > Pb^{2+} > Al^{3+} > Zn^{2+} > Mn^{2+} > Ca^{2+} > Mg^{2+}$. Chemical addition is done to adjust to lower the pH to remove the different metals and increase the pH again by chemical addition to separate the chelating agents.

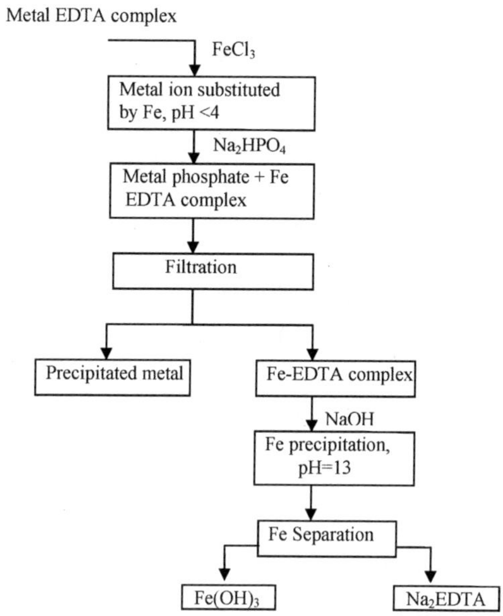

Figure 5.4 Flow chart of recycling process of EDTA (adapted from Kim and Ong, 1999).

The process based on stability of metal-EDTA is presented in Figure 5.4 (Kim and Ong, 1999). According to the sequence described above, Fe^{3+} can substitute most of the metals from the EDTA and make Fe-EDTA complex. Through addition of sulphates or phosphates metal can be separated from the Fe-EDTA complex. The final recovery of EDTA is not reported and Fe-EDTA complex solution is used as recycle of EDTA. The similar process was used by Chang et al. (2007a) and Chang et al. (2007b) and DTPA was also recovered by this process. The process suggested by Chang (1995) used the sulphuric acid to lower the pH rather than $FeCl_3$, after which ferrous sulphate was added to ensure the complete cation replacement with metal/Fe ratio of 3. Hong et al. (1998) used Na_2S (3-50mM) with and without $Ca(OH)_2$. EDTA was precipitated and used at slight excess on a molar basis at moderately alkaline conditions (pH=10) and reused over several cycles of operation. Lim et al. (2005) replaced $FeCl_3$ by $Fe(NO_3)_3$ to recover the metals Pb, Cd, and Ni. Similar process was adopted by Zeng et al. (2005), in which the metal-EDTA complex solution was treated with $Ca(OH)_2$, then a 0.5 M Na_2S solution with gentle stirring. The pH of the solution was elevated up to 10.5 and kept overnight to precipitate the metals. The pH after metal precipitation was lowered by 10% nitric acid to 4-5. This solution was recycled rather than obtaining solid EDTA.

The precipitation of EDTA in protonated form was done by Palma et al. (2003a). Before lowering the pH by acidification solution metal-EDTA was evaporated and reduced up to 75% by volume. Palma et al. (2003b) proposed reverse osmosis to reduce the volume of extractant. The sulphuric acid was used for acidification and EDTA in solid form was separated by filtration. A 93% of the EDTA was recovered by this process and transformation into sodium salt was done by an alkaline agent. The similar process was adopted by Goel et al. (2009) without evaporation, acidification was done in total volume of metal-EDTA complex solution. Cooling was done during addition of sulphuric acid to support the precipitation of EDTA. A 98% of the EDTA was recovered as solid form and reused again in protonated form. The transformation of protonated EDTA into sodium salt was done by stirring the protonated EDTA in Na_2S solution for few hours. The fresh and recycled EDTA efficiency to recover metal is shown in Figure 5.5. The efficiency of recycled EDTA was observed slightly less compared to fresh EDTA with different catalyst to extractant solution ratio. Vuyyuru et al. (2010) used nitric acid for the acidification without evaporation of solution to reduce the sulphur contamination in recovered EDTA.

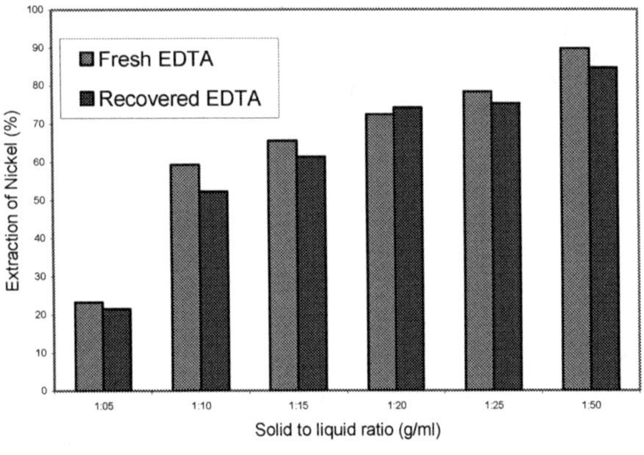

Figure 5.5 Efficiency of fresh and recovered EDTA with increasing catalyst to EDTA solution ratio to extract the nickel at atmospheric pressure, 100^0 C and 8 h of extraction time (Goel et al., 2009). Copyright Elsevier 2009 reproduced with permission.

The precipitation of other chelating agents such as DTPA was not observed useful by Goel and Gautam (2010). It was observed that DTPA could precipitate in smaller amount as compared to EDTA. After filtration of the extractant solution, the solution was kept in centrifuge for 30 min at 6000 rpm and kept for another 10 h at room temperature to precipitate the DTPA. A 90% of the DTPA was recovered by acidification and centrifugation. Efficiency of EDTA and DTPA after every cycle of Ni extraction is shown in Figure 5.6. The efficiency was reduced slightly but the capacity to extract Ni was greater of DTPA than EDTA. That may be the molecular structure of DTPA has eight binding sites for metal as it has two ethylene, three amine and five acetic acid while EDTA has only six sites to bind the metals. These sites enhance the capacity of DTPA to extract more metals than EDTA and takes longer and complex method to break the complex to recover DTPA.

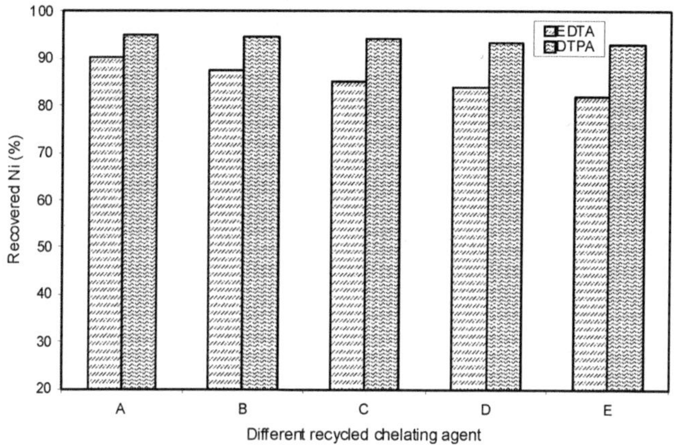

Figure 5.6 Reduced recovery of nickel after each recycle of EDTA and DTPA, A first recycle, B second recycle, C third recycle, D fourth recycle, E fifth recycle (Goel and Gautam, 2010). Copyright Elsevier 2010 reproduced with permission.

Chemical addition for recovery of metal and chelating agents is useful when there is only one metal present in the solution. Problem arises when there are more than one metal present in the solution and the pH of these metals for separation are very close. Without addition of chemicals, electrochemical recovery of metal and EDTA was proposed by Allen and Chen (1993). A two chamber cell separated by a cation exchange membrane, to prevent migration to the anode and the oxidative destruction of negatively charged metal-EDTA complexes, was used for this separation. The recovery of metal and EDTA was more than 95%. In electrochemical and reverse osmosis membranes can be clogged after continuous usage and thus diminish the performance and shorten the lifetime of the membranes. To separate the metal-EDTA complex some anion exchange resins are used by Tejowulan and Hendershot (1998). The chelex and AER resins were used to extract Cd, Cu, Pb and Zn up to 99% from the complex with the mixture with different ratios of the resin to the EDTA complex with metal. The chemical destruction of EDTA and its complexes using advanced oxidation processes (AOP) has been proposed (Korhonen et al., 2000; Munoz and Von Sonntag, 2000). AOP involves the use of ozone, H_2O_2, ultrasonic waves, UV irradiation, Fenton's reagent (Fe^{2+} and H_2O_2), alone or in combination, and electrochemical methods, to generate free hydroxyl radicals that are powerful, effective and non specific oxidizing agents.

5.7 Characterization of Recycled EDTA

Nuclear magnetic resonance (NMR) spectroscopy [Bruker AC 300 nuclear magnetic resonance spectrometer (300 MHz FT-NMR)] was used to determine its structure of the EDTA. SEM EDX analysis was used for its elemental composition. The system used was RONTEC's EDX system (model QuanTax 200), which provides an energy resolution of 127 eV for Mn KR. The samples were placed onto a metallic support and covered with a thin silver film. The electron micrographs were obtained at 15 kV. Elemental dot maps of sections were made by scanning the surface of EDTA with an electron beam to generate characteristic X-radiation from elements excited in the sample (Goel et al., 2009).

5.7.1 NMR (Nuclear Magnetic Resonance)

Proton NMR of fresh and recycled EDTA revealed only two types of hydrogen as shown in the Figure 5.7. The solvent used for the proton NMR was D_2O. Hydrogen impurities were present in this sample as shown by a peak in the spectrum (peak 3). There are three types of hydrogen in the EDTA structure.

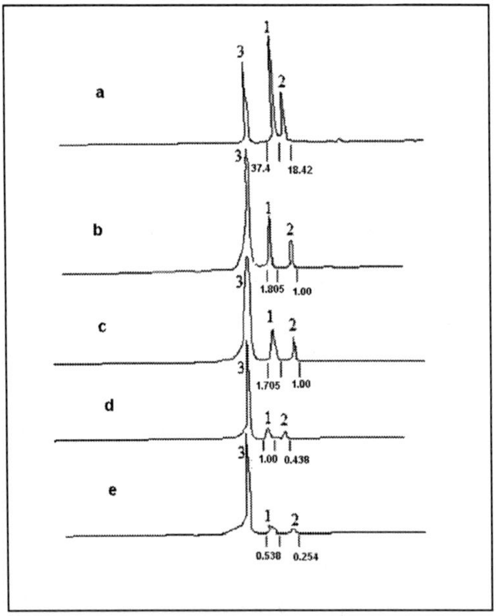

Figure 5.7 Confirmation of hydrogen bonding by Proton NMR spectrum of (a) fresh and recycled EDTA, (b) first recycle EDTA, (c) second recycled EDTA, (d) third recycled EDTA, (e) fourth recycled EDTA (Goel et al., 2009). Copyright Elsevier 2009 reproduced with permission.

Due to the reaction with deuterium, the acidic hydrogen was replaced by the deuterium and the other two hydrogen atoms, ethylene in acetic acid and ethylene with amine are labeled 1 and 2, respectively. The structure of the EDTA is symmetric if divided down the middle. It should show a total of three peaks in the spectrum. The hydrogen atoms present at ethylene of acetic acid are twice of ethylene with amine. The area of peak 2 and 1 are in the ratio of 1:2 due to the number of hydrogen atoms in the EDTA structure. The fresh and recycled EDTA shows the same behavior and did not change during repetitive usage. The area underneath the peaks of fresh EDTA is 18.42 and 37.4 in Figure 5.7a. For the fourth recycle of EDTA the area is 0.0438 and 0.0238 in Figure 5.7e. Thus the ratio is maintained at 1:2 up to the fourth recycle of EDTA. There is no dissociation of the structure of the hydrogen atoms.

5.7.2 FTIR (Fourier Transform Infrared Spectroscopy)

Infrared spectroscopy has long been used to identify structural characteristics of metal-ligands. EDTA salt compounds and recovered EDTA have previously been analysed by traditional transmission methods. The spectrum is shown in Figure 5.8 (a-c). The solid salt of EDTA was mixed with KBr and pellets were made for analysis.

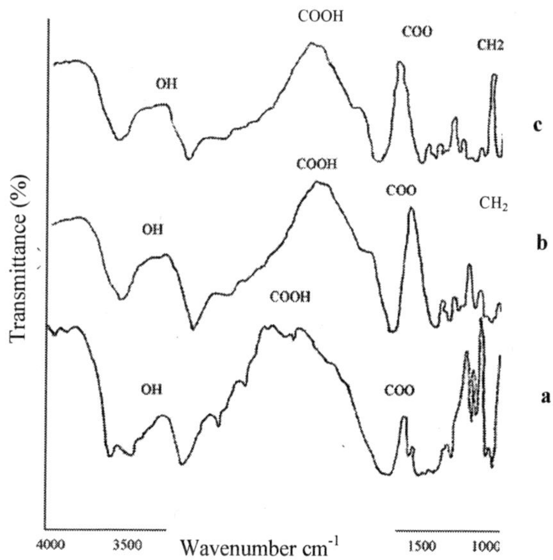

Figure 5.8 Confirmation of different group bonding present within EDTA by FTIR spectrum of (a) fresh EDTA, (b) first recycled EDTA, (c) second recycled EDTA (Goel et al., 2009). Copyright Elsevier 2009 reproduced with permission.

5.7.2.1 Region 3600- 3000 cm^{-1}

The region between 3600 and 3200 cm^{-1} is generally attributed to O-H and N-H stretching vibrations. For hydrogen bonded dimeric carboxylic acids, a downward shift of the O-H stretching band to near 3000 cm^{-1} is typical, this is shown by fresh and recycled EDTA. Bands between 3600 and 3000 cm^{-1} indicate the presence of hydroxyl groups as a result of hydrogen bonding and complexed water molecules as well as N-H stretching. All three of these stretching vibrations are observed in each of the spectrums shown Figure 5.8.

5.7.2.2 Region 3000-2500 cm^{-1}

Broad vibrational bands found between 2700 and 2500 cm^{-1} are characteristic of carboxylic acid dimmers and are assigned to overtone and combination bands of C-O stretching and O-H bending.

5.7.2.3 Region 1700-1500 cm^{-1}

Intermolecular interactions make the crystalline spectra more complicated than the aqueous spectra. The spectrum was taken in crystalline form of KBr pellets. Hydrogen bonding tends to decrease the frequency of the carboxyl groups and the carbonyl vibration of free carboxylic acid groups. It also increases the frequency of the antisymetric carboxylate stretching vibration. The typical range for saturated aliphatic carboxylic acids is 1725-1705 cm^{-1}. The band at 1711 cm^{-1} is likely the result of fresh and recovered EDTA acid groups.

Table 5.2 Assignments of vibrational bands (cm^{-1}) obtained from FTIR (Goel et al., 2009).

Molecular → structures / EDTA ↓	ν_{as} CO_2^-	ν_{sym} CO_2^-	δNH^+	$-COO^-$	-COOH	S=O
Fresh	1625	1400	1360	1320	1190	-
Recycled (I)	1643	1425	1371	1314	1256	1044
Recycled (II)	1648	1411	1368	1315	1257	1039

5.7.2.4 Region 1500-1900 cm^{-1}

For carboxylic acids and carboxylates, this region includes bands for O-H bending, C-O stretching, and C-H bending vibrations. These bands tend to have weaker intensities compared to the carboxyl group bands. In the region that have strong absorption can enhance the band intensity for other functional groups (Lanigan and Pidsosny, 2007). The bands are assigned in Table 5.2.

5.7.3 EDX (Energy Dispersive X-Ray Analysis)

Energy dispersive x-ray studies of the EDTA were performed on the EVO 50 apparatus. The samples were placed onto a metallic support and covered with a thin silver film. The electron micrographs were obtained at 15 kV.

Elemental dot maps of sections were made by scanning the surface of EDTA with an electron beam to generate characteristic X-ray radiation from elements excited in the sample. The scanned graphs show the different elements present in EDTA. Figure 5.9a shows the fresh EDTA elements in the graph and shows carbon, oxygen, nitrogen and sodium of the di sodium salt, sulphur is also present in traces. Figure 5.9b-e shows an increasing amount of sulfur present as the EDTA is further recycled. The sulfur in repeated cycles of EDTA is increased because the sulfuric acid was used for the dechelation to break the complex of EDTA and Ni metal.

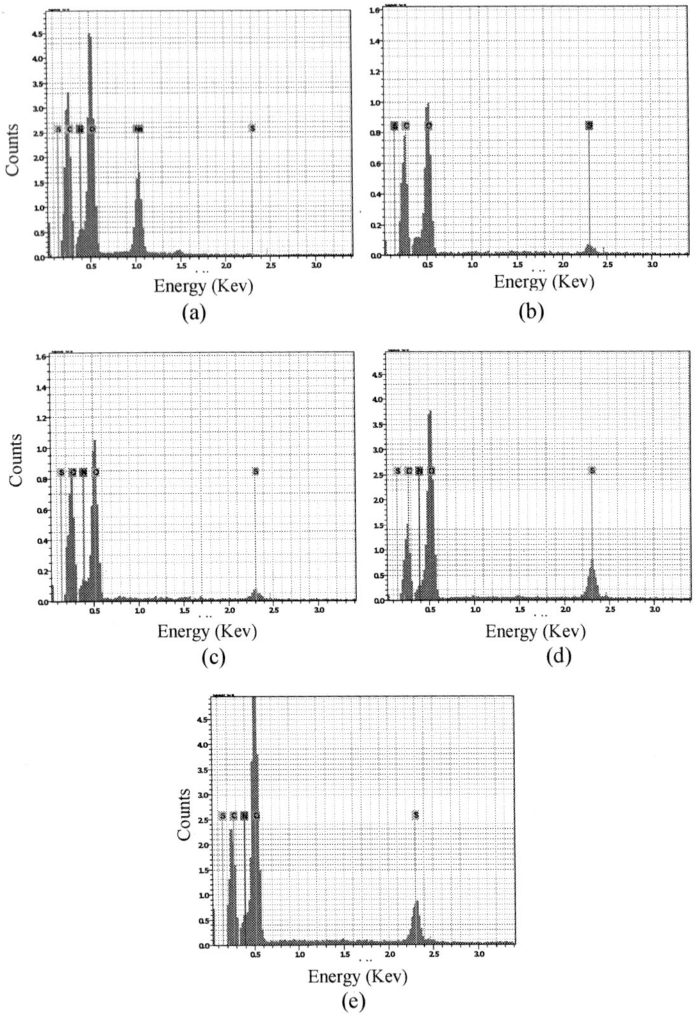

Figure 5.9 EDX of (a) fresh EDTA, (b) first recycled EDTA, (c) second recycled EDTA, (d) third recycled EDTA (e) fourth recycled EDTA (Goel et al., 2009). Copyright Elsevier 2009 reproduced with permission.

5.7.4 TGA (Thermogravimetry Analysis)

Thermogravimetry analysis was carried out in the range of 50^{0}C-450^{0}C using a Standard Deviation Thermogravimetry analyzer, model Q600 from TA instruments. During TGA the purge gas used was air. Figure 5.10b shows the weight loss up to 250^{0}C, after that no more losses occur in the fresh EDTA sample because the melting point of EDTA is 251^{0}C. Further losses are likely due to be associated with the EDTA salt.

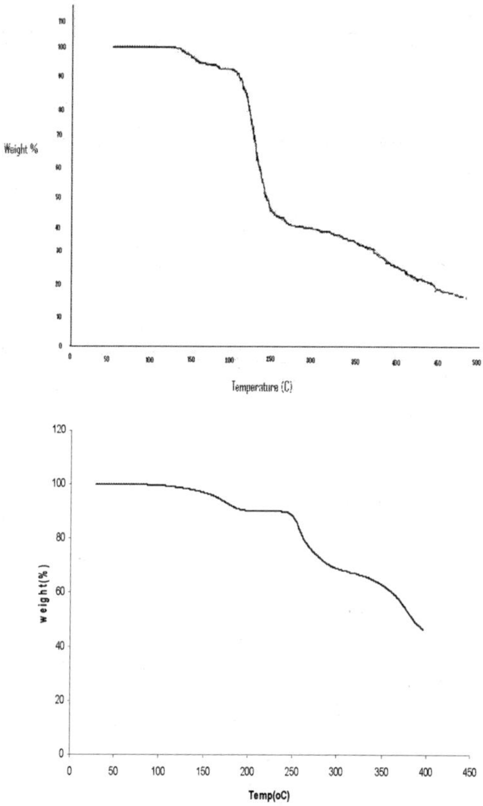

(a)

(b)

Figure 5.10 Weight loss of (a) first recycled EDTA, (b) fresh EDTA with increasing temperature at the rate of 10^0 C/min by thermogravimetry analysis (Goel et al., 2009) Copyright Elsevier 2009 reproduced with permission.

Figure 5.10a shows the weight loss with increasing temperature of the first time recovered EDTA. The figure shows higher weight losses than the fresh EDTA. As identified from EDX analysis a small amount of sulfur is present in the recovered EDTA. It is therefore expected that a higher weight loss occured due to the formation of sulfur oxide during the combustion of EDTA (due to dechelation with H_2SO_4). Therefore, the combined loss of water and sulfur is much higher than in the fresh EDTA sample.

5.8 Reusability of Spent Catalyst after Metal Extraction

The spent catalyst after metal removal was utilized by Marafi and Stanislaus (2003b). Spent catalyst was successfully utilized in cement production. In USA, cement kilns process about 60,000 ton/year of spent catalysts (Chen et al., 2006). Using the spent catalyst for the cement production seems an attractive option from the environmental and economical point of view. The reuse of spent catalyst to form new catalyst compositions has been reported in few studies (Gutnikov, 1971). The study investigated the possibility of preparation of an active hydrodesulphurization catalyst from spent HDS catalysts by mixing of alumina containing materials and shaping into compacted extrudates (Lopez et al., 1988). Decoked spent catalyst was ground and mixed with an unspecified additive and the resulting mixture was shaped to form particles of new hydroprocessing catalyst. High temperature sintering was used for pore enlargement to make the catalyst sites more active.

Although the utilization of the spent catalysts in the preparation of fresh or new active catalysts can help to reduce the spent catalyst problem to some extent, it does not solve the problem completely. All catalysts deactivate eventually to a point where further regeneration and recycling becomes uneconomical and they are discarded as wastes. Processes currently available for making spent catalysts non-leachable for safe disposal are very expensive. The cost involved in the treatment of the spent catalyst to make them non–leachable could be offset if the non-leachable material produced in the process were used in some other applications.

A process for making highly stabilized non-leachable anorthite glass ceramic materials from spent hydrotreating catalysts have been reported by Sun et al. (2001). The authors described a process to reduce non-leachable materials of high compressive strength such as a synthetic aggregate from spent hydroprocessing catalysts. The process involved mixing the spent catalyst in the form of a fine powder with clay, gatch, sand and water and shaping the wet mix into small balls of diameter 20 mm. After drying these balls at 110^0 C for 12 h, balls are reheated at $1150\text{-}1300^0$ C to make a strong bond with clay. To test the possibility of using the synthetic aggregate materials produced from spent catalysts, about 2 kg of aggregate was mixed with cement and sand and concrete cubes were prepared. After seven days concrete compressive strength was determined to be only 5 % less than the natural aggregate. This shows that the spent catalysts in the form of concrete are stable and non-leachable and they can be used in the construction industry incorporating in a cement industry.

5.9 Extraction Kinetic Models for Chelating Agents

The overall sorption rate depends on the path leading from the initial to the final state. In the catalyst where a porous media is considered, pathway includes events that are controlled either chemically or by molecular level mass transportation. The theoretical aspects of the sorption processes, the mathematics of mass transfer and of sorption kinetics, as well as the applicability of several different approaches to real samples are described by Weber et al. (1991). A simplest way to workout experimental data is by selecting a probable rate determining step, which controls the kinetics of the overall sorption process. However working with complex system such as multiple metals and other elements like Ca, Mg and other promoters added to the catalyst during catalyst production implies to work with very complex mathematical equations like partial differential equations. If there is free chelating agent present and to take this in account partial differential equation for each element diffusion and porosity of the catalyst should be considered (Tsang et al., 2007).

A number of different models have been proposed in the literature to describe the extraction kinetics from soil but very few are available for the catalyst. For extraction of metal with the chelating agents from spent catalyst, there are three basic physicochemical processes assumed (Kedziorek et al., 1998).

(a) Chelant transport advection dispersion equation;
(b) Solubilisation through complexing with heavy metals bound with the catalyst; and
(c) Transport of chelant metal complex in solution

5.9.1 Multiple First Order Reactions Model

In previous studies (Yu and Klarup, 1994; Lin and Chen, 1998; Fangueiro et al., 2005; Labanowski et al., 2008; Kuo and Mikkelsen, 1980; Yip et al., 2009), multiple first order reactions model is used by several soil chemists to describe the extraction of different trace metals like Zn, Fe, Mn, Pb and Cu. This model gives real physicochemical meaning but it needs several independent parameters which are different for every combinations of chemical and solids. In this model different first order reactions are ascribed to different types of discrete binding sites of catalyst to metal. There are two assumptions made:

(1) There are different first order reactions taking place and their rates of the reactions are independent of each other.
(2) The fast reaction completes before start of the slow reaction.

Thus, metals are bound to fractions associated to specific smaller portion of the catalyst. Multiple first order reactions can be described as:

$$\frac{dC_i}{dt} = k_i\left(C_i^0 - C_i\right) \tag{23}$$

Where C_i represents the quantity of desorbed metal from compartment i per gram of the catalyst at time t, C_i^0 represents the quantity of desorbed metal per gram of catalyst in compartment i per gram of catalyst at equilibrium and k_i is the rate constant of first order reaction for each compartment i. The fast reaction is considered to be the beginning of the extraction process and 0-3 h of the extraction time has been proposed by different authors. The slow reaction is considered for the rest of the time until reaching equilibrium. In practice, it is difficult to obtain equilibrium because extraction in the presence of free chelating agent takes place for infinite time; chelating agents are always present in excess compared to metal.

The first order reaction kinetics for both stages is given as:

$$\frac{dC_1}{dt} = k_1\left(C_1^0 - C_1\right) \tag{24}$$

$$\frac{dC_2}{dt} = k_2\left(C_2^0 - C_2\right) \tag{25}$$

According to the assumptions, metal desorption reactions are independent from each other. After integration of equation (24) and (25), taking initial condition as $C_1 = 0$ and $C_2 = 0$ at $t = 0$, and rearranging, the equation will become

$$C = C_1^0\left(1 - \exp\left(-k_1 t\right)\right) + C_2^0\left(1 - \exp\left(-k_2 t\right)\right) \tag{26}$$

C_1^0 (mg/g) represents the amount of metal readily extractable by the fast reaction and C_2^0 (mg/g) represents the amount of metal less extractable by the slow reaction at equilibrium.

5.9.2 Diffusion Model

This kinetic model shows a good physical meaning, the rate limiting step in this model is intraparticle diffusion or surface diffusion. This model assumes that among the sediment particulates, the complexing reactions of metals with chelating agents are fast and their product diffusion are slow. This model is based on the Fick's law of diffusion (Sivasubramaniam and Talibudeen, 1972; Jardine and Sparks, 1984). In most studies, ground particles or sediment particles are considered to be spherical. The diffusion model is applied to the mass transportation of the metal chelating agent complex to the surface of the particles and finally in the bulk of the solution by stirring of the solution.

For a spherical particle system, Fick's law is mathematically expressed as Crank (1976):

$$\frac{\partial C}{\partial t} = -D\left(\frac{\partial^2 C}{\partial r^2} + \frac{2}{r}\frac{\partial C}{\partial r}\right) \tag{27}$$

Where C is the concentration of the diffusive substance, D is the corresponding diffusion coefficient, r the radius of the particle and t is the time. According to Crank (1976), the total amount M_t of the complex entering or leaving a sphere of radius a at time t needs to be determined. Equation (28):

$$\frac{M_t}{M_\infty} = 1 - \frac{6}{\pi^2}\sum_{n=1}^{\infty}\frac{1}{n^2}\exp\left(\frac{-Dn^2\pi^2 t}{a^2}\right) \tag{28}$$

Where M_∞ is the amount of complex entering or leaving the sphere after infinite time. To apply these model equations in extraction rate data, the values of D, n, and a needs to be find out. The equation (28) is an infinite series of order n. The value of (D/a^2) is considered 1.5×10^{-5} min^{-1} for $n = 10$ and radius $a = 200\mu m$ by Yu and Klarup, (1994).

For cylindrical particles the model is defined as the parabolic model. According to Crank (1976), the equation which describes the diffusion in and out of the cylinder is:

$$\frac{M_t}{M_\infty} = 1 - \sum_{n=1}^{\infty}\frac{4}{a^2\alpha_n^2}\exp\left(-D\alpha_n^2 t\right) \tag{29}$$

Where α_n represents the positive roots of $J_0(a\alpha_n) = 0$, with $J_0(x)$ being the Bessel function of the first kind of zero order. For short time interval, this equation can be simplified to:

$$\frac{M_t}{M_\infty} = \frac{4}{\pi^{1/2}}\left(\frac{Dt}{a^2}\right)^{1/2} - \left(\frac{Dt}{a^2}\right) \tag{30}$$

Lin and Chen (1997) considered that in the initial phase of diffusion, there would be a linear relationship:

$$\frac{M_t}{M_\infty} = k(t)^{1/2} \tag{31}$$

Jardine and Sparks (1984) also used equation (31) for the study potassium-calcium exchange in multireactive soil system. The value of diffusion coefficient was assumed constant but the time range of valid use of this equation was not specified.

5.9.3 Empirical Models

The two constants equation was used by many researchers to fit metal extraction kinetic data (Lee et al., 2005; Yu and Klarup, 1994; Lin and Chen, 1998; Fangueiro et al., 2005; Kuo and Mikkelsen, 1980; Yip et al., 2009). The equation is described as:

$$C = At^B \tag{32}$$

Where *C* is the concentration of metal, *t* is the time and *A* and *B* are the two constants. After simplification, equation (32) becomes:

$$\ln C = B \ln t + \ln A \tag{33}$$

Kuo and Mikkelsen, 1980, considered the equation (33) as a multiple order equation by taking the constant *B* as the order of reaction, but there is still no physical significance of the constants. Experimental data fits well with the equation but the model has a lacks of significance. The two constants model is useful to predict the extraction kinetics, however, it is not consistent with the fact that *C* approaches the limit as *t* tends to infinity, and hence this model applies only within a certain time scale.

The other empirical model is the Elovich equation applied to the extraction of metals (Halin et al., 1985). This is expressed as:

$$\frac{dC}{dt} = ae^{-bc} \tag{34}$$

Where *C* is the concentration of metal in solution at time *t*, and *a* and *b* are constants. This equation is based on a continuous and specific range of site reactivity hypothesis. In fitting the equation (34) the real physical meaning is unclear being an empirical equation.

5.10 Conclusions

Over the past few year, emphasis has been given to the management of spent catalyst due to increase in demand of petroleum and fertilizer products. Disposal of spent catalyst by encapsulation and addition duing cement production can reduce the risk of exposure to environment. The compressive strength of concrete made from the mix of catalyst is only 5 % less than the natural aggregate mix. After recovery of metals the volume of catalyst was reduced by 5-35 % based on the metals present in the catalyst. Recovery of valuable metals from spent catalyst becomes necessary not only for lowering the catalyst cost but also for reducing the catalyst waste to prevent the environmental pollution. Acid leaching was the general procedure for the metal removal from spent catalyst, due to handling purposes of large quantity acid should be replaced by some other extractants. Alkaline reagents create no problem but corrosive in nature needs replacement of process plant equipment after some time and increase overall cost of the process. Both acid and alkaline reagents cannot be recycled.

Chelating agents proved better extractants than acids because of their recoverable property and safer use. This property makes this process closed loop and more economical and cost effective. Use of chelating agents was limited to metal extraction from soils, Goel et al. (2009) attempted first time to make use of chelating agents for the spent catalyst. The conditions of operation are such that higher temperature processes are carried out using EDTA solution, which is noncorrosive, and the use of mineral acids for dechelation is limited to very low temperatures. Therefore, materials of construction are expected to be cheaper, and the handling and storage are also expected to be less hazardous than those involving strong acids/oxidizing agents. This method promises to be a good industrial process for handling 1-2 t per batch of spent catalyst, within the infrastructure of a major nitrogenous fertilizer industry. The prospects of recovering and recycling Ni as solution are thus very high and will eventually contribute to lowering catalyst cost while benefiting the environment. The Characterization of EDTA shows the stability of chemical structure that enables reusability after four recycles. The extracted metals from spent catalyst are in the form of sulphates. These extracted metals can be used to make new catalyst by wet impregnation method (one of the method to load metals on alumina or silica surface). Another use of nickel sulphate is electroplating of batteries.

5.11 References

Al-Dalama, K., and Stanislaus, A. (2006). "Comparison between deactivation patterns of catalyst in fixed bed and ebullating bed residue hydroprocessing units." *Chemical Engineering Journal and the Biochemical Engineering Journal*, 120, 33-42.

Abdel-Aal, E. A., and Rashad, M. M. (2004). "Kinetic study on leaching of spent nickel oxide catalyst with sulphuric acid." *Hydrometallurgy*, 74, 189-194.

Al-Mansi, N. M., and Abdel Monem, N. M. (2002). "Recovery of nickel from spent catalyst." *Waste Management*, 22, 85-90.

Allen, H. E., and Chen, P. H. (1993). "Remediation of metal contaminated soil by EDTA incorporating electrochemical recovery of metal and EDTA." *Environmental Progress*, 12, 284-293.

Baghalla, M., Gh Khosravian, H., and Mortaheb, H. R. (2009). "Kinetics of platinum extraction from spent reforming catalysts in aqua-regia solutions." *Hydrometallurgy*, 95, 247-253.

Bartholomew, C. H. (1984). Catalyst deactivation in hydrotreating of residues, in: M. Obella. S.S. Shih (Eds.) Chemical Engineering.

Bell, C. F. (1977). Principles and Metal Chelation, Claredon Press, Oxford, U.K.

Brown, G. A., and Elliot, H. A. (1992). "Influence of electrolytes on EDTA extraction of Pb from polluted soil." *Water, Air, and Soil Pollution*, 62, 157-165.

Chang, L. Y. (1995). "A waste minimization study of a chelated copper complex in wastewater- treatability and process analysis." Waste Manage., 15, 209-220.

Chang, F. C., Lo, S. L., and Ko, C. H. (2007a). "A copper removal process for printed circuit board wastewater sludge applying extraction and cementation with chelating agents recovery." *Environmental Engineering Science*, 24, 1006-1016.

Chang, F. C., Lo, S. L., and Ko, C. H. (2007b). "Recovery of copper and chelating agents from sludge extracting solutions." *Separation and Purification Technology*, 53, 49-56.

Chaudhary, A. J., Donaldson, J. D., Boddington, S. C., and Grimes, S. M. (1993). "Heavy metals in the environment. Part II: a hydrochloric acid leaching process for the recovery of nickel value from a spent catalyst." *Hydrometallurgy*, 34, 137-150.

Chavan, D. V., and Dhadke, P. M. (2002). "Extraction separation of Ir(III) and Rh(III) with cyanex 923 from chloride media: a possible recovery from spent catalysts." *Journal of Chemical Technology and Biotechnology*, 77, 925-932.

Chen, Y., Feng, Q., Shao, Y., Zhang, G., Qu, L., and Lu, Y. (2006). "Research on the recycling of valuable metals in spent Al_2O_3- based catalyst." *Minerals Engineering*, 19, 94-97.

Chinchen, G. C. (1985). Selected Developments in Catalysis Critical Reports on Applied Chemistry, J.R. Jennings (Ed.), Vol. 12.

Crank J. (1976). The Mathematics of Diffusion, Oxford University Press, New York.

Dowden, D. A., (1978). Catalysis, Specialist Periodical Report of the Chemical Society, C. Emball (Ed.), Vol. 12, London.

Eichler, A., Hafner, J., and Kresse, G. (1996). "Hydrogen adsorption on the (1 0 0) surfaces of rhodium and palladium: the influence of non-local exchange–correlation interactions." *Journal of Physics: Condensed Matter*, 8, 7659-7670.

El-Nadi, Y. A., Awwad, N. S., and Nayl, A. A. (2009). "A comparative study of Vanadium extraction by Aliquat-336 from acidic and alkaline media with application to spent catalyst." *International Journal of Mineral Processing*, 90, 115-120.

Faisal, M., Atsuta, Y., Daimon, H., and Fujie, K. (2008). "Recovery of precious metals from spent automobile catalytic converters using supercritical carbon dioxide." *Asia-Pacific Journal of Chemical Engineering*, 3, 364-367.

Fangueiro, D., Bermond, A., Santos, E., Carapuca, H., and Duarte, A. (2005). "Kinetic approach to heavy metal mobilization assessment in sediment: choose of kinetic equations and models to achieve maximum information." *Talanta*, 66, 844-857.

Fisher, K., Bipp, H. P., Riemschnieder, P., Leidmann, P., Bieniek, D., Kettrup, and A. (1998). "Utilization of biomass residue for the remediation of metal-polluted soils." *Environmental Science & Technology*, 32, 2154-2161.

Furimsky, E. (1996). "Spent refinery catalysts: environment safety and utilization." *Catalysis Today*, 30, 223–286.

Furimsky, E., and Massoth, F. E., (1999). "Deactivation of hydroprocessing catalysts." *Catalysis Today*, 52, 381-495.

Garrabrants, A. C., and Kasson, D. S. (2000). "Use of a chelating agent to determine the metal availability for leaching from soils and wastes." *Waste Management*, 20, 155-165.

Goel, S., Pant, K. K., and Nigam, K. D. P. (2009). "Extraction of nickel from spent catalyst using fresh and recovered EDTA." *Journal of Hazardous Materials*, 171, 253-261.

Goel, S., and Gautam, A. (2010). "Effect of chelating agents on mobilization of metal from waste catalyst." *Hydrometallurgy*, 101, 120-125.

Groot, B. D., Bergers, P. J. M., and Groot, A. C. D. (1994). "The analysis of EDTA in water by HPLC." *Water Resources*, 28, 639-642.

Gutnikov, G. (1971). "Method of recovering metals from spent hydro treating catalysts." US Patent 3,567,433.

Ho, P. F., and Chi, K. M. (2004). "Size-controlled synthesis of palladium nano-particles from- diketonato complexes of palladium." *Nanotechnology*, 15, 1059-1067.

Hong., A. K. P., Li, C., Banerji, S. K., and Regmi, T. (1998). "Extraction, recovery, and biostability of EDTA for remediation of heavy metal contaminated soil." *Journal of Soil Contamination*, 8, 81-103.

Hong,. A. K. P., Cai, X., and Cha, Z. (2008). "Pressure assisted chelation extraction of lead from contaminated soil." *Environmental Pollution*, 153, 14-21.

Invascanu, S., and Roman, O. (1975). "Nickel recovery from spent catalyst: I. Solvation process." *BullInst Politeh Iasi Sect 2.*, 2(21), 47.

Iwao, S., El-Fatah, S. A., Furukawa, K., Seki, T., Sasaki, M., and Goto, M. (2007). "Recovery of palladium from spent catalyst with supercritical CO_2 and chelating agent." *Journal of Supercritical Fluids*, 42, 200-204.

Jardine, P. M., and Sparks , D. L. (1984). "Potassium-calcium exchange in multivariative soil system: I kinetics." *Soil Science Society of America Journal*, 48, 39-45.

Kedziorek, M. A. M., Dupuy, A., Bourg, A. C. M., and Compere, F. (1998). "Leaching of Cd and Pd from a polluted soil during the percolation of EDTA: laboratory column experiments modelled with a non equilibrium solubilisation step." *Environmental Science & Technology*, 32, 1609-1614.

Kar, B. B., Murthy B. V. R., and Misra V. N. (2005). "Extraction of molybdenum from spent catalyst by salt roasting." *International Journal of Mineral Processing*, 76, 143-147.

Kari, F. G., and Giger, W. (1995). "Modelling of photochemical degradation of ethylenediaminetetraacetate in the river Glatt." *Environmental Science & Technology*, 29, 2814-2827.

Lanigan, K. C., and Pidsosny, K. (2007). "Reflectance FTIR spectroscopic analysis of metal complexation to EDTA and EDDS." *Vibrational Spectroscopy*, 45, 2-9.

Kim, C., and Ong, S. K. (1999). "Recycling of lead-contaminated EDTA wastewater." *Journal of Hazardous Materials*, B69, 273–286.

Korhonen, M. S., Metsarinne, S. E., and Tuhkanen, T. A. (2000). "Removal of ethylenediaminetetraacetic acid (EDTA) from pulp mill effluents by ozonation." *Ozone Science and Engineering.*, 22, 279-286.

Kuo, S., and Mikkelsen, D. S. (1980). "Kinetics of zinc desorption from soils." *Plant Soil*, 56, 355-364.

Labanowski, J., Monna, F., Bermond, A., Cambier, P., Frenandez, C., Lamy, I., and Oort, F. V. (2008). "Kinetic extractions to assess mobilization of Zn, Pb, Cu, and Cd in a metal- contaminated soil: EDTA vs citrate." *Environmental Pollution*, 152, 693-701.

Lim, T .T., Chui, P. C., and Goh, K. H. (2005). "Process evaluation for optimization of EDTA use and recovery for heavy metal removal from a contaminated soil." *Chemosphere*, 58, 1031-1040.

Lee, I. H., Wang, Y. J., and Chern, J. M. (2005). "Extraction kinetics of heavy metal containing sludge." *Journal of Hazardous Materials*, B123, 112-119.

Lin, J. G., and Chen, S. Y. (1998). "Mobilization kinetics of heavy metals from contaminated river sediment of northern Taiwan by complexing agent." *Toxicological and Environmental Chemistry*, 67, 511-529.

Lopez, J. B., Eugene, A. P., and Anthony, V. C. "Heavy oil hydroprocessing including recovery of molybdenum catalyst." US Patent, 4,762,812.

Marafi, M., and Stanislaus, A. (2003a). "Studies on rejuvenation of spent residue hydroprocessing catalysts by leaching of metal foulants." *Journal of Molecular Catalysis A: Chemical*, 202, 117-125.

Marafi, M., and Stanislaus, A. (2003b). "Options and processes for spent catalyst handling and utilization." *Journal of Hazardous Materials*, B101, 123-132.

Marafi, A., Hauser, A., and Stanislaus, A. (2007). "Deactivation patterns of Mo/Al2O3, Ni–Mo/Al2O3 and Ni–MoP/Al2O3 catalysts in atmospheric residue hydro desulphurization." *Catalysis Today*, 125, 192-202.

Marafi, M., and Stanislaus, A. (2007). "Studies on recycling and utilization of spent catalysts: Preparation of active hydrodemetallization catalyst compositions from spent residue hydroprocessing catalysts." *Applied Catalysis B: Environmental*, 71, 199-206.

Matjie, R. H., Scurrell, M. S., and Bunt, J. (2005). "The selective dissolution of alumina, cobalt, and platinum from a calcined spent catalyst using different lixivants." *Minerals Engineering.*, 18, 801-810.

McCulloch, D.C. (1983). Applied Industrial Catalysis, Academic Press, New York, 103-110.

Mishra, D., Chaudhary, G. R., Kim, D. J., and Ahn, J. G. (2010). "Recovery of metal Values from spent petroleum catalyst using leaching-solvent extraction technique." *Hydrometallurgy*, 101, 35-40.

Munoz, F., and Von Sonntag, C. J. (2000). " The reaction of ozone with tertiary amines including the complexing agents nitrilotriacetic acid (NTA) and ethyleneidaminetetraacetic acid (EDTA) in aqueous solution." *Journal of the Chemical Society, Perkin Transactions 1*, 2, 2029-2033.

Nagib, S., Inoue, K., Yamaguchi, T., and Tamaru, T. (1999). "Recovery of Ni from a large excess of Al generated from spent hydrodesulfurization catalyst using picolyamine type chelating resin and complexane types of chemically modified chitosan." *Hydrometallurgy*, 51, 73-85.

Nowack, B. (2002). "Reactions of biogenic ligands in the presence of metals– what can we learn from studying anthropogenic chelating agents?" *Geochimica et Cosmochimica Acta*, 66, A560.

Noyes, R. (1991). Handbook of Pollution Control Processes. USA.

Palma, L. D., Ferrantelli, P., Merli, C., and Biancifiori, F. (2003a). "Recovery of EDTA and metal precipitation from soil flushing solutions." *Journal of Hazardous Materials*, B103, 153-168.

Palma, L. D., Ferrantelli, P., Merli, C., and Petrucci, E., (2003b). "Treatment of the solution extracted from metal contaminated soils by reverse osmosis and chemical precipitation." *Annali di Chimica*, 93, 1005-1011.

Papassiopi, N., Tambouris, S., and Kontopoulos, A. (1999). "Removal of heavy metals from calcareous contaminated soil by edta leaching." *Water, Air, and Soil Pollution*, 109, 1-15.

Park, K. H., Mohapatra, D., and Nam, C. W. (2007). "Two stage leaching of activated spent HDS catalyst and solvent extraction of aluminium using organo-phosphinic extractant, Cyanex 272." *Journal of Hazardous Materials*, 148, 287-295.

Peters, R. W. (1992). "Chelant extraction of heavy metals from contaminated soils." *Journal of Hazardous Materials*, 66, 151-210.

Pourmortazavi, S. M., Hajimirsadeghi, S. S., Kohsari, I., and Hosseini, S. G. (2004). "Orthogonal array design for the optimization of supercritical carbon dioxide extraction of different metals from a solid matrix with cyanex 301 as a ligand." *Journal of Chemical Engineering*, 49(6), 1530-1538.

Ringbom, A., and Wanninen, E. (1979). Treatise on Analytical Chemistry, Part I, Vol. 2, 2nd Ed, Wiley, New York.

Sahu, K. K., Agarwal, A., and Pandey, B. D. (2005). "Nickel recovery from spent catalyst." *Waste Management and Research*, 23, 148-154.

Santhiya, D., and Ting, Y. P. (2005). " Bioleaching of spent refinery processing catalyst using Aspergillus niger with high-yield oxalic acid." *Journal of Biotechnology*, 116, 171-184.

Sivasubramaniam, S., and Talibudeen, O. (1972). "Potassium-aluminum exchange in acid soils: I. kinetics." *Journal of Soil Science*, 23, 163-176.

Steele, M. C., and Pichtel, J. (1998). "Ex-situ remediation of a metal contaminated superfund soil using selective extractants." *Journal of Environmental Engineering*, 124, 639-645.

Sun, D. D., Tay, J. H., Cheong, H. K., Leung, D. L. K., and Qian, G. (2001). "Recovery of heavy metals and stabilization of spent catalyst using a glass-ceramic matrix." *Journal of Hazardous Materials*, B87, 213-223.

Tejowulan, R. S., and Hendershot, W. H. (1998). "Removal of trace metals from contaminated soils using EDTA incorporating resin trapping techniques." *Environmental Pollution*, 103, 135-142.

Trimm, D. L. (1990). "Deactivation, regeneration and disposal of hydroprocessing catalysts." *Studies in Surface Science and Catalysis*, 53, 41-60.

Tsang, D. C. W., Lo, I. M. C., and Chan, K. L. (2007). "Modeling the transport of metals with rate-limited EDTA-promoted extraction and dissolution during EDTA-flushing of copper-contaminated soils." *Environmental Science & Technology*, 41, 3660-3666.

Vuyyuru, K. R., Pant K. K., Krishnan V. V., and Nigam K. D. P. (2010). "Recovery of nickel from spent industrial catalysts using chelating agents." *Industrial and Engineering Chemistry Research*, 49, 2014-2024.

Wagner, A. L., Osborne, R. S., and Wagner, J. P. (2003). "Prediction of deactivation rates and mechanisms of reforming catalysts." *American Chemical Society, Division of Fuel Chemistry*, 48, 748-749.

Wasay, S. A., Barrington, S., and Tokunaga, S., (2001). "Organic acids for the in situ remediation of soils polluted by heavy metals: soil flushing columns." *Water, Air, and Soil Pollution*, 127, 301-314.

Weber, W. J., McGinley, P. M., and Katz, L. (1991). "Sorption phenomena in subsurface systems: concepts, models and effects on contaminant fate and transport." *Water Research*, 25, 499-528

Yip, T. C. M., Tsang, D. C. W., Ng, K. T. W., and Lo, I. M. C. (2009). "Empirical modelling of heavy metal extraction by EDDS from single metal and multi metal contaminated soils." *Chemosphere*, 74, 301-307.

Yu, J., and Klarup, D. (1994). "Extraction kinetics of copper, zinc, iron and Manganese from contaminated sediment using disodium ethylenediaminetetraacetate." *Water, Air, and Soil Pollution*, 75, 205-225.

Zeng, Q. R., Sauve, S., Allen, H. E., and Hendershot, W. H. (2005). "Recycling EDTA solutions used to remediate metal-polluted soils." *Environmental Pollution*, 133, 225-231.

CHAPTER 6

Enhanced Soil Flushing and Washing of Contaminated Soil and Sediments

Catherine N. Mulligan

6.1 Introduction

In the urban context, surface and subsurface soils and sediments may be contaminated and degraded. Derelict sites pose unique problems. Urban land can become degraded chemically and physically. They have the capacity to maintain their biodiversity, ecological functions, soil quality, and groundwater renewal. Brownfields, however, have been degraded by contamination from various sources such as refineries, rail yards, gas stations, warehouses, dry cleaners and other commercial enterprises using or storing hazardous chemicals. Abandoned urban lands (brownfields) are clearly contrary to the principles of sustainable development detailed in the Brundtland Report (World Commission on Environment and Development, 1987). Subsurface contamination by nonaqueous phase liquids (NAPLs), particularly petroleum, has been a major and widespread problem.

Numerous economic and health benefits can be achieved by restoring contaminated urban and brownfield sites. They include providing tax revenue, improving land and public health by improving air quality, removing threats to safety and reducing greenhouse gas emissions (NRTEE, 1998). In addition, transportation costs can be reduced by up to $66,000 per hectare per year if brownfields are redeveloped compared to greenfields by reducing urban sprawl (NRTEE, 2004). For every hectare of brownfield restored, 4.5 hectares of greenland can be preserved.

Regardless of the origin of the contaminants and pollutants in the area, an evaluation of the threats to human health and the environment must be undertaken before the remediation process. Both the potential exposure time and level must be considered. Both the wastes and the products they produce (such as leachates and emissions) are health and geoenvironmental threats. Disposal of wastes in the ground, illicit dumping, leaking underground storage tanks and others are also causes for concern in the urban environment. A recent sampling of backyards in Montreal,

Canada, where wastes had been previously dumped, indicated elevated levels of the heavy metals lead and zinc (Huang, 2005).

To evaluate the threats to human health and the environment, tests must be undertaken before the remediation process. Contaminant concentration, solubility, partition coefficients, and leaching potential should be determined. Techniques such as selective sequential extraction are useful in determining the likelihood that the heavy metals are mobile. Selective sequential extraction studies were performed on nine soil samples (Huang, 2005). Lead and zinc were shown to have different affinities towards different soil fractions. Both Pb and Zn have higher affinities towards the soil fractions of organic matter and oxides. Only a small fraction of both metals is associated with the exchangeable fraction. Metals bound to the exchangeable fraction of soil are mostly physically adsorbed (by electrostatic force) to the soil surfaces, and thus the bonding is weaker compared to other binding mechanisms. The moderate to high degree of leaching by rainfall and the competition from other cations present in the leachate solution could explain why only a limited amount of Pb and Zn were retained by this soil fraction. There is a high degree of association of Pb and Zn with soil oxides and organic matter. The metals associated with oxides are particularly susceptible to oxidation-reduction reactions and solubilization upon a decrease in pH by acid rain.

A variety of in situ and ex situ remediation techniques exist. For evaluation of the most appropriate technique, the procedure in Figure 6.1 should be followed. Ex situ techniques include excavation, contaminant fixation or isolation, incineration or vitrification, washing and biological treatment processes. In situ processes such as: (a) bioremediation, air or steam stripping or thermal treatment for volatile compounds, (b) extraction methods for soluble components, (c) chemical treatments for oxidation or detoxification, and (d) stabilization/ solidification with cements, limes, resins for heavy metal contaminants. Phytoremediation is less developed. The most suitable types of plants must be selected based on pollutant type and recovery techniques for disposal of the contaminated plants.

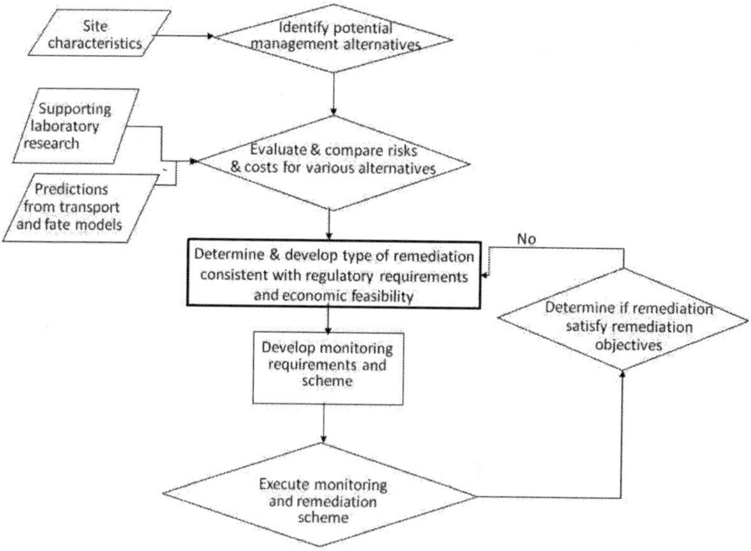

Figure 6.1 Process for evaluating soil remediation processes.

Most in situ remediation techniques are potentially less expensive and disruptive than ex situ ones, particularly for large contaminated areas. Natural or synthetic additives can be utilized to enhance precipitation, ion exchange, sorption and redox reactions (Mench et al., 2000). The sustainability of reducing and maintaining reduced solubility conditions is key to the long term success of the treatment. Ex situ techniques are expensive and can disrupt the ecosystem and the landscape. For shallow contamination, remediation costs, worker exposure and environmental disruption can be reduced by using in situ remediation techniques. In this chapter we will examine various agents used to enhance in situ flushing and soil washing processes for soil and sediment remediation. The main focus will be on metal removal by biologically produced surfactants due to their biodegradability and effectiveness.

6.2 Soil Flushing

To remove non-aqueous phase liquids (NAPLs) from the groundwater, extraction of the groundwater can be performed by extraction pumping of the contaminated dissolved phase and/or free phase NAPL zone. Drinking water standards can be achieved after treatment with activated carbon, ion exchange and other methods. However, substantial periods of time can be required before this occurs. To enhance the removal rates of the contaminants, extraction solutions can be infiltrated into the soil using surface flooding, sprinklers, leach fields, horizontal or vertical drains. Water with or without additives (surfactants or solvents) is employed to solubilize and extract the contaminants as shown in Figure 6.2 in soil flushing. Chemical additives include organic or inorganic acids, bases, water soluble solvents, complexing agents and surfactants.

Soil flushing is appropriate for highly permeable soils. The washing solution is pumped through the soil by injection wells, surface sprinklers or other means of infiltration. The washing solution should be treated to remove the contaminants and reuse the water. Control of the infiltrating agent may be difficult, particularly if the site hydraulic characteristics are not well understood.

As some residuals may remain in the soil or groundwater, the agents must be non-toxic and biodegradable. Contaminant removal efficiencies are related to, and affected by, soil pH, soil type, porosity and moisture content, cation exchange capacity, particle size, organic content, permeability and the type of contaminants. High soil permeabilities (greater than $1x10^{-3}$ cm/sec) are considered to be beneficial for such procedures. Depth to groundwater can increase costs. However, the spreading of contaminants and the fluids must be contained and recaptured. Emissions of volatile organic compounds (VOCs) should be monitored and treated if required. Recycling of additives is desirable. Metals, VOCs, polychlorinated biphenyls (PCBs), fuels and pesticides can be removed through soil flushing.

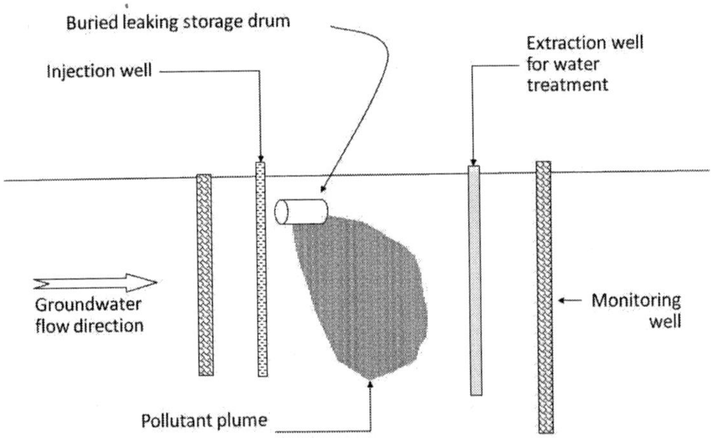

Figure 6.2 Schematic of Soil Flushing.

In choosing the remediation technologies to treat this problem, it is necessary to factor in the targets, exposure routes, future land use, acceptable risks, legislation, and resultant emissions. Laboratory and field treatability tests should be performed to obtain site specific information. Soil flushing is a proven technology and has been demonstrated as 12 Superfund sites. Costs can vary from $18 to $50 per cubic meter for large easy to small difficult sites. A schematic illustration of the criteria and tools for evaluating technologies and protocols for environmental management of contaminated soils and sediments is shown in Figure 6.3.

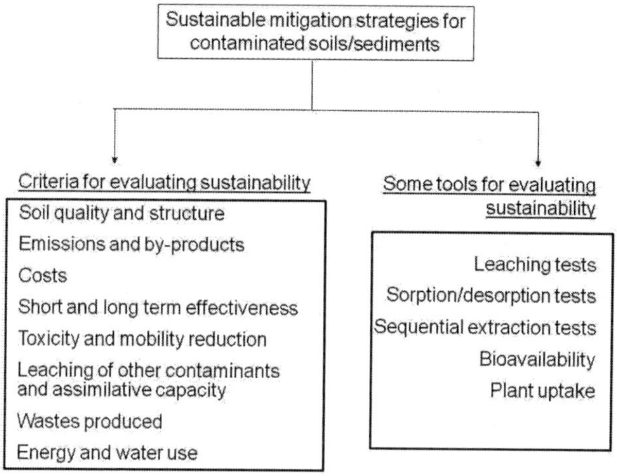

Figure 6.3 Schematic of evaluation of technologies.

6.3 Soil Washing

Soil washing has been suggested for a variety of soils contaminated with metals, mixed contaminants, and organic contaminants (El-Shafey and Canepa, 2003). Soil washing is a process that uses water to removal contaminants from soil and sediments by physical and/or chemical techniques. Contaminated sediments are problematic as they can potentially release contaminants severely impacting water quality. Soil washing involves the addition of a solution with the contaminated soil to transfer the contaminants to the wash solution. It is most appropriate for weaker bound metals in the form of hydroxides, oxides and carbonates. Mercury, lead, cadmium, copper, nickel, zinc and chromium can be recovered by electro-chemical processes if the levels of organic compounds are not significant. Metals can also be removed by precipitation or ion exchange. Precipitation is not applicable for metal sulfides. Pretreatment to remove uncontaminated coarser fractions can be used. Various additives can be employed such as bases, surfactants, acids, or chelating agents. Nitric, hydrochloric and sulfuric acids can be used. However, if sulfuric acid is used, 50% of the amount is required compared to hydrochloric acid (Papadopoulos et al., 1997).

Figure 6.4 illustrates a typical soil washing process where the separation consists of washing, rinsing, size separation, and other technologies similar to those used in the mineral processing industry. The Bergman process is an example of a commercial soil washing system. In general, it is recommended that the soil should contain at least 60% of the coarse fraction and less than 20% of an organic fraction by volume (USEPA, 1995). Surfactants may be added to the washing water. Larger particles are separated from the smaller ones as the smaller they are the higher the contamination level. The smaller volumes of soil can be treated more easily. A soil washing demonstration indicated that 18% of the received sediment can be classified as contaminated (USEPA, 1992). The optimal size range is 0.24 to 2 mm due to the surface charges of the soil clay particles that attract anionic metal contaminants and the organic fraction that binds organic contaminants. Wash water and additives should be recycled, regenerated or treated prior to disposal. The dewatering of particles is subsequently needed. The mechanical dewatering such as by a filter press, conveyer filtration, centrifugal separation, etc., is available. Froth flotation (the introduction of air bubbles in a slurry) may also be included and has been used to remove zinc from contaminated soil in Germany (Venghuis and Werther, 1998). The disposal of fine particles is different, depending on the type and levels of the contaminants. For contaminated sediments, the separated clean coarse sandy fraction has been used for construction fill while after blending with clay, and heat treatment in a kiln and the concentrated contaminant fraction has also been used in an aggregate feedstock (Zagular and Beitnger, 1996).

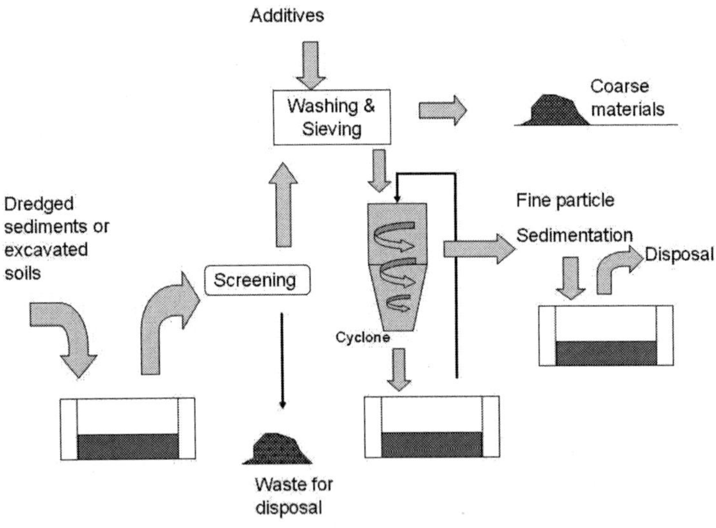

Figure 6.4 Schematic of soil washing.

Mixtures of metals and organic contaminants may require sequential washing with different additives to target the various contaminants. Soil washing processes generally use hot water. The viscosity of hydrocarbons is influenced by temperature and the increase in temperature reduces the viscosity. Since the increase in temperature of water increases the kinetic energy of water molecules, the diffuse double layer of soil particles becomes thinner. Then, surface attractive forces on the particles are reduced. The increasing temperature increases the solubilities of metal salts.

The treated soil can then be washed to remove any residual wash solution prior to disposal. Ideally the wash solution should be recycled. Costs of soil washing are usually in the order of \$70 to 190 US. per m^3 depending on site size and complexity (Racer software, Remedial action plan 2006). Although extensively used in Europe, full scale processes are not well known in the U.S. Extraction tests should be conducted to determine the optimal conditions (chemical type and dosage, contact time, agitation, temperature and extraction steps to meet regulatory requirements). Spent washwater can be a mixture of soluble contaminants and fine particles. Treatment is thus required to meet reuse or disposal requirements. Full scale demonstrations may be required to determine the feasibility of the selected treatment conditions. Soil washing and vitrification have shown costs of \$100,000 and \$1,000,000 per ha, respectively (Russel et al., 1991).

6.4 Use of Acids and Chelating Agents

A typical washing method may be acid leaching which refers to the soil remediation which extracting the metals with sulfuric acid. It is frequently ineffective for cadmium. Leaching can be abiotic or microbial (Löser et al., 2006; 2007). In abiotic leaching, the H_2SO_4 is supplied to the soil with circulating water. However, in microbial leaching, elemental sulfur is added to the sediment, and is oxidized to sulfuric acid (Tsai et al., 2003a; 2003b). Another approach is to produce organic acids (such as citric acid produced by the fungus *Aspergillus niger*) for complexation of heavy metals (Mulligan and Kamali, 2003). Both methods achieved removal efficiencies of greater than 90% for total extractable heavy metals. The pH ranges are controlled depending on the heavy metal species. For example, Al is markedly solubilized at pH< 4, and Fe at pH< 2.4.

Chelants desorb metals from soil by forming complexes with the metals. The complexes must be stable and soluble and enhance desorption of the metals. The chelant should not affect the environment and not destroy the physical characteristics of the soils. The complexing agent should be stable over a wide pH range, selective for the desired metal, be recoverable, have a low adsorption capacity, low toxicity and be cost effective. Ethylene diaminetetraacetic acid (EDTA) has been the most frequently used to enhance plant uptake and as a chelating agent in washing processes but the chemical is not highly degradable. Other chelants have been recently evaluated such as nitrilotriacetate (NTA), S,S-ethylenediaminedisuccinic acid (S,S-EDDS) and others. Usually lower pH levels extract more metals (van Benschoten and Matsumoto, 1997). Higher pH values (slightly alkaline) have been optimal for removal with S,S-EDDS of lead, zinc and cadmium. Frequent application of small dosages seems to be more beneficial than a single large dose (Finzgar and Lestan, 2007). The best method for choosing a chelating agent is to perform tests for each site to optimize the conditions of pH, concentrations and reaction times as there are many potential interfering agents and reactions.

Not all fractions are amenable to soil washing. Metal speciation and fractionation studies are useful in determining which fractions can potentially be removed from the soil by the chelant. Barona and Romero (1996) showed that the amount of lead removed by EDTA corresponded to the Fe- and Mn-oxide and organic fractions. Finzgar et al. (2005) showed that lead was removed from the carbonate and organic fractions by S,S-EDDS. Therefore the remaining metal fractions should be more stable and less mobile. However, climate, hydrological conditions and natural biota may influence the metal forms. Earthworms (*Eisenia fetida*) have been capable of enhancing Pb mobility by a factor of 6.2 times.

Kantar and Honeyman (2006) evaluated the use of citric acid for remediation of uranium-contaminated sandy soil. Citrate was able to remove U(VI) via complexation and extraction of Fe coatings and Fe-citrate complexes. U(VI) is the most common form. 98% removal of uranium could be achieved with 10^{-3} M citric acid in batch systems. In soil flushing experiments, more citric acid (4 pore volumes of 0.1 M citric acid) was required than for the batch experiments to remove approximately the same amount of uranium. Field implementation is feasible but uranium-citrate complexes must not migrate to undesirable areas.

Further studies have been performed to develop a feasible and economical technique to treat and microbially recover metals in low-grade oxide ores (Mulligan, 2002). Significant quantities of metals that are contained in low - grade ores and mining residues need to be extracted economically and to avoid endangering the environment. Pyrometallurgical and hydrometallurgical techniques are either very expensive, energy intensive or detrimental to the environment. For these reasons, biohydrometallurgical techniques employing the fungus *Aspergillus niger* are potentially more sustainable. *A. niger* has exhibited good potential in generating a variety of organic acids effective for metal solubilization (Mulligan and Kamali, 2003). Organic acid effectiveness was enhanced when sulphuric acid was added to the medium. Different agricultural wastes such as potato peels were tested. In addition, different auxiliary processes were evaluated in order to either elevate the efficiency or reduce costs. Finally, maximum solubilization levels of 68%, 46% and 34% were achieved for copper, zinc and nickel, respectively. Also iron dissolution was minimized to 7% which allows for further metal purification.

A later study was performed by Ouattara (2008). Leaching tests were performed to treat mining residues. Seven food wastes were used to produce organic acids by *A. niger* and *P. simplicissum*. Two steps were used, production followed by leaching. Whey permeate was the most promising substrate for production of a mixture of gluconic, malic, oxalic and citric acids. *A. niger* was superior to *P. simplicissum*. Cu, Zn, and Pb were leached from one type of residue while Ni and Zn were solubilized from the other. Mn was solubilized the most for both residues. The solubilized metals were mainly in the form of carbonates and oxides (which occurs at a pH less than 5). An important fraction was also solubilized in the organic fractions for Mn and Cu for both types of residues and Ni, Pb and Zn in one type of residue. The economic feasibility of this process depends on the value of the metals recovered.

Another research study was to elucidate the mechanisms of arsenic (As) mobilization from mine tailings in the presence of chelating natural organic matter (NOM). Humic acid (HA) was chosen as a model for NOM (Wang and Mulligan, 2009a). The introduction of HA at a low mass ratio (below 2 mg HA/g mine tailings) under acidic conditions inhibited As mobilization. As mobilization increased with increasing mass ratios. Under alkaline conditions, HA enhanced As mobilization significantly. A mobilization isotherm was developed to predict As mobilization from the mine tailings in the presence of HA. It was indicated that HA sorption to the mine tailings was essential for As mobilization. FTIR analyses indicated the carboxylic and amino functional groups in HA might be involved in As mobilization and complexation. Capillary electrophoretic analyses indicated that As redox reactions might not have a significant effect on As mobilization in this study. The mobilization of co-existing metals might enhance As mobilization by helping incorporate it into soluble complexes in the presence of HA.

6.5 Surfactant Washing and Flushing

6.5.1 Surfactants

Surface active agents (surfactants) are used to enhance soil flushing and washing procedures due to their amphiphilic properties. The hydrophilic and hydrophobic portions (Tsujii, 1998) enhance the solubility of organic or inorganic components. Through reduction of interfacial tension and micelle formation, surfactants or biosurfactants have shown many environmental applications including heavy metal and hydrocarbon removal from contaminated soils (Mulligan and Eftekhari, 2003; Mulligan and Wang, 2006; Urum et al., 2003). Low critical micelle concentrations (CMC) minimize the amount of agent required. Nonionic surfactants such as polyethoxylate nonyl phenol have shown their ability to enhance recovery of NAPLs such as gasoline (Chevalier et al., 1997). Nonylphenyl ethoxylate (Medina Agriculture Products) has been used for in situ and ex situ remediation. It is highly biodegradable under aerobic conditions and is approved for oil spill cleanups by the U.S. federal government.

After pretreatment to remove large debris and larger uncontaminated particles, the soil can be treated by mixing as a slurry with the chelating solution (Vandevivere et al., 2001). As the treatment can be severe, soil properties including soil flora, soil structure, hydraulic conductivity, and water holding capacity should be examined after treatment (Finzgar and Lestan, 2006a). This is not a sustainable practice. Surfactant-enhancement of remediation techniques is still developing and will be examined in detail in this chapter.

6.5.2 Biosurfactants

Yeast and bacteria are able to produce biological surfactants from various substrates including sugars, oils, alkanes and wastes (Lin, 1996). The CMCs of the biosurfactants typically range from 1 to 200 mg/L with molecular masses from 500 to 1500 Daltons (Lang and Wagner, 1986). They can be potentially as effective with some distinct advantages over the highly used synthetic surfactants including high specificity, biodegradability and biocompatibility (Cooper, 1986). Some classifications of biosurfactants include glycolipids, lipopeptides, phospholipids, fatty acids, neutral lipids, polymeric and particulate compounds (Biermann et al., 1987). Most are either anionic or neutral, while only a few with amine groups are cationic. The hydrophobic part of the molecule is based on long-chain fatty acids, hydroxy fatty acids or α-alkyl-β-hydroxy fatty acids. The hydrophilic portion can be a carbohydrate, amino acid, cyclic peptide, phosphate, carboxylic acid or alcohol.

Rhamnolipids are the most studied biosurfactants. They are produced by *Pseudomonas aeruginosa* (Hitsatsuki et al., 1971; Guerra Santos et al., 1984; Mulligan et al., 1999) and are commercially available (Jeneil Biosurfactant). Surface tensions of water are reduced to 29 mN/m by these biosurfactants. CMCs range from 10 to 230 mg/L, depending on the structure. Two types of rhamnolipids contain two rhamnoses attached to β-hydroxydecanoic acid or one rhamnose connected to the identical fatty acid (Figure 6.5). Type I, (R1) is L-rhamnosyl-β-hydroxydecanoyl-β-hydroxydecanoate, molecular mass = 504 Da; Type II, (R2) is L– rhamnosyl – β – L– rhamnosyl – β – hydroxydecanoyl – β –hydroxydecanoyl – β– hydroxydecanoate, molecular mass = 660 Da; Type III (R3) is one rhamnose attached to β–hydroxydecanoic acid and Type IV (R4) is two rhamnoses attached to β–hydroxydecanoic acid. *Pseudomonas aeruginosa* can utilize a wide variety of substrates including C_{11} and C_{12} alkanes, succinate, pyruvate, citrate, fructose, glycerol, olive oil, glucose and mannitol (Robert et al., 1989) which influences the type and yields of rhamnolipids produced. Fermentor design, pH, nutrient composition, and temperature are also influential (Mulligan and Gibbs, 1993). Other low molecular weight biosurfactants include surfactin (a lipopeptide), saponin (a plant derived biosurfactant) and others are discussed.

6.5.3 Soil Washing

6.5.3.1 Rhamnolipid

The feasibility of using biodegradable biosurfactants to remove heavy metals from an oil-contaminated soil (890 mg/kg of zinc, 420 mg/kg of copper with a 12.6%

oil and grease content) from a harbour area was recently demonstrated by batch washes with surfactin, a rhamnolipid and a sophorolipid (Mulligan et al., 1999). Five serial batch washes with 0.1% surfactin/1% NaOH removed 70% of the copper while 4% sophorolipid/0.7% HCl was able to remove 100% of the zinc. The results clearly indicated the feasibility of removing metals with the anionic biosurfactants tested even though the exchangeable metal fractions were very low. Since these agents are biodegradable, they can enhance hydrocarbon removal and can potentially be produced in the sediments.

Various studies have been performed to evaluate the use of biosurfactants for enhancement of soil washing. Anionic surfactants have been particularly effective for cation removal. Asci et al. (2007) evaluated the potential to remove Cd(II) from kaolinite. The desorption effects of pH, rhamnolipid concentration, and sorbed Cd(II) concentration were determined. The optimal conditions were pH 6.8, an initial concentration of 0.87 mM, and a rhamnolipid concentration of 80 mM. A removal of 71.9% of Cd(II) was achieved. Various sorption models were evaluated for Cd(II). The Kolbe-Corrigan model fitted best. Asci et al. (2008a) then examined the removal of zinc from Na-feldspar (another soil component) by a rhamnolipid biosurfactant. Significant sorption of zinc was shown. The optimal pH for removal was also 6.8 due to the small vesicles and micelles at a pH > 6.0. A concentration of 25 mM was optimal for removal of 2.2 mM of zinc (a 12.2 :1 molar ratio) or 98.8%.

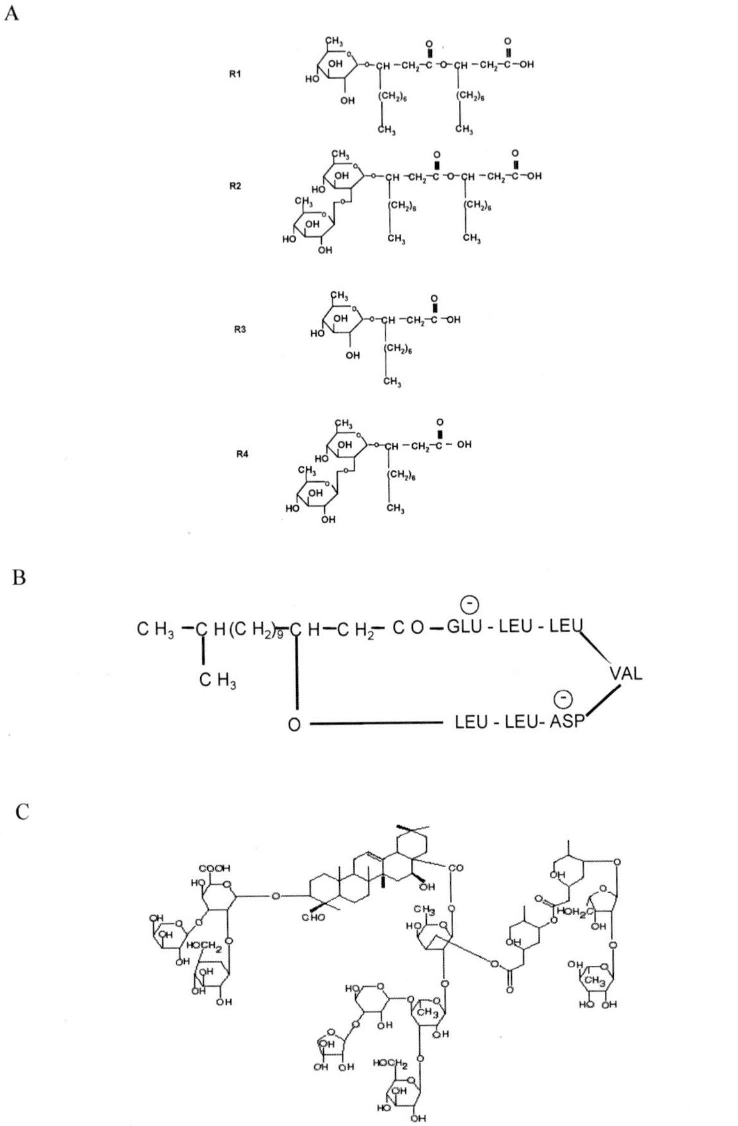

Figure 6.5 Chemical structure of (A) rhamnolipids (B) surfactin and (C) saponin.

As sorption of biosurfactants can reduce removal of contaminants from various components of the soil (Ochoa-Loza et al., 2007), sorption tests were performed for various soil components. Monorhamnolipid (R1) sorption depended on its concentration. The sorption followed the order of hematite > kaolinite > MnO_2~illite~Ca-montmorillonite > gibbsite > humic acid coated silica for low R1 concentrations for organic degradation. For higher concentrations, sorption by illite >> humic acid coated silica > Ca-montmorillonite > hematite > MnO_2 > gibbsite~ kaolinite. In spite of the stronger sorption by R1, it was more efficient of metal removal. This information will enable predictions to occur regarding the feasibility of rhamnolipid treatment and the quantity of rhamnolipid for rhamnolipid treatment. Use of the rhamnolipid as an R1/R2 mixture increases the R1 concentration available in solution for remediation.

Kim and Vipulanandan (2006) evaluated the removal of lead from water and contaminated soil (kaolinite). A linear isotherm was used to represent lead desorption from kaolinite. A biosurfactant to lead ratio of 100:1 showed optimal removal. The carboxyl group of the biosurfactant was found by FTIR spectroscopy to play a role in metal removal. Micelle partitioning could also be represented by Langmuir and Freundlich models. The biosurfactant micelle partitioning was more favorable than the synthetic surfactants, sodium dodecyl sulfate (SDS) and Triton X-100. The biosurfactant produced from vegetable oil at a concentration 10 x CMC could remove lead by over 75% from 100 mg/L contaminated water.

The effect of negatively-charged biosurfactant (rhamnolipid) addition on chromium contaminated kaolinite was studied (Massara et al., 2007). The results showed that the rhamnolipids have the capability of extracting a 25% portion of the stable form of chromium, Cr(III), from the kaolinite, under optimal conditions. The removal of hexavalent chromium was also enhanced compared to water by a factor of 2. Results from the sequential extraction procedure showed that rhamnolipids removed Cr(III) mainly from the carbonate, and oxide/hydroxide portions of the kaolinite. Close to 100% of the extracted Cr(VI) was reduced to Cr(III) by the biosurfactant over a period of 24 days. This study indicated that rhamnolipids could be beneficial for the removal of or long–term conversion of Cr(VI) to Cr(III).

As a continuation of this work to evaluate the feasibility of using rhamnolipid for the removal and reduction of anionic hexavalent chromium from contaminated soil and water, batch experiments were performed by Ara and Mulligan (2008). The effect of initial chromium concentration, rhamnolipid concentration, pH and temperature were evaluated and found to affect the reduction efficiency. The rhamnolipid can reduce 100% of initial Cr (VI) in water at optimum conditions (pH 6, 2% rhamnolipid concentration, 25°C) if the concentration is low (10 ppm). For higher initial concentrations (400 ppm), it took 24 h to reduce Cr by 24.4%). In the case of soil, rhamnolipid only removed the soluble part of the chromium present in the soil. The extraction increased with an increase in the initial concentration in soil but decreased slightly with the increase of temperature above 30°C. A sequential extraction study was used on soil before and after washing to determine from what fraction the rhamnolipid removed the chromium. The exchangeable and carbonate fractions accounted for 24% and 10% of the total chromium, respectively. The oxide and hydroxide portion retained 44% of chromium present in the soil. On the other hand, 10% and 12% of the chromium was associated with the organic and residual fractions. Rhamnolipid can remove most of the exchangeable (96%) and carbonate (90%) portions and some of the oxide and hydroxide portion (22%) but cannot remove chromium from the other fractions. This information is important in designing the appropriate conditions for soil washing.

The removal of another anionic contaminant, arsenic, was investigated the removal of arsenic from mining tailings by rhamnolipids (Wang and Mulligan, 2009b). As(V) was the only form extracted from the tailings, particularly at high pH. Significant removal of Cu, Zn, and Pb simultaneously occurred and was positively correlated. Either organic complexes or metal-bridging mechanisms resulted in the arsenic mobilization. Further studies (Wang and Mulligan, 2009c) focused on the development of mobilization isotherms to predict the As transport. Selective extraction was used determine which fractions were removable by the biosurfactant. Easily and moderately extractable fractions could be removed but redox or methylation reactions did not occur to any significant effect. It thus might be potentially useful for removal of As from mining tailings.

6.5.3.2 Saponin

A plant-based biosurfactant named saponin was evaluated by Song et al. (2008) for the removal of mixed contaminants, phenanthrene and cadmium, from soil. Phenanthrene is removed by solubilization while cadmium is complexed by the carboxyl groups of saponin. Removal rates were 87.7% and 76.2% for phenanthrene and cadmium respectively. Therefore, combined removal can be obtained by competition with each other.

Further saponin studies were performed by Chen et al. (2008). They measured the ability of saponin to extract copper and nickel from kaolin. It was found that 2000 mg/L of saponin could remove 83% of the copper and 85% of the nickel. Extraction was performed at pH 6.5. Metal desorption was in the following order ethylenediaminetetraacetic acid (EDTA) > saponin >> SDS. A three step washing mechanism was postulated: Adsorption followed by formation of ion pairs with adsorbed metal and then rearrangement to desorb the metal.

A variety of biosurfactants (rhamnolipids, saponin and mannosyl-erytritol lipids (MEL)) were evaluated by batch washes for metal removal from soil from a construction site in Canada and a lake sediment from Japan (Mulligan et al., 2008). The soil contained 890 mg/kg of zinc, 260 mg/kg copper, 170 mg/kg nickel and 230 mg/kg total petroleum hydrocarbons. After five washings of the soil with saponin (30 g/L), the highest levels of zinc removal (88%, pH 3) and nickel removal (76%, pH 5) were obtained. Copper removal (46%) was maximal with 2% rhamnolipids (pH 6.5). Multiple washings of the soil with 4% MEL (pH 5.6) provided lower levels of removal (17 % of the zinc and nickel and 36 % of the copper). The sediment contained 4440 mg/kg zinc, 94 mg/kg copper and 474 mg/kg of lead. From the sediment, the highest level of zinc (33%) and lead removal (24%) were achieved with 30g/L saponin (pH 5). Highest copper removal (84%) was achieved with 2% rhamnolipids (pH 6.5). The rhamnolipids seem to have more affinity for copper than for zinc and nickel as shown by the high removal rate. This phenomenon was also observed by Dahrazma (2005) where rhamnolipids removed more copper than zinc and nickel from sediment samples in a batch washing test. Multiple washings appeared to improve the removal of the metals significantly, especially the removal of copper. The same trend was seen for the sediment sample where up to 84% of the copper and 13% of the zinc were removed. Sequential extraction showed that the oxide fraction of zinc and organic fraction of copper were substantially reduced by the biosurfactants. The feasibility of reducing the zinc, copper, lead and nickel contents of contaminated soil and sediment with the anionic biosurfactants tested was thus demonstrated.

Saponin at a concentration of 30 g/L (pH 5) was used in a series of washes.The saponin at pH 5 was able to remove 79% of zinc from the soil; however the case was different for copper where the removal was 28% after 5 washings. The control (water) removed minimal amounts of all three metals (3.2% of zinc, <0.1% of copper at pH 5. Saponin seems to have a stronger affinity for zinc than for copper from the soil. However in terms of percentage for the sediment, it is different where 43% removal was achieved for copper compared to 33% for zinc. The significantly higher initial amount of zinc (4,441 mg/kg) in the sediment compared to 894 mg/kg in the soil could account for the difference for zinc. The removal trend is that more zinc is removed by sequential washes whereas copper seems to reach a static level after two or three washes.

Sequential extraction tests were performed on the sediments before and after washing with the controls and the washing agents. It was shown that copper could be removed mostly from the organic-bound fraction from the sediment and zinc and nickel from the oxide and carbonate-bound fraction by 2% rhamnolipid (pH 6.5). Saponin was effective for removal of heavy metals from all fractions with the exception of the residual. Residual fractions, the most difficult to remove, were not affected during the surfactant washing studies and thus could be considered stable and unlikely to leach metals and unnecessary to remediate. For copper, the organic fraction was the only significant fraction. The rhamnolipid was able to completely remove the copper from this fraction. The saponin removed only some of the organic fraction. The controls did not have any significant effect. Therefore the sequential extraction tests were useful for determining the metal binding fractions removed during washing.

6.5.3.3 Case study of a Washing Process

INRS developed the ORGANOMETOX process for removal of inorganic and organic contaminants (Mercier et al., 2008; Dragage Verreault, 2008) for soils and sediments. The contaminants were mainly concentrated in the fine fraction. The process is the subject of US, Canadian and European patents (7 countries). For the project initiated in 1995, the pilot plant treated at a rate of 8 tonnes/h more than 40 tonnes of soils/sediments. The first step (Figure 6.6) included metallurgical processes such as screening, for obtaining the fine fraction. Surfactants were then added for organic contaminant removal followed by various gravimetric separation processes. The surfactant added to the flotation columns was cocamidopropyl hydroxysultaine (CAS). This step was followed by centrifugation for solid/liquid separation. The pilot tests can then be used to determine the technical economic feasibility of a commercial scale plant of 50 tonnes/h. Commercialization is difficult since the results must be guaranteed by the contractor. The cost is usually relatively high (in the range of $100 to 300 per tonne). For sediments from Montreal with initial zinc and copper contents of 2682 and 117 mg/kg, respectively, removal of each was 88 and 70%.

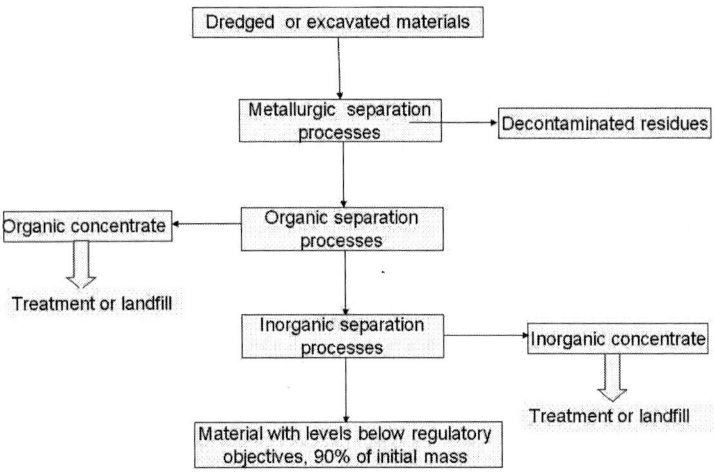

Figure 6.6 Schematic of ORGANOMETOX process (based on Mercier et al., 2008; Dragage Verreault, 2008).

6.5.4 Soil Flushing

6.5.4.1 Metal Removal

Other studies have been performed using soil columns to simulate in situ conditions. As previously shown the anionic characteristics of the biosurfactants enable metal removal from soil by biosurfactants. The removal of cadmium and lead by a biosurfactant produced by *Pseudomonas aeruginosa* BS2 was investigated in column experiments (Juwarkar et al., 2007). More than 92% of Cd and 88% of Pb was removed by the rhamnolipid (0.1%) within 36 h. The rhamnolipid was also able to decrease toxicity of the soils and allow microbial activity (*Azotobacter* and *Rhizobium*) to take place and does not degrade soil quality, indicating the sustainability of the process. The economics of the process, however, must be evaluated.

Rhamnolipid was evaluated for heavy metal removal (copper, zinc, and nickel) from the sediments taken from Lachine Canal, Canada (Dahrazma and Mulligan, 2006). Rhamnolipid solution with a constant rate was pumped continuously through a column containing the sediment sample. The washing tests were performed in a continuous flow configuration with a flow rate of 0.5 mL/min to evaluate the relationship between hydraulic conductivity and removal for heavy metals. The tests show that the wetted surface area is among the parameters that control the mechanism of metal removal and is an important issue in continuous flushing process. In addition, the removal of copper is more sensitive to wetted surface area as the majority of copper in this sediment exists in the organic fraction. Organic materials have the largest surface area among all the fractions in the sediment. A decrease in the wetted surface area affects this fraction more than the others and consequently reduces the copper removal from the sediment. The concentration of rhamnolipid and the additives, time and the flow rate were investigated. The removal of heavy metals from sediments was up to 37% of Cu, 13% of Zn, and 27% of Ni when rhamnolipid was applied. Addition of 1% NaOH to 0.5% rhamnolipid enhanced copper removal up to 4 times compared with 0.5% rhamnolipid alone.

To examine the effect of the flushing solution, scanning electron microscopy (SEM) was performed for four samples and showed that the use of the rhamnolipid does not affect the natural size distribution of the sediment. This is an advantage of rhamnolipid as a washing agent. It also can be added that removal by rhamnolipid is an environmentally safe sediment treatment technique in both *ex situ* and *in situ* soil remediation. The sediment after washing can be returned to the environment with minimal damage to its natural texture in *ex situ* remediation while the soil can remain in its natural place and structure for *in situ* sediment treatment. Addition of 1% NaOH decreased the particle size of the sediment due to dissolution of the organic matter which complexes the heavy metals.

The size and morphology of rhamnolipid micelles were evaluated by Dahrazma et al. (2008) using a small angle neutron scattering (SANS) technique. It was found that pH had a significant effect on the morphology. Small aggregates and micelles in the order of 1.7 nm were found at pH 11, whereas larger vesicles were formed under acidic conditions (50-60 nm diameter). As soil pores are typically in the order of 200 nm, filtering during soil flushing should not be a factor. Larger molecules, however, such as exopolymers could cause plugging of the pores. Complexation of the micelles with metals did not have any significant effect on the size of the micelles.

Yuan et al. (2008) tested the removal of heavy metals by a tea saponin by ion flotation. The biosurfactant functioned as both collector and frother. The complexed ions via carboxylate groups adsorb onto the air bubble surface. Lead (90%) removal was greater than copper (81%) and cadmium (71%). Increasing ionic strength slightly decreased removal efficiencies.

In a related technique, metal removal by a foam produced by 0.5% rhamnolipid solution from a sandy soil contaminated with 1,710 ppm of Cd and 2,010 ppm of Ni was then evaluated (Mulligan and Wang, 2004). Maximum removal was obtained after 20 pore volumes. Removal efficiency for the biosurfactant foam was 73.2 % of Cd and 68.1% of Ni. To evaluate the effect of the foam, a biosurfactant liquid solution was compared and 61.7% Cd and 51.0 % Ni were removed. This was superior to the nonionic Triton X-100 foam which removed 64.7% Cd and 57.3% Ni and liquid Triton X-100 which removed 52.8% Cd and 45.2% Ni. Distilled water removed only 18 % of both Cd and Ni. Concentrations of 0.5, 1.0 and 1.5% rhamnolipid at pH values of 6.8, 8 and 10 were also evaluated but did not show significant effects.

The hydraulic conductivity depends on the foam quality. Increasing foam quality decreased substantially the hydraulic conductivity from 2.9×10^{-3} cm/sec for 99% foam quality to 4.1×10^{-4} cm/sec for a 90% foam quality. All these values are lower than the conductivity of water at 0.02 cm/sec. This higher viscosity will allow better control of the surfactant mobility during *in situ* use. Therefore, rhamnolipid foam may be an effective and non-toxic method of remediating heavy metal, hydrocarbon or mixed contaminated soils. Further efforts will be required to enable its use at field scale. Abiotic and biotic factors on the soil after extraction need to be studied.

6.5.4.2 Hydrocarbon Removal

Hydrocarbon removal by surfactants has also been studied. A mixed anionic-nonionic surfactant solution was utilized for removal of phenanthrene (Zhou and Zhu, 2008). The solution consisted of anionic sodium dodecyl sulfate (SDS) and nonionic Triton-X100 surfactants. The mixed solution enhanced phenanthrene dissolution and decreased Triton sorption compared to the individual surfactants in column flushing experiments. Increasing the mass ratio of SDS to Triton from 1:8 to 1:1 increased the solubilization of phenanthrene. The presence of the SDS formed mixed micelles and led to a decrease in the critical micelle concentration (CMC) of Triton from 181 mg/L to 100 mg/L. This mixture has great potential for soil flushing.

Tsai et al. (2009) investigated a two stage remediation process, surfactant washing followed by Fenton-like oxidation. A biodegradable nonionic surfactant, Simple Green, was used at a concentration of 50 g/L for fuel-oil contaminated soil with a total hydrocarbon content of 50,000 mg/kg. After 45 pore volumes of flushing with the surfactant and then 25 pore volumes of deionized water, the TPH content decreased to 4,850 mg/kg. The water flushing was performed to decrease the levels of surfactant prior to oxidation as it would reduce oxidation efficiency. Oxidation efficiency increased with higher hydrogen peroxide concentrations (up to 1.765 mM) up to 7% removal. A solution of 2 mM of KH_2PO_4 was added to increase the half-life of hydrogen peroxide from 110 to 495 min. Complete oxidation of the fuel oil to non-toxic by-products was not possible. Molecular weights of the by-products were similar to the initial compounds before oxidation. Several parameters such as Fe content and phosphate addition need to be optimized.

Pilot scale work was performed for a polychlorinated biphenyl (PCB) contaminated soil flushing process. The pilot column contained 1.7 m^3 of contaminated sandy soil. The level of PCB needed to be reduced from 34.3 mg/kg to 10 mg/kg. An anionic surfactant (40 g/L Spolapon AOS) was passed through the column for 2.5 months. The surfactant contains linear sodium alkene sulfonates and hydroxyalkanesulfonates (C12-C16). The CMC was 2-3 g/L. After 2.5 months, 56% removal of the PCBs was reached. However, it was expected that 6 to 12 months would be needed. The cost was estimated to be 250 to 350 euro/tome. Water recycle was possible due to the low PCB concentration in the leachates (25 x 10^{-6} mg/L).

Perchloroethylene (PCE)-contaminated clayey soil was treated with UH-biosurfactant (Harenda and Vipulanandan, 2008). The soil was 82.5% sand with 17.5% kaolinite clay. Batch and column studies were carried out. Ionic strength of 0.1 g/L in the surfactant solution increased convection, diffusion and retardation. Water was able to remove only 1.18% PCE whereas the biosurfactant could remove an additional 0.27%.

Martel et al. (2005) investigated the removal of PCBs from under a building using soil washing. An anionic surfactant (Nansa HS 85, a dodecylbenzene sulfonate) and an alcohol (n-butanol) were evaluated. In the laboratory tests, 99% of the PCBs could be removed after 10 pore volumes of flushing. Only 25 % could be removed in the field however. Before the experiment, there was a previous spill of a surfactant that formed a gel and blocked the pores. Ethanol was then chosen to solubilize the surfactant and modify the surfactant to alcohol ratio. 99% of the PCBs were then removed. This application shows that the presence of other chemicals at a site can alter remediation efficiencies.

Batch and continuous flow washing experiments were used to evaluate the feasibility of using surfactants for the removal of styrene from contaminated soil (Guo, 2009). Higher styrene removal was achieved from batch experiments compared to column experiments. The highest styrene removal (98.4%) was achieved at 1% rhamnolipid in the batch test. Maximum styrene removal occurred at the highest rhamnolipid concentration in both batch and continuous washing tests. After rhamnolipid adsorption to soil, the two mechanisms of surfactant-aided soil washing are mobilization and solubilisation. The mobilization mechanism occurs at a rhamnolipid concentration below critical micelle concentration (CMC). In this situation, rhamnolipid reduced the surface and interfacial tension between air/water, oil/water, and oil/soil systems. When rhamnolipid concentration is above CMC, the solubility of styrene increased dramatically due to the aggregation of surfactant micelles. Solubilization was the dominant mechanism since most styrene existed within soil samples as pure phase. Rhamnolipid in this study has shown great potential in styrene removal due to its low CMC and high sorption to soil.

The capability of a rhamnolipid was evaluated for its ability to remove pentachlorophenol (PCP) from soil by using it in the form of foam (Mulligan and Eftekhari, 2003). The stability and quality (99%) of the rhamnolipid foam were excellent. Foam injection into a contaminated soil with a 1000 mg/kg level of PCP enabled removal of 60 and 61% of the PCP from a fine sand and sandy-silt soil, respectively. The high quality foams, such as those generated in this study, contain large amounts of air and thus large bubbles with thin liquid films. The foam then collapses easily and is thus less resistant to the soil upon introduction, resulting in lower soil pressures and avoidance of problems like soil heaving.

6.5.5 Treatment of Soil Washing Solutions

After soil treatment with washing solutions, a metal-chelant or surfactant solution is generated. Recovery of the chelating agent is difficult by settling, chemical precipitation or activated carbon. Processes for dealing with the solutions are limited. Reverse osmosis can reduce the volume of water for treatment. Colloidal materials can clog the membranes. Chelating agents are expensive and thus recycling can enhance the economic feasibility. If one metal is present, recycling is more feasible than a mixture of metals. EDTA is inexpensive so it could be destroyed ($4.3 per kg) (Chaney et al., 2000). Advanced oxidation processes can release the metal from the chelating agent. The metals can then be adsorbed by a resin (Finzgar and Lestan, 2006b). Only small amounts of water were used in the process. Recycling of toxic metals and chelating agents has proven difficult but would decrease costs.

A biodegradation process with a UASB reactor was evaluated for treatment of the leachate from soil washing experiments before release to the environment (Guo, 2009). Results from batch and continuous experiments were compared and led to the conclusion that anaerobic biomass could biodegrade styrene and styrene with rhamnolipid under anaerobic conditions. More than 96% of styrene (with acetic acid as the co-substrate and an initial styrene concentration of 75 mg/L, 150 mg/L or 250 mg/L) was biodegraded. The retention time was 16 hours in a UASB reactor with 82% COD removal. Therefore this process could potentially be used for leachate treatment after soil flushing as shown in Figure 6.7.

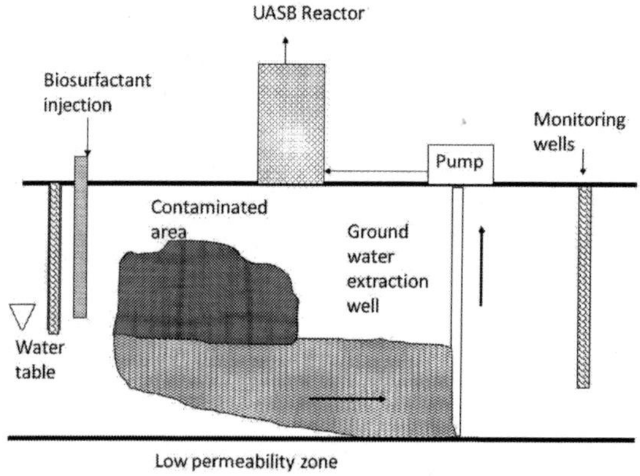

Figure 6.7 Anaerobic treatment of soil leachates.

Another process could be ultrafiltration for the removal of heavy metals from contaminated waters (Elzeftawy and Mulligan, 2011). The effects of different major operating conditions on the system performance were investigated for copper, zinc, nickel and cadmium using two membranes (5,000 and 10,000 dalton molecular weight cutoff). The optimal conditions were successfully applied to treat six contaminated wastewaters from metal refining industries using the two membranes (> 99% rejection ratio). Fouling was the major disadvantage of this process.

6.5.6 Other Enhanced Techniques

Chelants and surfactants have also been combined with other techniques for enhancement. During electrokinetic treatment, hydrogen (H^+) ions are produced during electrolysis. They then migrate into the soil. The low pH desorbs metals from the soil and precipitates at the cathode (Acar and Alshawabkeh, 1993). EDTA has been added to enhance electrokinetic extraction (Yeung et al., 1996). Electrokinetics of soils of high organic or clay contents can be enhanced with chelants for soils that are not amenable to soil washing.

Maturi et al. (2009) examined surfactant-enhanced electrokinetic remediation of mixed contaminants in a low permeability soil. For removal of a mixture of heavy metals and organic compounds, few technologies exist. Electrokinetics has been able to remove PAHs and heavy metals separately. Addition of 5% Igepal at the anode enhanced the removal of phenanthrene in the presence nickel, whereas Tween 80 inhibited electroosmotic flow and subsequently phenanthrene solubilization and transport.

Darko-Kagya et al. (2010) examined the addition of lactate-modified nanoscale iron particles with 2,4 dinitrotoluene (DNT) in soils. Aluminum lactate was selected as a green compound for enhancing the dispersion of nanoscale iron particles. The nanoscale iron particles (NIP) are of high specific surface areas but they aggregate which limits the ability to transport in in situ remediation cases. The DNT degradation ranged from 65 to 99% in kaolin and 59 to 98% in sand with modified nanoscale iron particles. Modification with aluminum lactate enhances transport in soil and thus its potential as a soil flushing agent. Polymers and surfactants are other approaches used to modify NIP (Saleh et al., 2007).

Reddy et al. (2010) studied the integration of hydraulic flushing and electrokinetics for the treatment of metal and organic contaminant mixtures. EDTA and Igepal CA 920 flushing was used in combination with electrokinetics. Zn (60%), Cu (30%) and Pb (80%) were removed by EDTA treatment. Igepal did not remove any metals but removed various PAHs. A sequential treatment could then be used to remove heavy metals and PAHs from manufactured gas plant silty sands.

Some solubilizing additives such as cyclodextrin are used to enhance biodegradation, solubility and microencapsulate (Leitgib et al., 2008; Boving et al., 2003). Phenols and PAHs have been removed by hydroxypropyl β-cyclodextrin. Randomly methylated β-cyclodextrins (RAMEB) reduced the toxicity of contaminants on microbes, plants and animals in the soil (Gruiz et al., 1996). Therefore RAMEB was investigated for the enhanced biodegradation of poorly soluble contaminants. An ex situ groundwater technique was combined with bioventing and flushing with RAMEB and nutrient solutions. Monitoring of soil gas and ground water is important followed by soil analyses. A mixture of diesel and engine oil was successfully treated.

Tetrachloroethylene (PCE) is a chlorinated solvent used in dry-cleaning industry. PCE is a toxic chemical and is hazardous to the human and environment. PCE is a dense non-aqueous phase liquid (DNAPL) and once leaked into the soil, moves downwards and contaminates the soil and groundwater. Common remediation techniques like soil washing would not easily remediate the soil contaminated with PCE, unless proper additives like surfactants are added to the washing solution. Surfactants reduce the surface and interfacial tension, and increase the solubility of PCE. Surfactants at concentrations above CMC form micelles and make PCE partition into the micelles, consequently PCE disperses in the aqueous phase and becomes more bioavailable.

A study by Hamidi and Mulligan (2010) evaluated the effectiveness of rhamnolipid and two surfactant-based products (ASAP™ and Degreaser™ from Avmor) on bioremediation of PCE. These two agents are non-toxic chemicals. Following the USEPA methods, and gas chromatography analysis, diverse 21 day experiments were conducted. ASAP™ and Degreaser™ at the ratio of 1:1 showed removal efficiencies of 45% and 52% respectively. Addition of rhamnolipid with a ratio to soil of 1:1 to these products at the same ratios enhanced the remediation by up to 55% and 58%. Conducting microbial analysis, a direct correlation was observed between microbial density and PCE removal, which proves that degradation had occurred (Figure 6.8). The removal trend was: biological agent + rhamnolipid > biological agent > rhamnolipid > control.

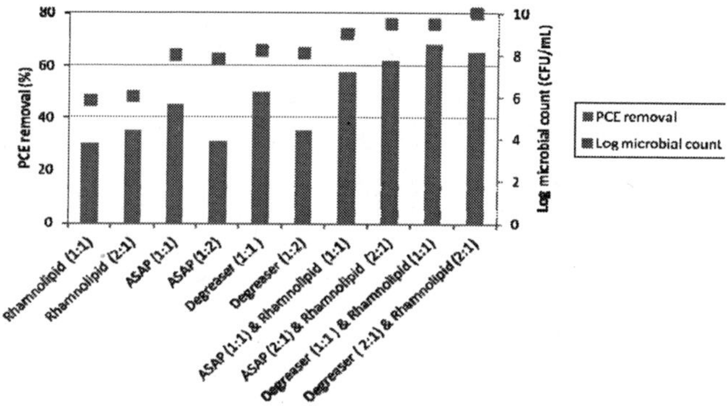

Figure 6.8 Enhanced biodegradation of PCE in the presence of rhamnolipid (Rhamn) and/or two other agents Degreaser (Deg) or ASAP (AS) (adapted from Hamidi and Mulligan, 2010)

6.5.7 Case Studies of Enhanced Biodegradation

Once the contaminants are desorbed, they are then more bioavailable to microorganisms and thus biodegradation is enhanced. The mechanism is indicated in Figure 6.9. The hydrophobic part of the surfactant is attracted to the hydrophobic contaminant on the soil while the hydrophilic is attracted to the water. These forces desorb the contaminant from the soil and thus making them available for biodegradation by microorganisms. Case studies have shown its effectiveness (Ivey, 2005). 200,000 kg of PAH contaminated soil were excavated and biopiled. Total hydrocarbon levels were greater than 10,000 ppm. Surfactant was added to the pile. After 12 weeks, levels were acceptable according to the standards.

In another case study, a brownfield was excavated. Bench scale tests were performed to evaluate the surfactant in 20:1 liquid to soil ratios. 19 PAHs were desorbed at an average of 90%. The range of removal was 84.3% for benzo(ghi)perylene to 97.3% for 2-methylnaphthalene. In 2007, a benzene toluene ethylbenzene xylene (BTEX) and DNAPL contaminated refinery site since 1950s. The site area was 20 acres. The soil was a glacial till. There was potential for ground water contamination. Pilot scale results indicated that DNAPL and BTEX removal rates increased by 549% and 303%. This was significant as three other technologies had failed.

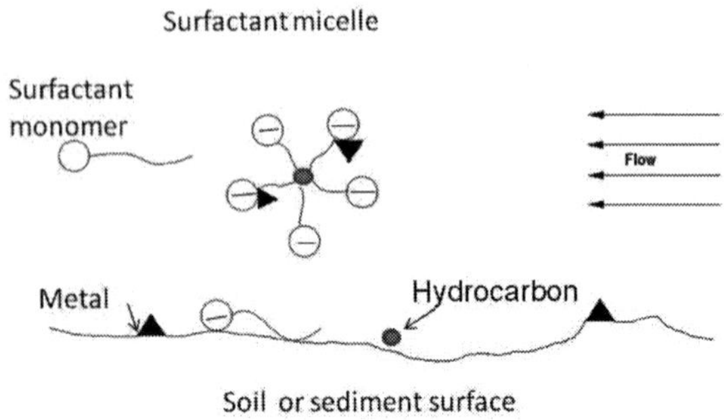

Figure 6.9 Mechanism of contaminant removal by a surfactant.

In 2006, Ivey-sol surfactant-enhanced bioremediation of F3 hydrocarbons was evaluated where there had been oil and gas activities. Ivey-sol is a nonionic surfactant for desorption of PAHs, PCBs, chlorinated hydrocarbons, MTBE, TBA and metals. The hydrocarbon levels decreased substantially over a 2 month period.

6.6 Conclusions

Chelating agents, surfactants and biosurfactants such as rhamnolipids and saponin have shown their potential for soil washing and flushing of contaminated soil and sediments. More biosurfactants need to be investigated. Both organic and inorganic contaminants can be treated through desorption or biodegradation processes. The biosurfactants seem to enhance the solubilization and emulsification of the contaminants via the mechanisms shown in Figure 6.8. Due to their biodegradability and low toxicity, biosurfactants such as rhamnolipids are very promising for use in remediation technologies. In addition, there is the potential for *in situ* production, a distinct advantage over synthetic surfactants. This needs to be studied further. Further research regarding prediction of their behaviour in the fate and transport of contaminants is required. More investigation into the solubilization and bioconversion mechanisms of both hydrocarbons and heavy metals by biosurfactants is required to enable model predictions for transport and remediation. New applications for the biosurfactants regarding nanoparticles and combinations with other techniques are developing. Future research should focus on the stabilization of the nanoparticles by biosurfactants before addition during remediation procedures. Sustainable remediation techniques must minimize energy and water requirements and leave a soil that is environmentally safe with its precontaminated properties (physical, chemical and biotic) restored. More field demonstrations are also required.

6.7 References

Acar, V.B. and Alshawabkeh, A. (1993). "Principles of electrokinetic remediation." *Environmental Science and Technology*, 27, 2638-2647.

Ara, I., and Mulligan, C. N. (2008). Conversion of Cr(VI) in water and soil using rhamnolipid. 61st Canadian Geotechnical Conference, Edmonton, AB, Sept. 20-24.

Asci, Y., Nurbas, M., Acikel, Y. S. (2007). "Sorption of Cd(II) onto kaolinin as a soil component and desorption of Cd(II) from kaolin using rhamnolipid biosurfactants." *Journal of Hazardous Materials*, B139: 50-56.

Asci, Y. Nurbas, M., and Acikel, Y. S. (2008a). "A comparative study for the sorption of Cd(II) by K-feldspar and sepiolite as soil components and the recovery of Cd(II) using rhamnolipid biosurfactants." *Journal of Environmental Management*, 88, 383-392.

Asci, Y., Nurbas, M. and Acikel, Y. S. (2008b). "Removal of zinc ions from a soil component Na-feldspar by a rhamnolipid biosurfactant." *Desalination*, 233, 361-365,

Barona, A., and Romero, F. (1996). "Fractionation of lead in soils and its influence on the extractive cleaning with EDTA." *Environmental Technology*, 17, 63-70.

Biermann, M., Lange, F., Piorr, R., Ploog, U. Rutzen, H., Schindler, J., Schmidt, R. (1987). Surfactants in Consumer Products, Theory, Technology and Application. Edited by Falbe, J., Heidelberg: Springer-Verlag.

Boving, T. B., McCray, J. E., Blanford, W. J., Brusseau, M. L. (2003). Cyclodextrin Enhanced In-situ Removal of Organic Contaminants from Groundwater at Department of Defense Sites. ESTCP Project Number: 200 113. Kingston, June 31, 2003.

Chaney, R. L., Brown, S. L., Li, Y-M., Angle, J. S., Stuczynsk, T. I., Daniels, W. L., Henry, C.L., Siebielic, G., Malik, M., Ryan, J.A. Compton, H., (2000). Progress in risk assessment for soil metals and in-situ remediation and phytoextraction if metals from hazardous contaminated soils. In: Proceedings of US-EPAs Conference Phytoremediation, Boston, MA.

Chen, W-J., Hsiao, L-C. Chen, K. K-Y. (2008). "Metal desorption from copper(II)/nickel(II)-spiked kaolin as a soil component using plant-derived saponic biosurfactant." *Process Biochemistry*, 43, 488-498.

Chevalier, L. R., Masten, S. J., Wiggert, D. C. and Walace, R. B. (1997). "Experimental investigation of surfactant enhanced dissolution of residual NAPL in saturated soil." *Ground Water Monitoring Review*, 17(4), 89-98.

Cooper, D. G. (1986). "Biosurfactants." *Microbiological Science*, 3,145-149.

Dahrazma, B., and Mulligan, C. N. (2007). "Investigation of the removal of heavy metals from sediments using rhamnolipid in a continuous flow configuration." *Chemosphere*, 69, 705-711.

Dahrazma, B., Mulligan, C. N. and Nieh, M. P. (2008). "Effects of additives on the structure of rhamnolipid (biosurfactant): a small-angle neutron scattering (SANS) study." *Journal of Colloid and Interface Science*, 319, 590-593.

Darko-Kagya, K., Khodadoust, A. P., and Reddy, K. R. (2010). "Reactivity of Lactate-Modified Nanoscale Iron Particles with 2,4-Dintrotoluene in Soils."

Journal of Hazardous Materials, 182(1-3), 177-183.

Dragage Verreault (2008). Technologies décontamination sols et sédiment métaux/hydrocarbures. Workshop on in situ remediation of contaminated sediment, Workshop at Concordia University, Montreal, Canada, March 10-11.

El-Shafey, E. L. and Canepa, P. (2003). "Remediation of a Cr (VI) contaminated soil : Soil washing followed by Cr (VI) reduction using a sorbent prepared from rice husk." *Journal de Physique IV*, 107 (1), 415-418.

El Zeftawy, M. A. M. and Mulligan, C. N. (2011). "Use of rhamnolipid to remove heavy metals from wastewater by micellar enhanced ultrafiltration (MEUF)" *Separation and Purification Technology*, 77 (1,2), 120-127.

Finzgar, N., Kos, B., and Lestan, D. (2005). "Heap leaching of lead contaminated soil using biodegradable chelator [S,S] ethylenediamine disuccinate." *Environmental Technology*, 26, 553-560.

Finzgar, N. and Lestan, D. (2006). "Advanced oxidation for treatment of aqueous extracts from EDTA extraction of Pb and Zn contaminated soil." *Journal of Environmental Engineering*, 132, 1376-1380.

Finzgar, N. and Lestan, D. (2007). "Multi-step leaching of Pb and Zn contaminated soils with EDTA." *Chemosphere,* 63, 824-832.

Franzetti, A., DiGennaro, P., Bevilacqua, A., Papacchini, M. and Bestetti, G. (2006). "Environmental features of two commercial surfactants widely used in soil remediation." *Chemosphere*, 62, 1474-1480.

Gruiz, K. Fenyvesi, É., Kriston, É,. Molnár M. and Horváth, B. (1996). "Potential use of cyclodextrins in soil bioremediation." *Journal of Inclusion Phenomena and Molecular Recognition in Chemistry*, 25, 233–236.

Guerra-Santos, L. H. Käppeli, O., and Feichter, A. (1984). "*Pseudomonas aeruginosa* biosurfactant production in continuous culture with glucose as carbon sources." *Applied and Environmental Microbiology*, 48, 301-305.

Guo, Y. (2009). Rhamnolipid-Enhanced Remediation of Styrene-Contaminated Soil Followed by Anaerobic Biodegradation, Ph.D. Thesis, Concordia University, Montreal, Quebec, Canada.

Hamidi, A. and Mulligan, C. N. (2010). Treatment of PCE-Contaminated Soil Using Biosurfactants and Other Additives, 11th International Environmental Specialty Conference, Winnipeg, Manitoba, June 9-12, 2010.

Harenda, S. and Vipulanandan, C. (2008). Biosurfactant flushing of PCE contaminated clayey soil, GeoCongress 2008, Geotechnics of Waste Management and Remediation, ASCE, Downloaded Aug. 6, 2010, pp 511-518.

Hitsatsuka, K., Nakahara, T. Sano, N., and Yamada, K. (1971). "Formation of a rhamnolipid by *Pseudomonas aeruginosa* and its function in hydrocarbon fermentation." *Agricultural Biology and Chemistry*, 35, 686-692.

Huang, Y-T., (2005). Heavy metals in urban soils. M.A.Sc Thesis, Concordia University, Montreal, Canada.

Ivey, G. A. (2005). Surfactant desorption is key to successful PAH bioremediation. *Environmental Science and Engineering*, September, 2 pages.

Juwarkar, A. A. ,Nair, A., Dubey, K. V., Singh, S. K., and Devotta, S. (2007). "Biosurfactant technology for remediation of cadmium and lead contaminated soils." *Chemosphere*, 68, 1996-2002.

Kantar, C. and Honeyman, B. D. (2006). "Citric acid enhanced bioremediation of soils contaminated with uranium by soil flushing and soil washing." *Journal of Environmental Engineering*, 132 (2), 247-255.

Kim, J. and Vipulanandan, C. (2006). "Removal of lead from contaminated water and clay soil using a biosurfactant." *Journal of Environmental Engineering*, 132(7), 777-786.

Lang, S. and Wagner, F. (1987). Structure and properties of biosurfactants. In: Biosurfactants and Biotechnology, Edited by Kosaric N, Cairns WL, Gray NCC, New York: Marcel Dekker, 21-45.

Leitgib, L., Gruiz, K., Fenyvesi, E., Balogh, G. and Muranyi, A. (2007). "Development of an innovative soil remediation: "cyclodextrin-enhanced combined technology." *Science of the Total Environment*, 192, 12-31.

Lestan, S., Luo, C-L., and Li, X-D. (2008). "The use of chelating agents in the remediation of metal-contaminated soils : A review." *Environmental Pollution*, 153, 3-13.

Lin, S. C. (1996). "Biosurfactant: Recent advances" *Journal of Chemical Technology and Biotechnology*, 63, 109-120.

Löser, C., Andreas, Z., Görsch, K. and Seidel, H. (2006). "Remediation of heavy metal polluted sediment in the solid bed: comparison of abiotic and microbial leaching." *Chemosphere*, 65, 9–16.

Löser, C., Andreas, Z., Petra, H. and Seidel, H. (2007). "Remediation of heavy metal polluted sediment by suspension and solid-bed leaching: estimate of metal removal efficiency." *Chemosphere*, 66, 1699–1705.

Martel, R., Foy, S., Saumure, L., Roy, A., Lefebvre, R., Therrien, R. Gabriel, U., and Gelinas, P. J. (2005). "Polychlorinated biphenyl (PCB) recovery under a building with an in situ technology using micellar solutions." *Canadian Geotechnical Journal*, 42, 932-948.

Massara, H., Mulligan, C. N., and Hadjinicolaou, J. (2007). "Effect of rhamnolipids on chromium contaminated soil." *Soil and Sediment Contamination*, 16, 1-14.

Maturi, K., Reddy, K. R. and Camselle, C. (2009). "Surfactant-enhanced electrokinetic remediation of mixed contamination in low permeability soil." *Separation Science and Technology*, 44, 2385-2409.

Mench, M. Vangronsveld, J., Clijsters, H., Lepp, N. W., and Edwards, R. (2000), *In situ* metal immobilization and phytostabilization of contaminated soils. In *Phytoremediation of Contaminated Soil and Water,* N. Terry, and G. Bauelos, eds. Lewis Publishers, Boca Raton, FL., 323-358.

Mercier, G. Blais, J. F., and Drogui, P. (2008). Groupe de recherche en assainissement et techologies environmentales, Workshop on in situ treatment of contaminated sediments, Concordia University, Montreal, Canada, March 10-11.

Mulligan, C. N. (2002). Environmental Biotreatment, Government Institutes, Rockville Maryland.

Mulligan, C. N. (2005). "Environmental applications for biosurfactants." *Environmental Pollution*, 133,183-198.

Mulligan, C. N. and Eftekhari, F. (2003). "Remediation with surfactant foam of PCP contaminated soil." *Engineering Geology*, 70, 269-279.

Mulligan, C. N., and Gibbs. B. F. (1993). Factors influencing the economics of biosurfactants. In Biosurfactants, Production, Properties, Applications. Kosaric, N. (Ed.): Marcel Dekker, New York, 329-37.

Mulligan, C. N. and Kamali, M. (2003). "Bioleaching of copper and other metals from low-grade oxidized mining ores by *A. niger*." *Journal of Chemical Technology and Biotechnology*, 78, 497-503.

Mulligan C. N., Oghenekevwe, C., Fukue, M. and Shimizu, Y., Biosurfactant Enhanced Remediation of a Mixed Contaminated Soil and Metal Contaminated Sediment, 7th Geoenvironmental Engineering Seminar, Japan-Korea-France, May 19-24, 2007, Grenoble, France.

Mulligan, C. N. and Wang S. (2006). "Remediation of a heavy metal contaminated soil by a rhamnolipid foam." *Engineering Geology*, 85, 75-81.

Mulligan, C. N. Yong, R. N. and Gibbs, B. F. (1999). "On the use of biosurfactants for the removal of heavy metals from oil-contaminated soil." *Engineering Progress*, 18, 50-54.

National Round Table for the Environment and Economy (NRTEE), (1998), Greening Canada's Brownfield Sites, NRTEE, Ottawa.

National Round Table for the Environment and Economy (NRTEE), (2004), Cleaning up the Past, Building the Future, A National Brownfield Redevelopment Strategy for Canada, NRTEE, Ottawa.

Ochoa-Loza, F. J. Noordman, W. H. Jannsen, D. B., Brusseau, M. L., and Miller. R. M. (2007). "Effects of clays, metal oxides, and organic matter on rhamnolipid sorption by soil." *Chemosphere,* 66, 1634-1642.

Ouattara, A. (2008). Lixiviation fongique des residus miniers par *A. niger* and *P. simplicissum*. Ph.D. thesis, U. Laval, Quebec, Canada.

Papadopoulos, D., Pantazi, C. Savvides C., Harlambous, K. J. Papadopoulos A., and Loizidou, M., (1997). "A Study on Heavy Metal Pollution in Marine Sediments and Their Removal from Dredged Material." *Journal of Environmental Health,* A32(2), 347-360.

Reddy, K .R., Cameselle, C., and Ala, P. R. (2010). "Integrated Electrokinetic-Flushing to Remove Mixed Organic and Metal Contaminants." *Journal of Applied Electrochemistry*, 40(6), 1269-1279.

Robert, M., Mercadé, M. E., Bosch, M. P., Parra, J. L., Espiny, M. J. Manresa, M. A. Guinea, J. (1989). "Effect of the carbon source on biosurfactant production by *Pseudomonas aeruginosa* 44T1." *Biotechnology Letters*, 11, 871-874.

Russel, M., Colgazier, E. W., and English, M. R. (1991). Hazardous Waste Remediation: The task ahead. Waste Management Research and Education Insitute, University of Tennessee, Knoxville, TN.

Saleh, N., Sirk, K. Liu, Y., Phenrat, T., Dufour, B., Matyjaszewski, K., Tilton, R. D., and Lowry, G. V. (2007). "Surface modifications enhance nanoiron transport and NAPL targeting in saturated porous media." *Environmental Engineering Science*, 24, 45–57.

Song, S. Zhu, L., and Zhou, W. (2008). "Simultaneous removal of phenanthrene and cadmium from contaminated soils by saponin, a plant-derived biosurfactant." *Environmental Pollution*, 156, 1368-1370.

Svab, M., Kubal, M., Mullerova, M., and Raschman, R. (2009). "Soil flushing by surfactant solution: Pilot scale demonstration of complete technology." *Journal of Hazardous Materials*, 163, 410-417.

Tsai, L. J., Yu, K. C., Chen, S. F., Kung, P. Y., Chang, C. Y. and Lin, C. H., (2003a), "Partitioning variation of heavy metals in contaminated river sediment via bioleaching: effect of sulfur added to total solids ratio." *Water Research,* 37, 4623–4630.

Tsai, L. J., Yu, K. C., Chen, S. F. and Kung, P. Y., (2003b), "Effect of temperature on removal of heavy metals from contaminated river sediments via bioleaching." *Water Research,* 37, 2449–2457.

Tsai, T. T., Kao, C. M., Surampalli, R. Y. and Liang, S. H. (2009). "Treatment of fuel-oil contaminated soils by biodegradable surfactant washing followed by Fenton-like oxidation." *Journal of Environmental Engineering*, 135(10), 1015-1024.

Tsujii, K. (1998). Surface Activity, Principles, Phenomena, and Applications. Academic Press, USA.

Urum, K., Pekdemir, T., and Gopur, M. (2003). "Optimum conditions for washing of crude oil-contaminated soil with biosurfactant solutions." *Transactions of the Institution of Chemical Engineers*, 81B, 203-209.

USEPA (1992). "Soil Sediment Washing System." EPA/540/MR-921075, USEPA, October .

USEPA (1995). "Bergman USA Soil Washing Technology: Applications Analysis Report" EPA/540/AR-92/075, September.

Van Benschoten, J. E. and Matsumoto, M. R. (1997). "Evaluation and analysis of soil washing for seven lead-contaminated soils." *Journal of Environmental Engineering,* 123, 217-224.

Venghuis, T. and Werther, J. (1998). "Flotation as an additional process step for the Washing of Soils Contaminated with heavy metals." Proceedings of the 6th International FZK/TNO Conference, ConSoil' 98, May 1998, Edinburgh, Thomas Telford Publishing, London, Vol.1, 479-480.

Wang, S. and Mulligan, C. N. (2009a). "Effect of natural organic matter on arsenic mobilization from mine tailings" *Journal of Hazardous Materials*, 168, 721-726.

Wang, S. and Mulligan, C. N. (2009b). "Rhamnolipid biosurfactant-enhanced soil flushing for the removal of arsenic and heavy metals from mine tailings." *Process Biochemistry*, 44(3), 296-301.

Wang, S., and Mulligan, C. N. (2009c). "Arsenic mobilization from mine tailings in the presence of a biosurfactant" *Applied Geochemistry*, 24, 938-935.

Yeung, A. T., Hsu, C. N., and Menon, R. M. (1996). "EDTA-enhanced electrokinetic extraction of lead." *Journal of Environmental Engineering*, 122(8), 666–672

Yuan, X. Z., Meng, Y. T., Zwng, G. M., Fang, Y. Y., and Shi, J. G. (2008). "Evaluation of tea-derived biosurfactant on removing heavy metal ions from dilute wastewater by ion flotation." *Colloids and Surfaces A: Physicochemical and Engineering Aspects*, 317, 256-261.

Zacharias, R. A. (2007). Ivey Technology accelerates cleanup, Reclaiming Canada Issue 1, http://www.ivey-sol.com.

Zagular, S. J. and Beitnger, E. W (1996). Effective remediation of contaminated sediment, Proceedings of the 2nd International Symposium on Environmental Contamination in Central and Eastern Europe, Technical University of Budapest, Budapest, September, 1996.

Zhou, W. and Zhu, L. "Enhanced soil flushing of phenanthrene by anionic-nonionic mixed surfactant." *Water Research*, 42, 101-108.

CHAPTER 7

Heavy Metal Leaching from Contaminated Soils during the Percolation of EDTA: Observations and Modeling

Monika A.M. Kedziorek, Fabrice Compère, Alain Dupuy, and Alain C.M. Bourg

7.1 Introduction and Outline

Amongst the diversity of pollutants associated with our industrial civilization, heavy metals are of great concern, not necessarily due to extremely high toxicity but because of the absence of degradation. In hydrologic cycles, natural and anthropogenic solids (soils, sediments, industrial residues, wastes...) act as reservoirs for heavy metals. Contaminants trapped in these solids present a risk to the environment because they can be remobilized depending on local hydrogeochemical conditions, be these naturally controlled or induced by human activity. A significant part of the heavy metal load of sediments and soils, especially under natural or weakly polluted conditions, is not in equilibrium with groundwater (i.e. permanently trapped in mineral structures). If present in great amounts and in potentially soluble form (e.g., Kedziorek and Bourg, 2007) (which is the case for many contaminated solids), they must be removed in order to prevent their potential dissemination in the environment.

The contamination of natural solids (agricultural soils, sediments and solids in active or derelict industrial sites) is one of the major challenges of our modern society. A great effort and many advances have been made recently in remediation techniques for degradable and volatile organic pollutants. For heavy metals, the magnitude of the problem is much greater (larger volumes, no degradation). Several chemical extraction methods have been investigated but few have potential applications due to numerous practical problems (e.g., Kowalick, 1992).

To our knowledge, few practical applications of in situ heavy metal decontamination have been reported. Urlings et al. (1990) aspersed a sandy soil contaminated by cadmium with an acidic solution and achieved a good decontamination after several months of treatment. However, although acids are relatively inexpensive, they have undesirable side effects such as the dissolution of the porous matrix leading to destruction of the solid, and poor chemical extractive efficiency because much of their solubilizing efficiency is consumed by the dissolution of non-toxic elements.

An alternative to using acids is using chelates, which have a strong metal complexing capacity to solubilise pollutant metals and entrain them out of the target soil. EDTA (ethylenediaminetetraacetic acid) has the best complexing ability and is the most well known for its use in analytical chemistry and various industrial processes. Its potential for decontaminating soils contaminated by metals has been studied by many researchers (e.g. Lestan et al., 2008). As will be seen in this book and discussed briefly later in this chapter, other chelates might now be preferred.

Batch Experiments. Numerous authors have investigated the extractive action of EDTA using batch experiments (e.g. Baghdady and Sippola, 1984; Samanidou and Fytianos, 1990; Yu and Klarup, 1994; Elliott and Brown, 1989; Tuin and Tels, 1990; Brown and Elliott, 1992; Hong and Pintauro, 1996; Li and Shuman, 1996; Bryce and Clark, 1996; Nowack and Sigg, 1997; Ghestem and Bermond, 1998; Puginier et al., 1998; Peters, 1999; Garrabrants and Kosson, 2000; Sun et al., 2001; Xie and Marshall, 2001; Fangueiro et al., 2002; Manouchehri et al., 2006; Tsang and Lo, 2006; Kedziorek and Bourg, 2007; Di Palma and Mecozzi, 2007; Vaxevanidou et al., 2008; Zhang et al., 2010).

The conclusions of these investigations are three-fold. (1) In situations where there is true chemical equilibrium, the extractive behavior of EDTA is a result of the competition of metal dissolution/desorption from the solid and metal complexation with the aqueous chelate. In some cases, metal extraction can be aided by a preliminary weak adsorption of the uncomplexed chelate. This adsorption is however restricted to positively charged solids such as aluminum oxides, iron oxyhydroxides and calcite (Stumm and Morgan, 1981). (2) Components other than toxic metals (such as Ca, Mg, Al, Fe) can be removed from the soil by EDTA, thus decreasing the extraction yield and soil fertility. Sufficient addition of EDTA is therefore necessary for an efficient pollutant extraction (e.g., Nowack and Sigg, 1997; Sun et al., 2001; Kedziorek and Bourg, 2007; Zhang et al., 2010). (3) Time is a very important factor as EDTA complexation with heavy metals is far from instantaneous and its reversibility is also time-dependent. The time factor for the most efficient pollutant extraction resides in the competition between fast desorption of pollutant metals and slower dissolution of metals from indigenous minerals (e.g., Puginier et al., 1998; Kedziorek and Bourg, 2000; Sun et al., 2001).

Laboratory Percolation Experiments. Few people have studied dynamic situations in laboratory column experiments (e.g., Jardine et al., 1993; Szecsody et al., 1994; Bourg et al., 1994; Kedziorek et al., 1998; Kedziorek and Bourg, 2000, Sun et al., 2001, Tsang et al., 2007a, 2007b). The planning and conceptual interpretations of these investigations are difficult because of the complexity of the processes involved in terms of kinetics, reversibility and competitive effects, as illustrated by the batch study observations. Some of the most pertinent experiments are discussed here.

We investigated the mobilization and movement of Cd and Pb in an EDTA flux both by laboratory experiments (in pulse and step modes) and mathematical modeling (Kedziorek et al., 1998). A numerical model linking solute transport of EDTA and EDTA-metal chelates to metal solubilization from contaminated solids was validated using experimental breakthrough curves (BTC). We used an implicit finite difference scheme to solve advection-dispersion equations for free and complexed EDTA with, for both, inclusion of a second-order kinetic law to express the extraction reaction. The hydrodispersive parameters of the column were calculated using the bromide BTCs. Our model simulated various EDTA injection modes (pulse or step) at different chelate concentrations (10^{-2} to 10^{-4} M), by fitting the kinetic rate constant K of solubilization with the observed Cd and Pb BTCs. Interestingly, pulse and step experiments at different EDTA concentrations gave similar K values, 2.4 10^{-6} s^{-1} for Cd and 2.1 10^{-5} s^{-1} for Pb. The model also accounted for the decreasing metal extraction efficiency as the metal in the solid is depleted or as the available EDTA concentration decreases. In this paper we presented methodological suggestions for necessary investigations prior to remediation operations.

Different leaching patterns in four contaminated soils were observed for Zn, Cd, Cu and Pb in a column experiment using 10^{-2} M EDTA (Sun et al. 2001). The relative mobility decreased from Cu (corresponding to the peak EDTA arrival in the leachate) to Pb with Zn and Cd as intermediate velocities. The elution of Pb increased after the Cu and Zn peaks. Considerable dissolution of Fe occurred in two of the four soils during EDTA leaching. The authors proposed that exchangeability of metals in soil, the kinetics of metal desorption/dissolution and the mode of EDTA addition were the main factors controlling the behaviour of metal leaching with EDTA.

EDTA solutions (10^{-2} or 10^{-3} M) flushed through artificially contaminated soils in laboratory column experiments induced Cu extraction but also mineral dissolution (Tsang et al., 2007). To explain the time-dependent depletion of extractable metals (both from contamination sources and by natural occurrence), the observed BTC of Cu, Fe, Al and Ca were simulated, as by Kedziorek et al. (1998), by second-order rate-limited reactions of EDTA-promoted desorption and dissolution. The fitted rate constants of Cu were about one order of magnitude greater than those of Fe and Al. This might be explained by the fact that Cu is extracted from weakly sorbed sites while Fe and Al are dissolved from amorphous oxides. The retardation of Fe, Al and Ca transport was attributed to surface complexation. The simultaneous simulation of Cu desorption and dissolution of soil Fe, Al, and Ca minerals is more realistic than separate modeling of each metal because the individual metal approach certainly overestimates the free EDTA available during transport and consequently underestimates the rate coefficients of EDTA-promoted dissolution (Tsang et al., 2007).

In summary, the percolation experiments (1) demonstrate the significance of kinetics in EDTA heterogeneous reactions (desorption, dissolution) and (2) confirm the competitive action of EDTA reactions between the dissolution of metals from minerals naturally present in the solids and the removal of surface contaminant metals, in terms of both thermodynamics and reaction rates.

Field Studies or Experiments. Natural organic ligands are widely considered to increase the solubility of metals and especially those with high organic ligand affinity such as Cu, Pb, Ni or Hg (e.g., Bourg and Schindler, 1985; Von Gunten et al. 1991; Bourg and Darmendrail, 1992, Di Palma et al., 2007). It is therefore assumed that they will compete with anthropogenic chelates (Nowack et al., 1997) to transport dissolved metals and remove them from contaminated solids.

Nowack et al. (1997) investigated, under natural conditions, the behavior of metals (Ca, Fe(III), Mn(II), Cu, Zn and Ni) and the anthropogenic chelate EDTA (concentration range 3 to 7 x 10^{-8} M) during the infiltration of river water into an adjacent alluvial aquifer (at the Glatt River field site, Switzerland). EDTA behaved conservatively regarding its total concentration in the infiltration flow path (as compared to borate used as a natural tracer), but not with respect to its speciation. In the river, 35% of the EDTA was bound to Fe(III) as opposed to only 11% in the aquifer. Nowack et al. (1997) explained this variation by exchange with metals either transported with the water flow or adsorbed on infiltrated aquifer sediments according to slow kinetics reactions described by the generic equation Fe(III)-EDTA + Me(aqueous or adsorbed) $\rightarrow$ M-EDTA + Fe(III). Desorption of metals is therefore possible. Natural organic matter was very competitive, especially for Cu and Zn, which is explained by its higher concentration, EDTA accounting for only 0.2 to 0.4% of the DOC (dissolved organic carbon).

A series of field experiments involving the injection of EDTA and metals, together with a water tracer, was carried out at the Cape Cod Toxic Substances Hydrology Research Site, near Falmouth, Massachusetts. In a field experiment (Davis et al., 1993), EDTA (0.5 mM, as K_2H_2EDTA) was injected together with the reactive metal, zinc (0.1 mM as $Zn(NO_3)_2.6H_2O$), and the tracer, bromide (1.9 mM as KBr), as a 380-L pulse in the oxic zone of this shallow, unconfined, sand and gravel aquifer. Two meters down-gradient from the injection well, the arrival of excess metals (i.e. Zn, Fe, Al, Mn and Ca above background level) assumed to be transported as EDTA complexes was retarded by a factor of 2 for Zn and by about 1.3 for the other metals as a bulk. The authors proposed that the slight, but significant, retardation of the transported Zn (as EDTA complexes) was due to weak adsorption of the complexed metal.

In a larger-scale experiment, nearly 10,000 L of groundwater containing Br as a water tracer and EDTA (in slight excess) complexed with Pb, Zn, Cu and Ni were injected into the Cape Cod site (Davis et al., 2000). The transport of metal EDTA complexes (some over more than 200 m) was explained by aqueous complexation, adsorption and dissolution/precipitation of solid Fe oxyhydroxides in the aquifer sediments. The behavior of the 4 heavy metals was quite different. Dissolved Pb-EDTA disappeared first from the EDTA plume, probably due to exchange reactions with Fe and adsorbed Zn (present prior to the experiment). Dissolved Cu-EDTA remained for a longer distance. Even though the thermodynamic stability of Pb-EDTA is greater that that of Cu-EDTA, stronger Pb adsorption on aquifer sediments destabilizes its EDTA complex. The mass of dissolved Zn-EDTA increased due to desorption of Zn in the aquifer sediments as mentioned earlier. The dissolved Ni-EDTA mass remained constant. Simulations by Kent et al. (2008) under conditions similar to those in the experiment above confirmed the mechanistic hypotheses discussed by Davis et al. (2000).

Friedly et al. (2002) compared simulations and results of small-scale field experiments at the Cape Cod site in which free EDTA and complexes with Ni, Zn and Ca were injected into three chemically different areas of this aquifer contaminated by domestic sewage. One zone contained large amounts of adsorbed, sewage-derived zinc, another had a large amount of adsorbed manganese due to oscillating, mildly reducing conditions created by the sewage plume, and the third had negligible adsorbed manganese and zinc contents compared to the natural levels. The injectate plume was near equilibrium with regard to Fe in the sediment after 11 m of transport in the Zn-contaminated area but remained far from equilibrium in the other two areas. Similarly, EDTA reactions with elements common in aquifers (Ca, Fe, Al) were often not at equilibrium.

Adsorption and dissolution of Fe(III) minerals strongly influence the transport of heavy metals complexed with EDTA. During metal-EDTA injection tests under mildly reducing conditions in the Cape Cod site, the rates of desorption of metals from aquifer sediments were observed to be much faster that the dissolution of Fe oxyhydroxides and consequently adsorbed pollutant metals could be efficiently removed from soils and sediments if the water flows rapidly enough (Kent et al., 2002).

In summary, extraction is metal-specific and competition between elements can displace metals bound and transported by EDTA. Consequently, the behavior of different metals is very site specific (i.e. chemical conditions, presence of surface species). Modeling of processes is complex because some reactions are at equilibrium while others are not.

Outline. This chapter presents two laboratory experiments of heavy metal leaching in an EDTA flux: a previously unpublished study of a soil contaminated by long-term spreading of sewage sludge/wastewater (near Paris, France) and a detailed description of a computer model simulating the long-term contamination of agricultural soils by smelting activities (Northern France).

7.2 Observations: Solubilization of Heavy Metals from a Soil Contaminated by Long-Term Spreading of Sewage Sludge/Wastewater—Reactivity and Kinetics

7.2.1 Soil Sample and Methods

The soil sample used in these leaching experiments was collected from the upper 20 cm of an agricultural plot at the Paris-Achères water treatment plant, 20 km NW of Paris (France). This plot was used for more than 100 years for spreading wastewater and sludge (initially with untreated wastewater and later with increasing amounts of treated water and sludge). The sample was dried at 50 °C for 24 hours and sieved through a 500 μm nylon sieve to eliminate the coarse fraction. The total geochemical content measured by ICP after fusion with sodium borate is given in Table 7.1.

Table 7.1 Selected characteristics of the Achères soil sample

org. matter	Si	Al	Ca	Mg	Fe	Mn	Cd	Co	Cu	Ni	Zn
(%)	(%)	(%)	(%)	(%)	(%)	(%)	μg/g	μg/g	μg/g	μg/g	μg/g
8.1	39.9	1.7	2.3	<0.6	1.0	0.015	3.5	7	296	34	1138

A PVC column (length: 9.0 cm; internal diameter: 2.1 cm) was packed with the dried solid material. At the bottom of the column, a fiberglass filter Whatman GF:D, with an average pore size of 2.5 μm, was inserted on a filter support. Eluted water was filtered through 0.45 μm membrane before analysis.

Three series of leaching experiments were carried out as gravity induced percolations with a 0.001 M EDTA solution:

Percolation A: one pore volume percolated every 30 min.
Percolation B: one pore volume percolated every hour with 2 interruptions of one night each - the first before the 10th pore volume and the second before the 20th pore volume.
Percolation C: one pore volume percolated every 3 days.

7.2.2 Results and Interpretation

The results of the three leaching experiments are presented in Figures 7.1 to 7.10, as leachate pH and dissolved metal concentrations for each pore volume eluted. The leaching efficiency (eluted *vs.* total metal content) is given in Table 7.3.

The pH of the eluted water is relatively constant for each percolation rate: 7.16 ± 0.26 for series A, 7.24 ± 0.26 for series B and 7.85 ± 0.25 for series C (Figure 7.1). The first two series indicate pH values in the same range. Series C presents slightly higher pH values, which is unexpected. These higher pH values are difficult to explain. They are in contradiction with CO_2 production due to the oxidative microbial degradation of organic matter (inducing more reducing conditions as evidenced in series C by enhanced iron dissolution (Figure 7.4)).

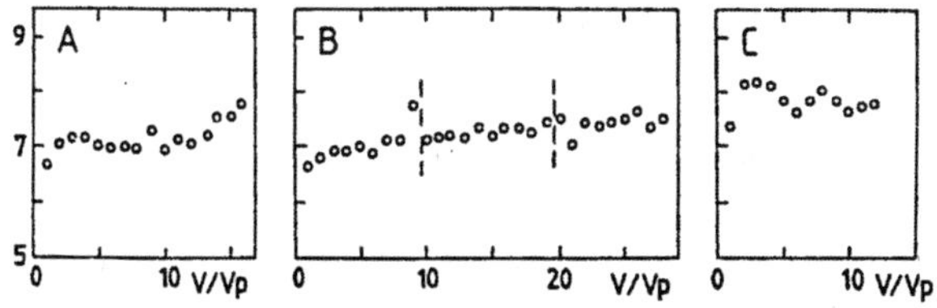

Figure 7.1 pH in eluates (V/Vp is the number of eluted pore volumes).

Magnesium and calcium show classical leaching curves, i.e. an exponential decrease in leached metal (Figures 7.2 and 7.3). After 7 or 8 pore volumes, a steady state is achieved (2 mg/L for series A and B and 3 mg/L for series C in the case of magnesium, and 40 to 50 mg/L for all series of calcium). The leaching yields are small for both alkaline earth metals (< 2 % for Mg and < 1 % for Ca, for 10 pore volumes eluted) (Table 7.3).

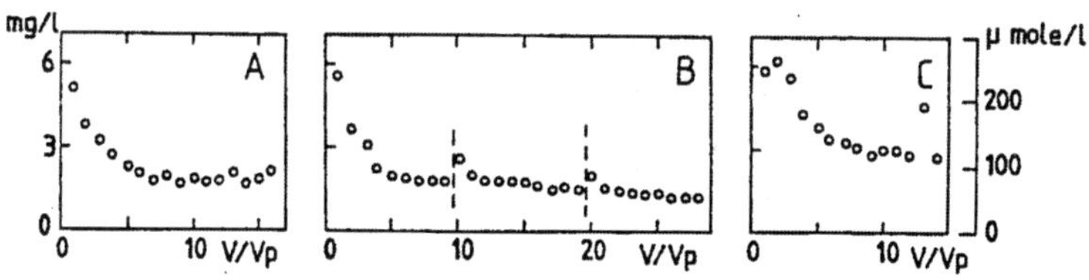

Figure 7.2 Magnesium in leachates (V/Vp is the number of eluted pore volumes).

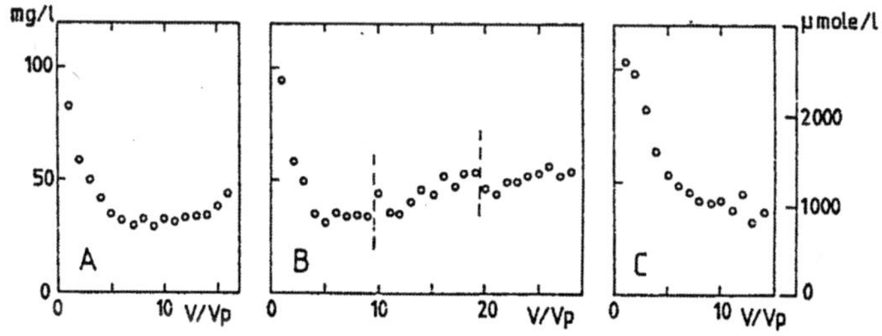

Figure 7.3 Calcium in leachates (V/Vp is the number of eluted pore volumes).

High concentrations of iron and manganese are observed in the eluted water (Figures 7.4 and 7.5). This is not surprising since the formation constants of Fe(III)-

EDTA and Mn(IV)-EDTA are much greater than those of Ca and Mg (Table 7.2). Moreover Fe and Mn are often present in soils as coatings and therefore much more available for reaction with a dissolved constituent such as EDTA. Leached Fe and Mn tend to increase with time, probably due to local conditions becoming more reducing, thus solubilising Fe and Mn as Fe(II) and Mn(II). This process is enhanced by complexation, albeit weaker for the divalent cationic forms (Table 7.2), by EDTA. The leached quantities are, however, small (< 1 % for Fe and 1 to 6 % for Mn for 10 pore volumes eluted) (Table 7.3).

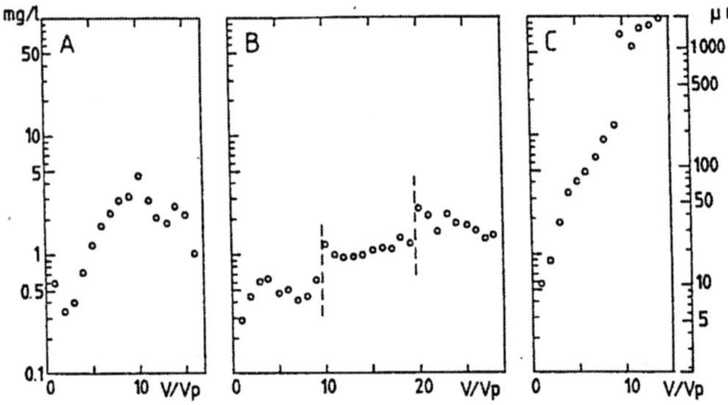

Figure 7.4 Iron in leachates (V/Vp is the number of eluted pore volumes).

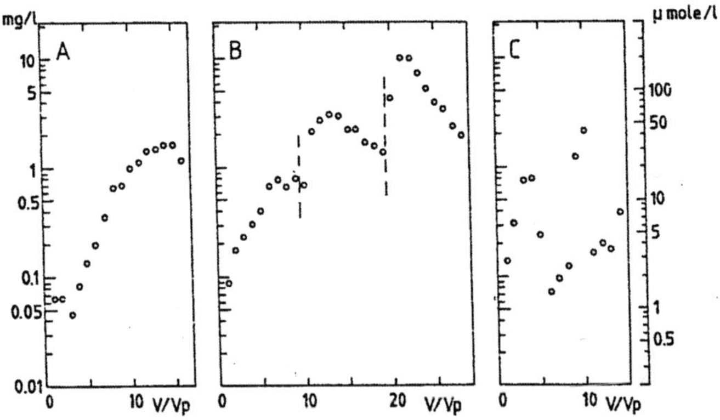

Figure 7.5 Manganese in leachates (V/Vp is the number of eluted pore volumes).

Zinc and cadmium leaching decreases sharply with the volume eluted (Figures 7.5 and 7.6). Contact time significantly increases leaching, especially in series C.

Solubilization also increases after a night of no flow (series B). However, for the same volume of eluted water, the leached metal quantity is quite similar (Table 7.3).

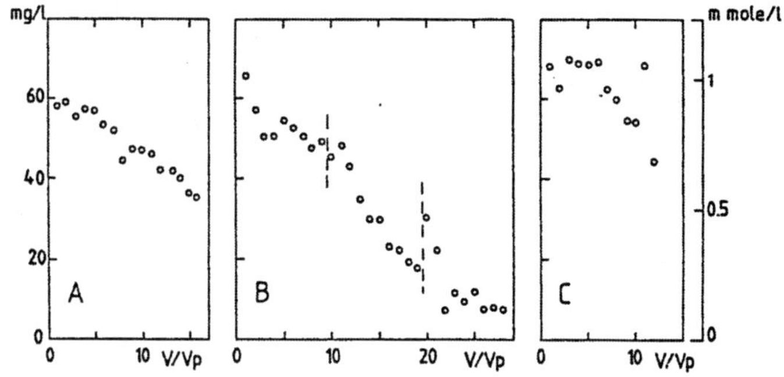

Figure 7.6 Zinc in leachates (V/Vp is the number of eluted pore volumes).

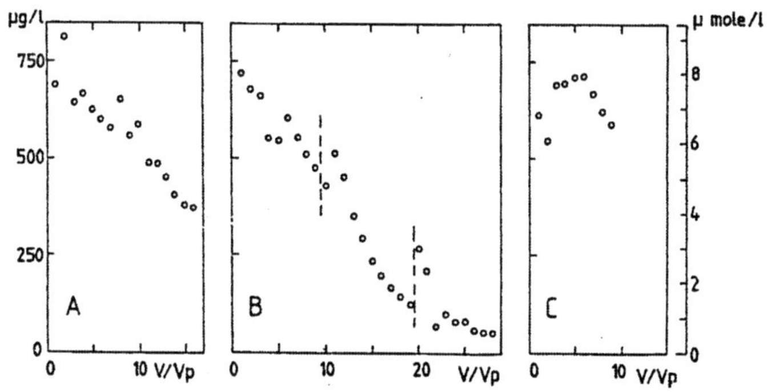

Figure 7.7 Cadmium in leachates (V/Vp is the number of eluted pore volumes).

The leaching of nickel, like that of calcium and magnesium, rapidly reaches equilibrium (100 µg/L for series A and B and 300 µg/L for series C) (Figure 7.8). A longer contact time favors leaching. This kinetic effect is even more marked for cobalt (Figure 7.9). Leached cobalt appears (above the detection limit of 10 µg/L) only after 3 days (series B) and 6 days (series C).

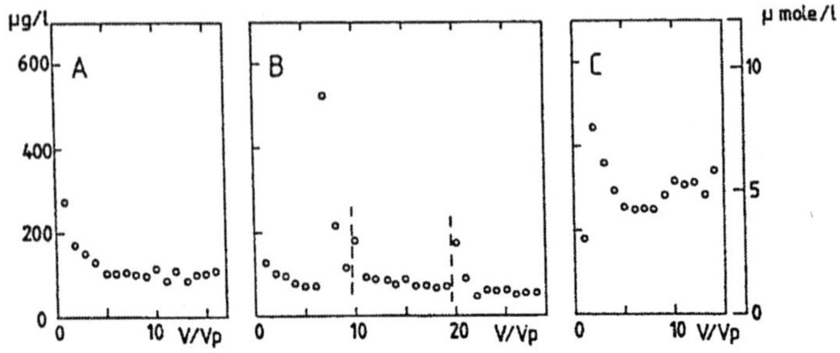

Figure 7.8 Nickel in leachates (V/Vp is the number of eluted pore volumes).

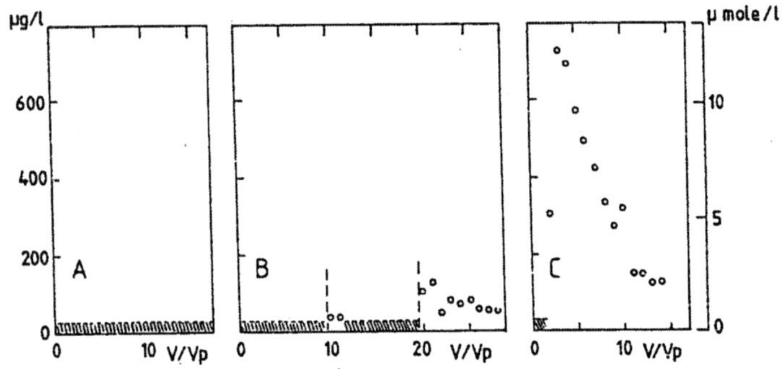

Figure 7.9 Cobalt in leachates (V/Vp is the number of eluted pore volumes).

The case of copper is quite complex (Figure 7.10). In series A, leaching increases with eluted volume and the same behavior is observed for the first day of series B. Thereafter, however, the elution curve shows a decrease in leached metal with time. Overnight contact time does increases elution (series C). The leaching behavior of copper during the first day is similar to that of iron and manganese and is, therefore, most likely adsorbed on, or co-precipitated with, Fe and/or Mn oxyhydroxides. As conditions become more reducing, Fe and Mn are dissolved and any associated metals (i.e. copper) are also. The slower elution rate of series C does not result in a significantly greater leaching of copper. The amount of eluted copper for 10 pore volumes is similar for the three experiments (Table 7.3).

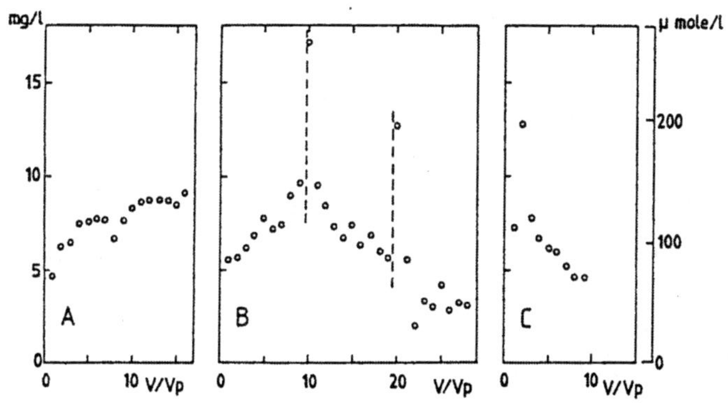

Figure 7.10 Copper in leachates (V/Vp is the number of eluted pore volumes).

Table 7.2 Complexation constants of metals with EDTA (literature values).

metal	Ca	Mg	Fe(III)	Fe(II)	Mn(IV)	Mn(II)	Cd	Co	Cu	Ni	Zn
$\log_{10}K$	13.1	9	27.7	1.2E14	15.6	6.2E13	18.2	2E16	19.7	20.4	17.2

Table 7.3. EDTA leaching efficiency.

metal	% of total metal leached for 10 pore volumes eluted*		
	series (a)	series (b)	series (c)
Ca	0.45	0.73	0.72
Mg	< 1.0	< 1.1	< 1.9
Fe	0.05	0.04	0.95
Mn	1.2	5.9	0.95
Cd	41	33	38
Co	< 1.4	< 1.3	14.4
Cu	6.7	7.9	6.1
Ni	0.9	1.1	2.6
Zn	10.8	10.0	15.8

* In order to better compare the leaching efficiency of the 3 series, the eluted quantities are averaged and the results given for 10 pore volumes.

7.2.3 Conclusion

EDTA 0.001 M is able to leach significant fractions of some of the metals contained in a soil contaminated by the long-term application of municipal waste (waste water and sludge). Cadmium is by far the most efficiently leached of all metals (about 40 % for 10 pore volumes), followed by zinc (Table 7.3). These extraction yields are high, even though the EDTA solution percolation rates are quite rapid (1 pore volume per hour corresponds to a water velocity of ca. 8 cm/hour, or 2 m/day). There is no direct correlation between leached yield and metal-EDTA association constant, which is not surprising as the metals are neither present under the same geochemical form nor under the same physical availability (coatings, included in solids…).

The kinetics of dissolution is evident for all heavy metals. In conclusion, modeling of EDTA leaching of heavy metals must include a kinetic term.

7.3 Observations and Modeling: Solubilization of Heavy Metals from a Soil Polluted by Long-Term Smelting Activities—A Finite Difference Model, including a Kinetic Leaching Term

7.3.1 Introduction

Here we describe the leaching by EDTA of a heavy metal (Cd) in an agricultural soil polluted by smelting activities. A simple numerical model is developed to simulate the solubilization and subsequent transport of dissolved Cd species. Simultaneous complexation and solubilization are considered to be a single process that depends on the available metal content in the solid and the dissolved ligand concentration. An implicit finite difference scheme is numerically implemented to calculate the movement of both free chelate and metal-chelate species. The model uses two advection-dispersion equations (one for the uncomplexed EDTA, the other for the EDTA-metal complexes) both including a second-order kinetic solubilization term. The model is validated with experimental data obtained during the leaching of cadmium in the presence of EDTA through percolation columns under different flow conditions.

7.3.2 The Model

Formulation. A model was developed to simulate (a) EDTA transport (advection-dispersion equation), (b) solubilization kinetics of metals bound to the soil, and (c) transport of EDTA-metal complexes in solution (advection-dispersion equation) (Figure 7.11) (symbols are defined and units are given in Table 7.4).

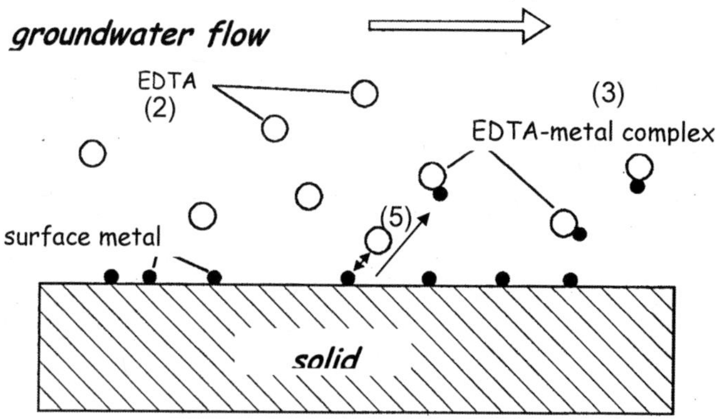

Figure 7.11 Schematic representation of the modeled system. Numbers in brackets correspond to the equation numbers in the text.

The transport of a simple conservative solute (bromide) through the column is described by the classic one-dimensional advection-dispersion equation (Beruch and Street, 1967; Hoopes and Harleman, 1967; Bear, 1979):

$$\frac{\partial C}{\partial t} = \alpha u \frac{\partial^2 C}{\partial x^2} - u \frac{\partial C}{\partial x} \qquad (1)$$

where C is the molar concentration of solute in the aqueous phase, α the longitudinal dispersivity, u the pore velocity (Darcy velocity divided by the kinematic porosity, ω_k), t the time and x the distance.

For the transport of the solute EDTA, a kinetic sink term must be added to equation (1) because the concentration of uncomplexed EDTA (C_E) in the aqueous phase decreases as complexation progresses:

$$\frac{\partial C_E}{\partial t} = \alpha u \frac{\partial^2 C_E}{\partial x^2} - u \frac{\partial C_E}{\partial x} + \frac{\partial C_M}{\partial t} \qquad (2)$$

where C_M is the number of moles per pore volume unit of potentially extractable metal in the soil. No sink term was included in equation (2) for EDTA adsorption because, at the pH of the experiments (7.5), non-complexed EDTA did not adsorb onto the soil matrix (Kedziorek, 1997).

The transport of EDTA-metal complexes is written with a source term as:

$$\frac{\partial C_{EM}}{\partial t} = \alpha\, u \frac{\partial^2 C_{EM}}{\partial x^2} - u \frac{\partial C_{EM}}{\partial x} - \frac{\partial C_M}{\partial t} \qquad (3)$$

where C_{EM} is the molar concentration of EDTA-metal complexes in the aqueous phase.

Only part of the metal load of solids might be in geochemical equilibrium with the surrounding aqueous phase. The value of C_M is obtained according to the following equation:

$$C_M = \frac{d_A \, .10^{-6}}{\omega_k \, .m} C_M^* \qquad (4)$$

where, C_M^* is the solubilizable metal content (µg of metal/g soil), d_A the bulk density of the soil (dimensionless), ω_k the kinematic porosity (dimensionless), and m the molecular weight of the metal (g).

The solubilization term of equations (2) and (3) is represented by a second-order kinetic equation:

$$\frac{\partial C_M}{\partial t} = -K\, C_E \frac{C_M}{C_{Mi}} \qquad (5)$$

where K is the kinetic solubilization coefficient, and C_{Mi} the initial number of moles of metal bound to soil per pore volume unit. The solubilization rate was assumed to be proportional to the EDTA concentration in the aqueous phase (Roucaché, 1997) and to the ratio of the concentration of metal present in the soil divided by the initial metal concentration. This ratio accounts for the decrease in the solubilization rate as the metal content available for solubilization in the column solids decreases. The solubilization reaction involves the complexation of one mole of cadmium with one mole of EDTA.

At the end of each time step of the resolution algorithm, equations (2) and (3) are solved and the concentration of extractable metal in the soil is corrected (equation (5)).

Table 7.4 Symbols and units

α	longitudinal dispersivity [L]
Δt	discrete time step [T]
Δx	discrete space step [L]
ω_k	kinematic porosity [dimensionless]
C_E	molar concentration of EDTA in aqueous phase [mol.L^{-3}]
C_{Ex}^{t}	molar concentration of EDTA in aqueous phase at distance x and time t [mol.L^{-3}]
C_{Ei}	injection concentration of EDTA [mol.L^{-3}]
C_{EM}	molar concentration of EDTA-Metal complexes in aqueous phase [mol.L^{-3}]
C_M^*	solubilizable metal content (µg of metal per g of soil) [M.M^{-1}]
C_M	number of solubilizable moles of metal bound to soil per pore volume unit [mol.L^{-3}]
C_{Mi}	initial number of solubilizable moles of metal bound to soil per pore volume unit [mol.L^{-3}]
d_A	bulk density of soil [dimensionless]
K	kinetic coefficient of solubilization [T^{-1}]
L	length of the column [L]
N	number of column discretization segments
m	metal molecular weight (in g) [M]
t	variable of time [T]
u	pore velocity [L.T^{-1}]
x	variable of space [L]

Resolution. The model calculates the solubilization of one metal species at a time. An implicit finite difference method was used to solve equations (2) and (3). Discrete space and time steps were labeled Δx and Δt, respectively. The column was discretized into N equal segments each with a length of Δx, giving N+1 discrete points for which all the equations are solved. C_{Ex}^{t} is the concentration of EDTA at distance x and time t. The resolution algorithm has 4 steps.

Step 1: Resolution of equation (2), transport by advection and dispersion of uncomplexed EDTA, with a solubilization kinetic term. Rewriting equation (2) with Taylor developments gives:

$$\frac{C_{Ex}^{t+\Delta t} - C_{Ex}^{t}}{\Delta t} = \alpha\, u\, \frac{C_{Ex+\Delta x}^{t+\Delta t} - 2\, C_{Ex}^{t+\Delta t} + C_{Ex-\Delta x}^{t+\Delta t}}{(\Delta x)^2} - u\, \frac{C_{Ex+\Delta x}^{t+\Delta t} - C_{Ex-\Delta x}^{t+\Delta t}}{2\,\Delta x} + K\, C_{Ex}^{t+\Delta t}\, \frac{C_{Mx}^{t}}{C_{Mi}} \tag{6}$$

The initial conditions are:

$$C_{Ex}^{t=0} = 0 \tag{7}$$

and C_{Mi} is constant for all values of x (homogeneous column).

Several boundary conditions must be accounted for:

(i) for a one-step injection occurring between times t1 and t2 at the inlet of the column:

$$C_{Ex=0}^{t<t1} = 0 \qquad C_{Ex=0}^{t1 \le t \le t2} = C_{Ei} \qquad C_{Ex=0}^{t>t2} = 0 \tag{8a, b, c}$$

(a narrow pulse injection corresponds to an instantaneous injection such as t=t1=t2 or, in other words, a perfect Dirac injection)

(ii) at the end of the column:

$$\left.\frac{\partial C_E}{\partial x}\right|_{(x=L)} = 0 \tag{8d}$$

where C_{Ei} is the concentration of a given volume of EDTA injected and L the column length.

Equation (6) involves the resolution of a classic tridiagonal matrix. The model was tested and was in close agreement with analytical solutions for a punctual injection with a uniform, one-dimensional flow in a semi-infinite medium (De Marsily, 1986).

Step 2: Resolution of equation (3), transport by advection and dispersion of dissolved EDTA-metal complexes, including a kinetic term for solubilization.

Rewriting equation (3) with Taylor developments gives:

$$\frac{C_{EM_x}^{t+\Delta t} - C_{EM_x}^{t}}{\Delta t} = \alpha\, u \frac{C_{EM_{x+\Delta x}}^{t+\Delta t} - 2\, C_{EM_x}^{t+\Delta t} + C_{EM_{x-\Delta x}}^{t+\Delta t}}{(\Delta x)^2} - u \frac{C_{EM_{x+\Delta x}}^{t+\Delta t} - C_{EM_{x-\Delta x}}^{t+\Delta t}}{2\,\Delta x} - K\, C_{E_x}^{t+\Delta t} \frac{C_{M_x}^{t}}{C_{Mi}} \tag{9}$$

Initial conditions are:

$$C_{EM_x}^{t=0} = 0 \tag{10a}$$

$$C_{M_x}^{t=0} = C_{Mi} \tag{10b}$$

Boundary conditions are:

(i) a null flux is imposed at the inlet of the column:

$$u\, C_{EM\, x=0}^{t+\Delta t} = \alpha\, u \frac{\partial C_{EM0}^{t+\Delta t}}{\partial x} \tag{11a}$$

where $C_{EM0}^{t+\Delta t}$ is the value of C_{EM} at x = 0 (beginning of the column) at time t+Δt : there is no EDTA-metal input in the column

(ii) at the end of the column:

$$\left.\frac{\partial C_{EM}}{\partial x}\right|_{(x=L)} = 0 \tag{11b}$$

Step 3: Equation (5) is used to calculate the evolution of the metal content remaining in the soil at each discrete point.

Rewriting equation (5) in finite difference form gives:

$$\frac{C_{Mx}^{t+\Delta t} - C_{Mx}^{t}}{\Delta t} = -K\, C_{Ex}^{t+\Delta t} \frac{C_{Mx}^{t}}{C_{Mi}} \tag{12}$$

C_M at time t + Δt is therefore:

$$C_{Mx}^{t+\Delta t} = C_{Mx}^{t}\left(1 - K\, \Delta t \frac{C_{Ex}^{t+\Delta t}}{C_{Mi}}\right) \tag{13}$$

Equation (13) implies that K has a maximum value of $\frac{C_{Mi}}{\Delta t\, C_{Ei}}$.

7.3.3 Experimental Section

The soil sample used in the percolation experiments was collected in the upper 30 cm of an agricultural plot near Lille (Nord-Pas-de-Calais, France) in the vicinity of smelting activities. It was dried at 40 °C for 24 hours and sieved through a 2 mm nylon sieve to eliminate the coarse sand fraction. The total Cd content in the solid, measured by ICP after fusion with sodium borate, was 0.08 µmole/g. The solubilizable metal in the soil was estimated by measuring, in batch experiments (10 g solid/L), the fraction extractable with 0.01 M EDTA (Kedziorek, 1997). All of the Cd was extractable from the soil used in the experiments carried out to test the model.

A borosilicate glass column (length: 25 cm; internal diameter: 2.5 cm) was packed with the dried solid material (ca. 220 g) so as to create uniform bulk densities. An electrolyte solution (0.01 M $NaNO_3$) was pumped upwards through the column until steady-state conditions (constant pH and conductivity) were reached. This required approximately four pore volumes (approximately 24 hours). The quantity of

Cd eluted during this equilibration stage was negligible (<0.1%) with respect to the total amount eluted during the percolation of EDTA. The flow rate in the column was constant (*ca.* 14 $cm^3.h^{-1}$).

Various EDTA concentrations were used in three separate experiments performed for different injection modes (pulse and step injections). Bromide, an ideal tracer (conservative and non-reactive), was injected with EDTA to provide information on the water movement.

In the first experiment (pulse mode), three 0.5-mL pulses, each containing 0.005 M of bromide, were injected. The pulses were separated by four pore volumes of a 0.01 M $NaNO_3$ electrolyte solution. The first pulse contained no EDTA. The second and third contained 0.001 M and 0.01 M of EDTA, respectively, which corresponds to a total of 0.5 and 5 μmoles of injected EDTA.

In the two step mode experiments, EDTA and bromide were injected at constant concentration in the feed solution. This feed solution contained bromide (usually 0.65 mM) and, depending on the experiment, total EDTA concentrations of 10^{-2} or 10^{-4} M.

Evolution of total EDTA, EDTA-cadmium chelates and the bromide concentration in the eluate were monitored by dissolved organic carbon measurement, atomic absorption spectrophotometry and ion chromatography, respectively. During the percolation experiments, the eluate flow was sampled continuously and the actual flow rate was determined by dividing each sample volume by the time required to obtain it.

The soil and experimental protocol are described in detail elsewhere (Kedziorek, 1997).

7.3.4 Model Simulations with Experimental Validation

To determine the kinematic, or effective, porosity (ω_k) and the dispersivity (α), the model was first run with a null kinetic parameter K. K was then determined by manually varying its value until simulated curves fit experimental curves. This was done for the two successive pulse injections of 10^{-3} M and 10^{-2} M of EDTA, and the two long-step injections (10^{-4} M and 10^{-2} M of EDTA).

Pulse EDTA Injections. In our short step injections, at a flow rate of 14.2 $cm^3.h^{-1}$, the bromide solution was injected for 126 s. This injection of 0.5 mL represents 1.25 % of the column pore volume and can therefore be considered to be a pulse injection. The fit between model calculations and observed values is excellent (Figure 7.12). The average values of the kinematic porosity and dispersivity are 33 % and 0.42 cm, respectively (Table 7.5).

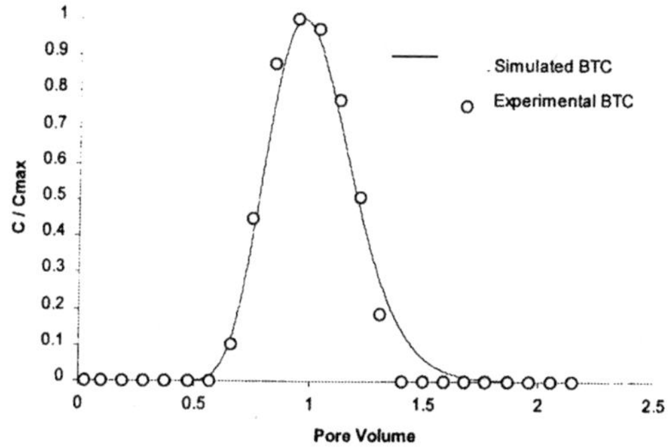

Figure 7.12 Example of simulated and measured breakthrough curves for bromide injection (values of fitting parameters are given in Table 7.5) (Kedziorek et al. 1998;Copyright American Chemical Society 1998, reproduced with permission).

Table 7.5 Fitted values of kinematic porosity and dispersivity for the pulse experiments

Pulse	kinematic porosity, ω_k [cm^3/cm^3]	dispersivity, α [cm]
Pulse 1	0.345	0.44
Pulse 2	0.32	0.45
Pulse 3	0.325	0.38

For the two successive pulses of EDTA (10^{-3} M and 10^{-2} M), the Cd-chelate is eluted with bromide, with a greater extraction as the injected concentration of EDTA is raised ten-fold (Figure 7.13). Fitting was done for each step separately (Pulse 2 and Pulse 3) and together (Pulse 2-3). The fitted K values are almost identical for the 3 simulations (from 1.9 to 2.5 10^{-6} s^{-1}) (Table 7.6). The agreement between calculated and observed values of the fraction of Cd extracted from the soil (ratio between the number of moles of Cd eluted as EDTA-Cd complexes and the initial number of moles of Cd present in the column) is good (Table 7.6). Using a single set of fitting parameters, the model simulates the EDTA-Cd chelate breakthrough curves (BTC) well for the two successive pulses (Figure 7.13).

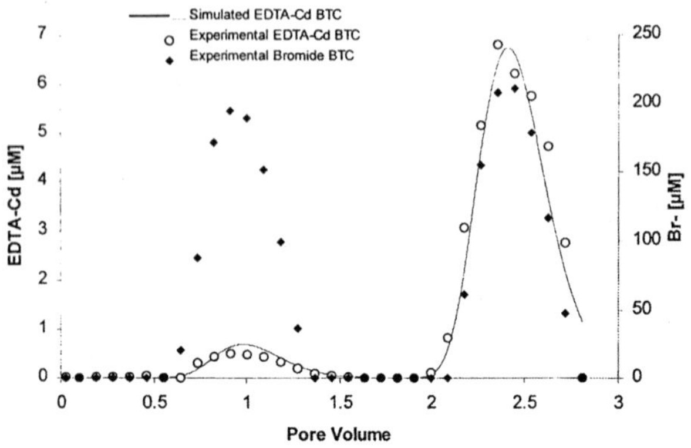

Figure 7.13 Experiment with two successive injections of EDTA of 10^{-3} M and 10^{-2} M (during 126 sec each). Observed and simulated cadmium breakthrough curves are presented with bromide restitution. (Flow rate = 14.2 $cm^3.h^{-1}$, $\omega_k = 0.32$, $\alpha = 0.42$ cm and $K = 2.5 \times 10^{-6}$ s^{-1}).

Table 7.6 Pulse injection experiments: fitting values and results of modeling

Pulse	C EDTA [M] (C_{Ei})	K [s^{-1}]	% Cd extracted	
			observed	simulated
Pulse 2	10^{-3}	1.9×10^{-6}	0.1	0.1
Pulse 3	10^{-2}	2.4×10^{-6}	1.4	1.3
Pulse 2-3*	10^{-3} and 10^{-2}	2.5×10^{-6}	1.5	1.4

*Fitting using both Cd pulses.

Long Step EDTA Injections. The kinematic porosity and dispersivity values obtained from the bromide BTCs were very similar to those of the pulse injections (33% and 0.44 cm, respectively) even though each step experiment was carried out with a column filled with a new soil sample. This demonstrates the excellent reproducibility in filling the column with the solid material. K values ranged from 1.7 to 4 10^{-6} s^{-1} with the highest value obtained with the lower EDTA concentration (10^{-4} M) (Table 7.7). For the low EDTA concentration, the model calculations are poor at the beginning of the BTC (Figure 7.14). The model predicts that the BTC will slope slightly downwards as the efficiency of the extraction process decreases along with that of the Cd content (to 88% of the original content at the end of the experiment).

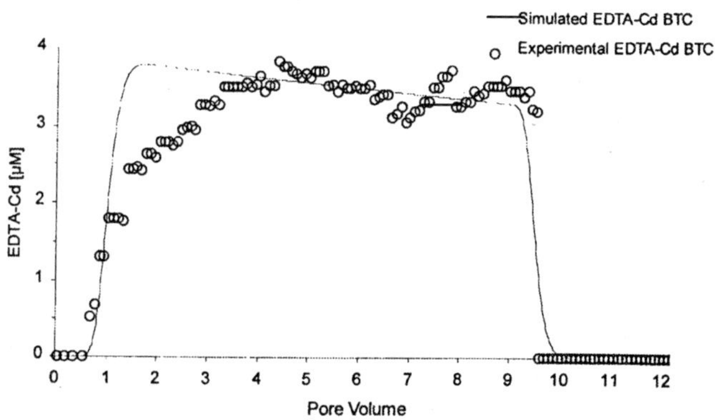

Figure 7.14 Step experiment (injection of 10^{-4} M EDTA injected): experimental and simulated EDTA-cadmium breakthrough curves (parameters are in given in Table 7.7) (Kedziorek et al. 1998. Copyright American Chemical Society 1998, reproduced with permission).

The Cd eluted in the presence of the higher EDTA concentration step injection (10^{-2} M) (Step 2, Figure 7.15) is quite well represented by the model. The eluted Cd concentration peaks rapidly and the BTC decreases as the amount of extractable Cd remaining in the soil is depleted. Simulated and observed percentages of extracted Cd are in good agreement (Table 7.7).

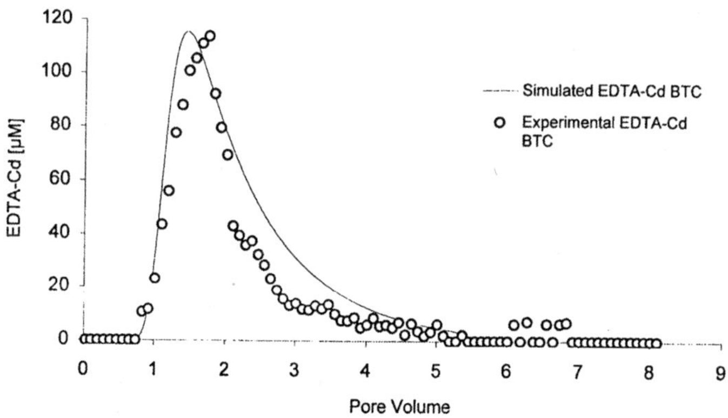

Figure 7.15 Step experiment (injection of 10^{-2} M EDTA): experimental and simulated EDTA-cadmium breakthrough curves (parameters are in given in Table 7.7) (Copyright American Chemical Society 1998, reproduced with permission).

Table 7.7 Step injection experiments: fitting values and results of modeling

Step	Flow rate [mL.h^{-1}]	C_{Ei} [M]	ωk	α [cm]	K [s^{-1}]	% Cd extracted observed	% Cd extracted simulated
Step 1	14.6	10^{-4}	0.33	0.44	4×10^{-6}	12	13
Step 2	14.7	10^{-2}	0.33	0.44	1.7×10^{-6}	66	76

7.3.5 Discussion and Summary

Fittings of short step (pulse) and long step experiments gave very similar kinetic coefficients (Tables 7.6 and 7.7). A mean value of 2.1×10^{-6} s^{-1} can be retained, if we disregard the highest value of 4×10^{-6} s^{-1} obtained with the lower concentration of injected EDTA. Simulation of succeeding pulses (pulse 2-3) required almost no modification in fitting of the kinetic coefficient (2.4×10^{-6} s^{-1} for pulse 3 alone and 2.5×1010^{-6} s^{-1} for pulse 2-3).

The kinetic coefficient is relatively independent of the EDTA concentration and of the mode of injection. It can therefore be considered to be a characteristic of the EDTA-Cd complexation reaction under our experimental conditions.

This section describes a model in which simultaneous transport of EDTA and EDTA-metal chelates is accounted for with advection-dispersion equations linked to a second order kinetic term describing the complexation/solubilization process. Fitting of experimental BTCs showed that the kinetic coefficient for cadmium solubilization is quite independent of EDTA concentration. Pulse and step injections gave similar values for this parameter. The model can simulate the complexation of one metal at a time with EDTA. In situations where there are several metals with high kinetic coefficients, the corresponding decrease in non-complexed EDTA should affect the extent of remobilization. In such cases, a multi-component approach is needed to model the simultaneous formation of different chelates (Tsang et al., 2007).

7.4 Potential for Actual Use of Complexing Agents for Soil Decontamination

Using chemical reagents such as chelates for *in situ* decontamination operations is, to the say the least, a tricky business. Four major points should be considered before such a technique is proposed and safely and economically implemented.

(1) Groundwater flow should be properly understood and controlled during operations in order to prevent the dissemination of dissolved heavy metal- and organic reagent-rich water in the environment. This is especially important near environmental receptors.

(2) Metal extraction procedures should be optimized (operating variables for EDTA soil washing) in a chemical engineering approach (Fangueiro et al., 2002; Kim et al., 2003; Finžgar and Leštan, 2007; Zou et al., 2009).

(3) Using a chemical reagent such as a chelate could be expensive. Cost-effective methods will depend on regeneration of this chelate during elimination of the contaminant heavy metals (Xie and Marshall, 2001; Di Palma et al., 2003; Lim et al., 2005). An alternative would be to use less-efficient organic reagents that are, however, inexpensive and readily available as by-products of the chemical industry.

(4) Due to the persistence of EDTA in the environment, efforts should be made to find and use less aggressive compounds and degradable substances such as aminopolycarboxylic acids or EDDS (an isomer of diaminedisuccinic acid) (Oviedo and Rodriguez, 2003; Tardy et al., 2004; Saifullah et al., 2009; Yip et al., 2009; Tsang et al., 2009).

7.5 Conclusions

Extraction of heavy metals from contaminated soils and sediments is still far from being an exact science. At present, it is still somewhat of a semi-empirical technology. Practical applications will require extensive laboratory column experiments and modeling prior to the implementation of field operations.

7.6 Acknowledgments

This work was supported in part by the Ministère de l'Environnement (grant SRETIE 84296), ADEME (French Agency for Environment and Energy Management) and BRGM (Bureau of Geological and Mining Research). We thank Magalie Roucaché (Univ. of Limoges) for carrying out some of the EDTA extractions and Anna Ross for editing our English.

7.7 References

Baghdady, N. H., and Sippola, J., (1984). "Extractability of polluting elements Cd, Cr, Ni and Pb with three methods." *Acta Agriculturae Scandinavica*, 34, 345-348.

Bear, J. (1979). Hydraulics of Groundwater, MacGraw-Hill, New York.

Beruch, J. C., and Street, R. A. (1967). "Two-dimensional dispersion." *Journal of Sanitary Engineering Division*, Amer. Soc. of Civil Engineers, 93, SA6, 17-39.

Bourg, A. C. M., Crouzet, C., and Bertin, C. (1994)." Influence of EDTA on the mobility of trace metals in aquifers." In: Dracos, T.H., Stauffer, F. (Eds.),

Transport and Reactive Processes in Aquifers, A.A. Balkema, Rotterdam, 79-81.

Bourg, A. C. M., and Darmendrail, D. (1992). "Effect of dissolved organic matter and pH on the migration of zinc through river bank sediments." *Environmental Technology*, 13, 695-700.

Bourg, A. C. M., and Schindler, P. W. (1985). "Control of trace metals in natural aquatic systems by the adsorptive properties of organic matter." *Comptes Rendus Conf. Internat. Métaux Lourds dans l'Environnement*, Athènes, CCE, 1, 97-99.

Brown, G. A., Elliott, H. A. (1992). "Influence of electrolytes on EDTA extraction of Pb from a polluted soil." *Water, Air, and Soil Pollution*, 62, 157-165.

Bryce, A. L., and Clark, S. B. (1996). "Nickel desorption kinetics from hydrous ferric oxide in the presence of EDTA." *Colloids and Surfaces A*, 107, 123-130.

Davis, J. A., Kent, D. B., Rea, B. A., Maest, A. S., Garabedian, S. P., 1993. Influence of redox environment and aqueous speciation on metal transport in groundwater. In: Allen, H.E., Perdu, E.M., Brown, D.S. (Eds.), Metals in Groundwater, Lewis Pub., Boca Raton, Florida, 223-271.

Davis, J. A., Kent, D. B., and Coston, J. A. (2000). " Multispecies reactive tracer test in an aquifer with spatially variable chemical conditions." Water Resources Research, 36, 119-134.

De Marsily, G. (1986). Quantitative Hydrogeology, Academic Press.

Di Palma, L., Ferrantelli, P., Merli, C., and Biancifiori, F. (2003). "Recovery of EDTA and metal precipitation from soil flushing solutions." *Journal of Hazardous Materials*, 103, 153–168.

Di Palma, L. and Mecozzi, R. (2007). "Heavy metals mobilization from harbour sediments using EDTA and citric acid as chelating agents." *Journal of Hazardous Materials*, 147, 768-775.

Di Palma, L., Ferrantelli, P., Merli, C., Petrucci, E., and Pitzolu, I. (2007). "Influence of Soil Organic Matter on Copper Extraction from Contaminated Soil." *Soil and Sediment Contamination*, 16, 323–335.

Elliott, H. A., and Brown, G. A. (1989). "Comparative evaluation of NTA and EDTA for extractive decontamination of Pb-polluted solids." *Water, Air and Soil Pollution*, 45, 361-369.

Fangueiro, D., Bermond, A., Santos, E., Carapuça, H., and Duarte, A. (2002). "Heavy metal mobility assessment in sediments based on a kinetic approach of the EDTA extraction: search for optimal experimental conditions." *Analytica Chimica Acta*, 459, 245-256

Finžgar; N. and Leštan; D. (2007). "Multi-step leaching of Pb and Zn contaminated soils with EDTA." *Chemosphere*, 66, 824-832.

Friedly, J. C., Kent, D. B., and Davis, J. A. (2002). "Simulation of the mobility of metal−EDTA complexes in groundwater: The influence of contaminant metals." *Environmental Science and Technology*, 36, 355–363.

Garrabrants, A. C., and Kosson, D. S. (2000). "Use of a chelating agent to determine the metal availability for leaching from soils and wastes." *Waste Management*, 20, 155-165.

Ghestem, J. P. and Bermond, A., (1998). "EDTA extractability of trace metals in polluted soils: a chemical-physical study." *Environmental Technology*, 19, 409-416.

Hong, J. and Pintauro, P. N. (1996). "Desorption-complexation-dissolution characteristics of adsorbed cadmium from kaolin by chelators." *Water, Air, and Soil Pollution*, 86, 35-50.

Hoopes, J. A., and Harleman, D. R. F. (1967). "Wastewater recharge and dispersion in porous media." *Journal of Hydraulics Division*, Amer. Soc. of Civil Engineers, 93, HY5, 51-71.

Jardine, P. M., Jacobs, G. K., and O'Dell, J. D. (1993). "Unsaturated transport processes in undisturbed heterogeneous porous media II. Co-Contaminants." *Soil Science Society of America Journal*, 57, 954-962.

Kedziorek, M. A. M. (1997). "Mobilité naturelle et provoquée de polluants métalliques dans les sols et les sediments." Doctoral thesis (in French), Univ. of Limoges, France, pp. 156.

Kedziorek, M. A. M. and Bourg, A. C. M. (2000). "Solubilization of lead and cadmium during the percolation of EDTA through a soil polluted by smelting activities." *Journal of Contaminant Hydrology*, 40, 381-392.

Kedziorek, M. A. M. and Bourg, A. C. M. (2007). "Extraction with EDTA to assess the global risk presented by heavy metals (Cd, Ni, Pb) in soils and sediments." *Soil Chemical Pollution, Risk Assesment, Remediation and Security*, Simeonov, L, and Sargsyan, V., eds., Springer, 127-141.

Kedziorek, M. A. M., Dupuy, A., Bourg, A. C. M., and Compère, F. (1998). "Leaching of Cd and Pb from a polluted soil during the percolation of EDTA: Laboratory column experiments modeled with a non-equilibrium solubilization step." *Environmental Science and Technology*, 32, 1609-1614.

Kent, D. G., Davis, J. A., Anderson, L. C. D., Rea, B. A., and Coston, J. A. (2002), "Effect of adsorbed metal ions on the transport of Zn- and Ni-EDTA complexes in a sand and gravel aquifer." *Geochimica et Cosmochimica Acta*, 66, 3017–3036.

Kent, D. B., Davis, J. A., Joye, J. L., and Curtis, G. P. (2008). "Influence of variable chemical conditions on EDTA-enhanced transport of metal ions in mildly acidic groundwater." *Environmental Pollution*, 153, 44-52.

Kim, C., Lee, Y., and Ong S., K. (2003). "Factors affecting EDTA extraction of lead from lead-contaminated soils." *Chemosphere*, 51, 845-853.

Kowalick, W. W. Jr. (1992). "Trends in innovative treatment technologies at contaminated sites." *Water Science and Technology*, 26, 99-106.

Lestan, D., Luo, C.-L., and Li, X.-D. (2008). "The use of chelating agents in the remediation of metal-contaminated soils: A review." *Environmental Pollution*, 153, 3-13.

Li, Z. and Shuman, L.M. (1996). "Extractability of zinc, cadmium, and nickel in soils amended with EDTA", *Soil Science*, 161, 226-232.

Lim, T.-T., Chui, P.-C., and Goh, K. H. (2005). "Process evaluation for optimization of EDTA use and recovery for heavy metal removal from a contaminated soil." *Chemosphere*, 58, 1031-1040.

Manouchehri, N., Besancon S., and Bermond, A. (2006). "Major and trace metal extraction from soil by EDTA: Equilibrium and kinetic studies." *Analytica Chimica Acta,* 559, 105-112.

Nowack, B., and Sigg, L. (1997). "Dissolution of Fe(III) (hydroxides by metal-EDTA complexes." *Geochimica et Cosmochimica Acta*, 61, 951-963.

Nowack, B., Xue, H., and Sigg, L. (1997). "Influence of natural and anthropogenic ligands on metal transport during infiltration of river water to groundwater." *Environmental Science and Technology*, 31, 866-872.

Oviedo, C., and Rodríguez, J. (2003). "EDTA: the chelating agent under environmental scrutiny." *Quimica Nova*, 26, 901-905.

Peters, R. W. (1999). "Chelant extraction of heavy metals from contaminated soils." *Journal of Hazardous Materials*, 66, 151-210.

Puginier, B., Kedziorek; M. A. M., and Bourg, A. C. M. (1998). "Adsorptive behavior of nickel with goethite in the presence of EDTA: Kinetics and reversibility." *Environmental Technology*, 19, 829-835.

Roucaché, M. (1997). "Effet de faibles concentrations d"EDTA sur la solubilisation de métaux lourds contenus dans des sols et sédiments pollués." D.E.A. Dissertation (in French), Univ. of Limoges, France.

Saifullah, E., Meers, M., Qadir, P., de Caritat, F. M. G., Tack, Du Laing, G., and Zia, M. H. (2009). "EDTA-assisted Pb phytoextraction." *Chemosphere*, 74, 1279-1291.

Samanidou, V. and Fytianos, K. (1990)." Mobilization of heavy metals from river sediments of northern Greece by complexing agents." *Water, Air and Soil Pollution*, 52, 217-225.

Stumm, W., and Morgan, J. J. (1981). Aquatic Chemistry, 2nd ed., Wiley, New York, 780 pp.

Sun, B., Zhao, F. J., Lombi, E., and McGrath, S. P. (2001). "Leaching of heavy metals from contaminated soils using EDTA." *Environmental Pollution*, 113, 111-120.

Szecsody, J. E., Zachara, J. M., and Bruckhart, P.L. (1994). "Adsorption-dissolution reactions affecting the distribution and stability of Co(II)EDTA in Fe-oxide coated sand." *Environmental Science and Technology*, 28, 1706-1716.

Tandy, S., Bossart, K., Mueller, R., Ritschel, J., Hauser, L., Schulin, R., and Nowack, B. (2004). "Extraction of heavy metals from soils using biodegradable chelating agents." *Environmental Science and Technology*, 38, 937-944.

Tsang, D. C. W., and Lo, I. M. C. (2006). "Competitive Cu and Cd Sorption and Transport in Soils: A Combined Batch Kinetics, Column, and Sequential Extraction Study." *Environmental Science and Technology*, 40, 6655–6661.

Tsang, D. C. W., Lo, I. M. C., and Chan, K. L. (2007a). "Modeling the transport of metals with rate-limited EDTA-promoted extraction and dissolution during EDTA-flushing of copper-contaminated soils." *Environmental Science and Technology*, 41, 3660-3668.

Tsang, D. C. W., Zhang, W., and Lo, I. M. C. (2007b). "Copper extraction effectiveness and soil dissolution issues of EDTA-flushing of artificially contaminated soils." *Chemosphere*, 68, 234-243.

Tsang, D. C. W., Yip, T. C. M., and Lo, I. M. C. (2009). "Kinetic interactions of EDDS with soils. 2. Metal-EDDS complexes in uncontaminated and metal-contaminated soils." *Environmental Science and Technology*, 43, 837-842.

Tuin, B.J.W., Tels, M., 1990. Removal of heavy metals from contaminated clay soils by extraction with hydrochloric acid, EDTA or hypochlorite solutions, Environ. Technol., 11, 1039-1052.

Urlings, L.G.C.M. (1990). "In situ cadmium removal - Full scale remedial action of contaminated soil." paper presented at Internat. Symp. on Hazardous Waste Treatment, Treatment of Contaminated Soils, Air & Waste Management Assoc. and US EPA Risk Reduction Laboratory, Cincinatti, Ohio, February 1990.

Vaxevanidou, K., Papassiopi, N., and Paspaliaris I. (2008). "Removal of heavy metals and arsenic from contaminated soils using bioremediation and chelant extraction techniques." *Chemosphere*, 70, 1329-1337.

von Gunten, H. R., Karametaxas, G., Krahenbühl, U., Kuslys, M. Giovanoli, R. Hoehn, E., and Keil,. R. (1991). "Seasonal biogeochemical cycles in riverborne groundwater." *Geochimica et Cosmochimica Acta*, 55, 3597-3609.

Xie, T., and Marshall, W. D. (2001). "Approaches to soil remediation by complexometric extraction of metal contaminants with regeneration of reagents." *Journal of Environmental Monitoring*, 3, 411-416.

Yip, T. C. M., Tsang, D. C. W., Ng, K. T. W. and Lo, I. M. C (2009). "Kinetic interactions of EDDS with soils.1. Metal resorption and competition under EDDS deficiency." *Environmental Science and Technology*, 43, 831-836.

Yu, J., and Klarup, D. (1992). "Extraction kinetics of copper, zinc, iron and manganese from contaminated sediments using disodium EDTA." *Water, Air, and Soil Pollution*, 75, 205-225.

Zou, Z., Qiu, R., Zhang, W., Dong, H., Zhao, Z., Zhang, T., Wei, X., and Cai, X. (2009). "The study of operating variables in soil washing with EDTA." *Environmental Pollution*, 157, 229-236.

Zhang, W., Huang, H., Tan, F., Wang, H., and Qiu, R. (2010). "Influence of EDTA washing on the species and mobility of heavy metals residual in soils." *Journal of Hazardous Materials*, 173, 369-376.

CHAPTER 8

Roles of Metal-(Hydr)oxides in Chelant-Enhanced (Phyto)extraction

Michael Komárek

8.1 Introduction

In the last two decades, chelating agents have been studied extensively in soil remediation procedures, either *in situ* (soil flushing, chemically-enhanced phytoextraction) or *ex situ* (soil washing, heap leaching) (Leštan et al., 2008; Evangelou et al., 2007). The chelants most investigated and used include EDTA (ethylenediaminetetraacetate) and biodegradable EDDS (ethylenediaminedisuccinate). The latter was presented as an "environmentally friendlier" alternative to the often criticized persistent chelant EDTA (Kos and Leštan, 2003; Nowack et al., 2006). The chelating agents are applied to soils in order to extract the contaminating metals (soil washing/flushing) from the soils or to enhance their mobility and uptake by plants (chelant-enhanced phytoextraction). However, the efficiency of the process is still under debate. One of the main drawbacks of the method is the competition of major soil cations (e.g., Al, Ca, Fe, Mn) for the ligand (chelant) which subsequently reduces the extraction efficiency of the chelant. It has been found that EDTA and EDDS are able to dissolve soil (hydr)oxides and subsequently complex Al, Fe and Mn, which reduces the extraction efficiency (Nowack and Sigg, 1997; Tsang et al., 2007; Koopmans et al., 2008; Komárek et al., 2009). The aim of this chapter is to describe the role of soil (hydr)oxides during metal extraction using chelants and chelant-enhanced phytoextraction of metals from contaminated soils.

8.2 Interactions of Soil (Hydr)oxides with Chelating Agents

8.2.1 Soil (Hydr)oxides – Important Soil Components

Soil (hydr)oxides (or oxides, hydroxides, hydrous metal oxides, oxyhydroxides etc.) play an important role in soil chemistry. They are present in soils as discrete crystals, coatings on other particles and as mixed gels (Sparks, 2003).

These ubiquitous accessory minerals are due to their high reactive surface areas, important constituents of the soil sorption complex (especially amorphous and poorly crystalline ones) and significantly influence the behavior of many substances in the environment (Essington, 2004). They are dominant sorbents capable of forming strong surface complexes with many different metals, ligands and molecular species, through the formation of either positive or negative surface charge depending on the composition of the soil solution (Dzombak and Morel, 1990; Essington, 2004). Soil (hydr)oxides are usually characterized by their small particle size and low solubility in the range of common pH values found in soils (Sposito, 2008). Common definable soil (hydr)oxides are summarized in Table 8.1.

Table 8.1 Selected metal (hydr)oxides commonly occurring in soils (adapted from Essington, 2004; Sposito, 2008).

(Hydr)oxide	Formula
Gibbsite	$\gamma\text{-}Al(OH)_3$
Boehmite	$\gamma\text{-}AlOOH$
Allophane	$Al_2O_3 \cdot (SiO_2)_{1\text{-}2} \cdot 2.5\text{-}3H_2O$
Hematite	$\alpha\text{-}Fe_2O_3$
Goethite	$\alpha\text{-}FeOOH$
Lepidocrocite	$\gamma\text{-}FeOOH$
Hematite	$\alpha\text{-}Fe_2O_3$
Maghemite	$\gamma\text{-} Fe_2O_3$
Ferrihydrite	$Fe_5HO_8 \cdot 4H_2O$ or $Fe_{10}O_{15} \cdot 9H_2O$
Birnessite	$(Na, K, Ca,Mn^{II})(Mn^{III}Mn^{IV})_7O_{14} \cdot 2.8H_2O$
Vernadite	$\delta\text{-}MnO_2$

Gibbsite is the most common Al-(hydr)oxide, especially in highly weathered and acidic soils. Other secondary Al oxides, such as boehmite, are less common. The surface area of $Al(OH)_3$ phases vary according to the crystallinity of the mineral from 20 to 600 $m^2\ g^{-1}$. The surface area of gibbsite usually ranges from 20 to 50 $m^2\ g^{-1}$ (Essington, 2004) and its pHzpc (pH at zero point of charge) is ~9; therefore, in common soil conditions, its surface bears a net positive charge (Sposito, 2008).

Goethite is the most common stable Fe-(hydr)oxide in soils, it exists in almost all types of soils with higher abundances in cool and wet climates. Hematite and goethite's polymorph, lepidocrocite, are usually found in association with goethite. Hematite is formed mainly in warm climatic regions (e.g., tropics and subtropics). The specific surface area of synthetically prepared goethite and hematite varies from 60 to 200 and 6 to 115 $m^2\ g^{-1}$, respectively. Substantial Al-Fe substitution is possible in both minerals (Sposito, 2008). Ferrihydrite is, with its specific surface area of 100–700 $m^2\ g^{-1}$, an important component of the soil sorption complex (Essington, 2004). It is often found in association with goethite and can be formed in soils when other organic ligands (e.g., humic substances) or soluble silica are present and the crystallization of goethite or hematite is inhibited (Sposito, 2008).

Manganese-(hydr)oxides are another group of secondary minerals commonly found in soils, especially in soil environments with intensive reducing/oxidizing conditions. They often occur as fine-grained coatings of other soil particles and as nodules (Post, 1999; Essington, 2004). Most Mn-(hydr)oxides are amorphous (Sparks, 2003). The most common Mn-(hydr)oxide is birnessite which typically forms poorly crystalline particles in soils. Birnessite precipitates in soils mainly as a result of Mn^{2+} oxidation promoted by bacteria and fungi, and similarly to bacteriogenic ferrihydrite, it forms highly reactive biofilms (Sposito, 2008). It was found that Mn-(hydr)oxides were much more efficient in adsorbing metals compared to Fe-(hydr)oxides (Dong et al., 2000); nevertheless, due to the fact that Fe oxides are more abundant in soils compared to Mn ones, the former present one of the most important sinks for metals in soils. Besides the mentioned minerals, other Fe, Mn and Al phases (and their mixtures) exist in soils. However, it is often impossible to precisely define their mineralogy and simple extractions, such as the acid oxalate extraction, should be thus used to determine reducible and extractable contents of Fe, Mn and Al.

8.2.2 Chelates – Complexes of Metals with Chelants

Because metal (hydr)oxides are ubiquitous soil components, their interactions with chelating agents need to be taken into account when considering the efficiency of chelant-induced (phyto)extraction of metals. The choice of chelating agents is mainly influenced by their affinity for the targeted contaminating metals, which can be predicted by the stability constants of the formed metal–chelant complexes (chelates) (Table 8.2). When the chelating agent enters the soil, it complexes available metals based on their concentrations and stability constants until new equilibrium has been attained. This process leads to solubilization and mobilization of metals in soils and subsequent leaching or plant uptake. Metals available for chelant complexation belong mainly to the non-residual fractions, i.e., exchangeable metals, metals bound to (hydr)oxides and to soil organic matter (Tandy et al., 2004). The importance of metal "lability" during their extraction by chelants has also been highlighted by Yan et al. (2010).

The application of chelating agents leads to intensive metal solubilization from soils and when plants are grown on the soil, subsequent diffusion down the concentration gradient into plants roots takes place (Shen et al., 2002). The formed chelates may enter the root system at disruptions of the endodermis and Casparian strip (Nowack et al., 2006). The translocation of these metals into shoots in the presence of increased concentrations of chelating agents is most likely through a passive uptake mechanism (apoplastic) of the formed chelates (Sarret et al., 2001; Nowack et al., 2006; Tandy et al., 2006).

Table 8.2 Stability constants (log K) of selected metal complexes with EDTA and EDDS at 0.1 M ionic strength (adapted from Martell et al., 1998).

Metal	EDTA	EDDS
Al	16.4	15.6 [a]
Ca	10.7	4.6
Cd	16.5	10.9
Cu	18.8	18.5
Fe(III)	25.1	22.0
Mn	13.9	9.0
Ni	18.4	18.0
Pb	18.0	12.7
Zn	16.5	13.4

[a] data from Koopmans et al. (2008) at 0 M ionic strength

The extraction efficiency of the synthetic chelants is influenced by several factors; e.g., type of extracted metal(s), metal/chelant ratio, soil physico-chemical characteristics, adsorption and biodegradability of the chelant etc. (Nowack et al., 2006; Leštan et al., 2008; Meers et al., 2008; Yip et al., 2009a). However, the subsequently formed soluble metal–chelant complexes together with free chelant (if present) can possibly dissolve soil (hydr)oxides and the solubilized metals (Al, Fe, Mn) can compete for the chelating agent. Besides Fe, Mn and Al from secondary soil oxides, Ca is another important cation competing for the ligand. This is especially true for calcareous soils and EDTA (Nowack et al., 2006). Several recent studies have demonstrated that the dissolution of soil phases containing Ca, Fe, Mn and Al influences the speciation of chelating agents and thus reduces the amount of available ligands for the targeted metals (Nowack et al., 2006; Koopmans et al., 2008). While Ca can be an important competitor during EDTA-enhanced (phyto)extraction, the competition of Ca for EDDS seems to have only a minor influence (Table 8.2; Nowack et al., 2006; Tsang et al., 2009; Komárek et al., 2010).

8.2.3 Sorption of Chelants and Chelates onto Soil (Hydr)oxides

When the free chelants or chelates (i.e., complexed with metals) come into contact with metal (hydr)oxides in soils, they are first adsorbed on the surfaces of the minerals. This adsorption process is an important precursor for the chelant-induced dissolution of the oxides. Free EDTA is adsorbed as a binuclear complex at low pH and as a mononuclear complex at high pH values. The process can be described as surface complexation with the formation of inner-sphere complexes (Nowack and Sigg, 1996). Adsorption of chelates onto metal oxides can be explained by the formation of stable ternary surface complexes (metal–ligand–reactive surface of the oxide). The complex is adsorbed to the oxide surface by ligand exchange with uncoordinated carboxylate groups. Bonding of the complex to the surface occurs either through the metal (Eq. 1) or the ligand (chelant) (Eq. 2) as described for a Fe-oxide by Nowack and Sigg (1996) and Nowack (2002):

Type A ternary complex:

$$n{\equiv}\text{Fe–OH} + \text{Me}^{n+} + \text{H}_x\text{Lig}^{m-} \leftrightarrow {\equiv}\text{Fe–OMe–Lig}^{(n-m-1)+} + (x+1)\text{H}^+ \quad (1)$$

Type B ternary complex:

$$n{\equiv}\text{Fe–OH} + \text{Me}^{n+} + \text{H}_x\text{Lig}^{m-} \leftrightarrow {\equiv}\text{Fe–Lig–Me}^{(n-m-1)+} + (x-1)\text{H}^+ + \text{H}_2\text{O} \quad (2)$$

where ≡FeOH is the hydroxyl functional group of an Fe-oxide surface (≡Fe), Lig is the ligand (chelant) and Me is the complexed metal.

The mechanism described in Eq. 1 is usually characterized by increasing adsorption with increasing pH. On the contrary, adsorption mechanisms described by Eq. 2 occurs preferably at low pH (Nowack, 2002; Bradl, 2004). Another important factor influencing the adsorption of chelates is the structure of the complex. Chelates with the same structure and charge (e.g., Cu(II)EDTA, Pb(II)EDTA) will have the same adsorption behavior (Yang and Davis, 1999). However, different adsorption mechanisms were observed for complexes with metals with different charges (e.g., Fe(III)EDTA) or with the same charge but different structure (e.g., Pd(II)EDTA). Therefore, not only the charge, but the structure of the complex influences the adsorption behavior (Nowack, 2002).

Tsang et al. (2009) found that the adsorption of metal–EDDS complexes to soils is more pronounced at higher pH values, corresponding to a ligand-like adsorption behavior. On the other hand, an opposite trend was found for CuEDDS sorption onto ferrihydrite (Komárek et al., 2009). Higher EDDS adsorption at lower pH values is in accordance with the work of Nowack and Sigg (1996) where EDTA adsorption onto goethite was the highest at pH < 7. While uncomplexed EDTA is adsorbed as an inner-sphere complex under acidic conditions, the adsorption of Fe(III)EDTA forms a nonspecific surface complex and is weaker compared to free EDTA (Nowack and Sigg, 1996). Due to the fact that EDDS is a structural isomer of EDTA, the adsorption behavior will probably be similar. Tsang et al. (2009) summarized the possible sorption mechanisms of metal–EDDS complexes onto soil constituents: (i) ternary surface complexes on oxide hydroxyl groups through inner-sphere complexation, (ii) outer-sphere complexation through electrostatic interactions and (iii) hydrogen bonding. The prevailing mechanism depends on several factors (e.g., pH, complex concentration, type of metal, metal/EDDS ratio, etc.).

8.2.4 Chelant-Induced Dissolution of Soil (Hydr)oxides

The dissolution of Fe-(hydr)oxides by free EDTA and metal–EDTA complexes has been already described some thirty years ago by Rubio and Matijević (1979) and other studies followed (e.g., Szecsody et al., 1994; Nowack and Sigg,

1997). Chelant-induced dissolution of soil minerals is mostly attributed to ligand exchange reactions where the surface metal–oxygen bonds are broken after the adsorption of the chelating agent on the mineral surface (Nowack, 2002). The sequence of chelate-promoted dissolution of metal (hydr)oxides has been comprehensively described by Nowack and Sigg (1997): (i) after the adsorption of the chelate a partial dissolution of the oxide takes place and an intermediate (e.g., Fe-EDTA-Pb) species is formed; (ii) the surface complex further dissociates and the complexed metal (e.g., Pb) is released; (iii) depending on the equilibrium solubility concentration of the chelate (e.g., Fe(III)EDTA), some free $EDTA^{4-}$ can be released from the complexes which is further available for (hydr)oxide dissolution.

While the type of the metal complexed with EDTA does not necessary influence the dissolution rate of crystalline goethite, the dissolution of Fe-(hydr)oxides depends on the metal–chelant speciation, with free (uncomplexed) EDTA being the most efficient. According to Nowack and Sigg (1997), the dissolution rate decreases in the following order: EDTA >> CaEDTA > PbEDTA > ZnEDTA > CuEDTA > Co(II)EDTA > NiEDTA. Tsang et al. (2009) observed that, in uncontaminated soils, free EDDS and metal complexes with EDDS were adsorbed onto soil particles and induced mineral dissolution, while in contaminated soils, metal–EDDS rather underwent exchange reactions with Cu, Pb and Zn sorbed on the soil surfaces. Nevertheless, the dissolution of Fe-(hydr)oxides is a relatively slow process depending mainly on the concentration of the added chelant, pH, amount of the oxides in the soil, etc. (Nowack et al., 2006). The rate-limiting processes of dissolution of Fe-(hydr)oxides in the presence of chelants include detachment of Fe(III) from the oxide and dissociation of the chelates (Nowack and Sigg, 1997).

The dissolution of soil (hydr)oxides in the presence of EDDS has also been observed (Komárek et al., 2009). The dissolution of selected Fe-(hydr)oxides in the presence of free EDDS and metal–EDDS complexes are presented in Figure 8.1. The higher dissolution of Fe from ferrihydrite at higher pH values can be explained by increased detachment of FeEDDS due to the higher stability of the complex at higher pH values (Tsang et al., 2009). The high surface area of ferrihydrite enables fast and efficient sorption of EDDS onto the mineral surface, which is the essential precursor of mineral dissolution by chelating agents (Nowack and Sigg, 1997).

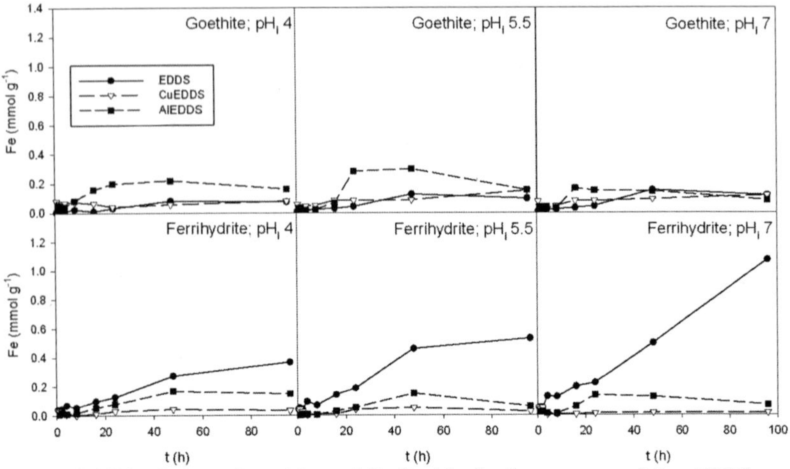

Figure 8.1 Dissolution of goethite and ferrihydrite in the presence of free EDDS, CuEDDS and AlEDDS at various pH values (adapted from Komárek et al., 2009).

Similar to EDTA, when EDDS is applied to soils, several processes occur: (i) EDDS forms complexes with readily available metals in the soil; (ii) metal-exchange reactions take place between the metal–EDDS complexes and the sorbed metallic contaminants. If there is an excess of EDDS, fast adsorption onto the soil minerals (especially onto amorphous or poorly crystalline ones) takes place and dissolution of these phases begins. The dissolution of amorphous minerals (e.g., ferrihydrite) seems to be the most important factor controlling dissolved Fe and Al concentrations in soils in the presence of metal–EDDS complexes and especially free EDDS (Nowack and Sigg, 1997; Koopmans et al., 2008; Komárek et al., 2009; Yip et al., 2009b). These metals subsequently compete for the chelating agent and the (phyto)extraction of the targeted metals is lowered. In Cu-contaminated soils, however, the formation of stable CuEDDS complexes, results in a lower dissolution of ferrihydrite and goethite compared to free EDDS and AlEDDS and the competition of Fe and Al could be less significant (Komárek et al., 2009). Such conclusions are backed by the results of Yip et al. (2009b) who observed higher dissolution of soil oxides in the presence of PbEDDS complexes compared to the more stable CuEDDS complexes. Therefore, the dissolution of the oxides (and competition of dissolved Fe, Mn, Al) depends also on the targeted contaminating metal(s) (i.e., higher dissolution with less stable metal–EDDS complexes).

Different dissolution patterns of Al in the presence of EDDS were observed by various authors. The EDDS-promoted dissolution of Al from synthetic gibbsite as presented by Komárek et al. (2009) is not as intensive as the increased dissolutions of Al from soils described in other works (Koopmans et al., 2008; Yip et al., 2009b). Such Al probably originated from other phases, i.e., less crystalline oxides and exchangeable Al (Yan et al., 2010). Komárek et al. (2009) further observed a comparable dissolution of goethite and gibbsite in the presence of EDDS which was significantly lower than the dissolution of ferrihydrite. This highlights the importance of (hydr)oxide crystallinity during chelant-enhanced (phyto)extraction.

8.2.5 Influence of Fe, Mn and Al Dissolution on (Phyto)extraction Efficiency

As mentioned above, increased solubilization of Al and Fe from soils in the presence of EDTA or EDDS has been documented. Koopmans et al. (2008) observed that the addition of 0.27–0.33 mM EDDS resulted into increased Al (32-fold) and Fe (109-fold) concentrations in soil solutions. The addition of EDDS thus solubilized up to 50% Fe and 37% Al from the soils. Similar findings were observed by Komárek et al. (2009) where EDDS-promoted dissolution of Fe-oxides led to the intensive formation of FeEDDS complexes. Yip et al. (2009a) studied the influence of EDDS concentrations on the competition of soil Fe, Mn, Al and found that under EDDS deficiency, Al is the major competing cation (although not as important at longer reaction times due to equilibrium changes). On the other hand, under EDDS excess, Fe became a significant competitor for the EDDS ligand (Yip et al., 2009a; Yan et al., 2010). Complexation of EDDS/EDTA with Fe is more pronounced at low pH (Tandy et al., 2004). The use of higher chelant concentrations leads to higher adsorbed concentrations of chelants and increased dissolution of soil (hydr)oxides (Nowack and Sigg, 1997; Yip et al., 2009b).

Figure 8.2 depicts the solubilization of Pb and Fe from a contaminated soil after the application of EDDS and EDTA. The observed decrease of water-soluble Pb with respect to time is evident and is most probably caused by exchange reactions between complexed Pb and Fe originating from slowly dissolving Fe-(hydr)oxides present in the soil. This fact was proved by increasing Fe concentration with respect to time after the addition of the chelants (especially EDTA), leading to a strong competition between Pb and Fe for the ligands. Lead extraction is therefore reduced with increasing time due to the co-extraction of Fe, which outcompetes other metals. Due to the lower stability of PbEDDS complexes (Table 8.2), EDDS was less effective in extracting Pb from the soil compared to EDTA.

Assuming that most of the dissolved Fe, Mn and Al originated from soil (hydr)oxides, these findings directly confirm the strong influence of these mineral phases on the extraction efficiency of the chelating agents. Nevertheless, due to the high stability of CuEDDS species (log K($CuHEDDS^-$) = 30.4; log K(CuH_2EDDS^0) = 34.5; Koopmans et al., 2008), the dissolution of (hydr)oxides present in soils with high Cu concentrations (e.g., Cu-polluted soil) is not as intensive, which was also proven by speciation modeling (Komárek et al., 2009). The preferential formation of CuEDDS complexes in systems where other "heavy metals" are present (e.g., Pb, Zn) through metal exchange has also been documented by Yip et al. (2009a). The high stability of the CuEDDS complexes has been attributed to its high ionic potential and four strong equatorial bonds between Cu and EDDS (Tsang et al., 2009).

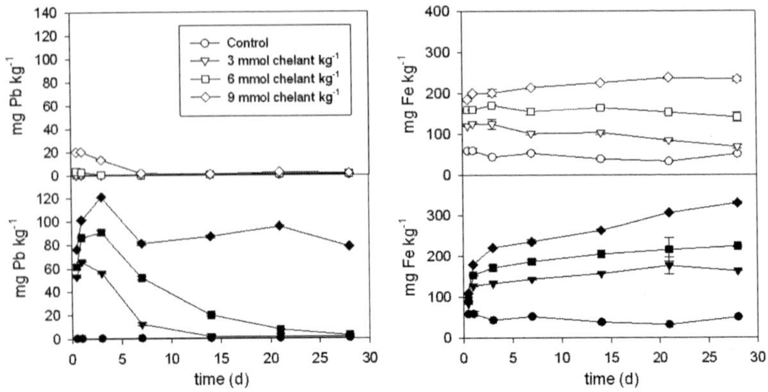

Figure 8.2 Dissolution of Pb and Fe from a contaminated soil in the presence of EDDS (white symbols) and EDTA (black symbols) (adapted from Komárek et al., 2007a).

Increased concentrations of Fe, Mn and Al originating mainly from the dissolution of soil (hydr)oxides were observed in plants grown on soils treated with EDDS (Figure 8.3). Strong stem–leaf translocation of the solubilized metals (probably as metal–EDDS complexes of high stability) can be observed in the tested plants. Nevertheless, the most pronounced change in metal uptake after the application of EDDS was observed in the case of the targeted metal (Cu) which again correlates with the high stability of the CuEDDS complexes. Nevertheless, the importance of soil (hydr)oxides during chelant-enhanced phytoextraction is undisputable.

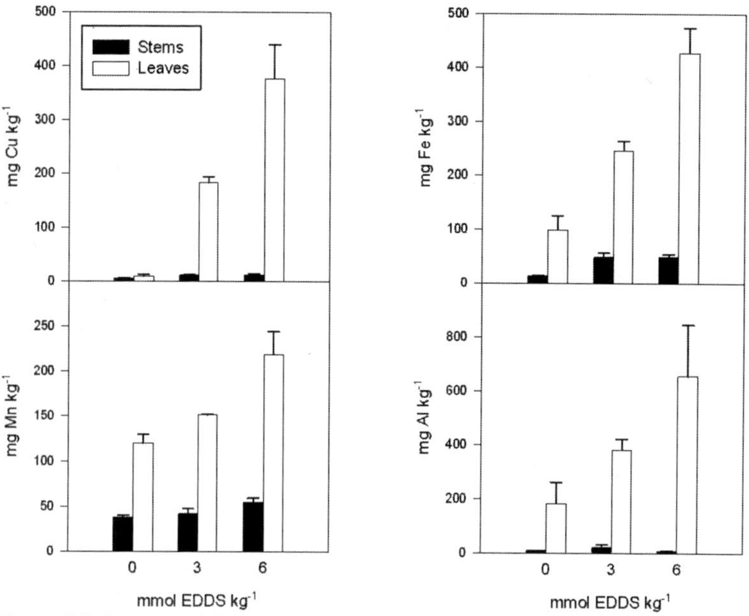

Figure 8.3 Contents of Cu, Fe, Mn and Al in poplar (*Populus nigra* L.) stems and leaves (dry weight) grown on a Cu-contaminated soil amended with EDDS (mmol EDDS kg^{-1} soil) ($n = 4$) (adapted from Komárek et al., 2010).

The process of chelant-induced dissolution of Fe-(hydr)oxides is more pronounced with increasing reaction time as more Fe, Mn and Al dissolves at later stages of the extraction process (Figures 8.1, 8.2; Tandy et al. 2004; Hauser et al., 2005; Komárek et al., 2007a,b; 2009; Yip et al., 2009a). Therefore, during chelant-enhanced phytoextraction, the competition of the dissolved major cations is more emphasized during subsequent years and the extraction efficiency is lowered and surpluses of chelating agents have to be applied to soils in order to overcome this drawback (Nowack et al., 2006; Komárek et al., 2008). This fact is associated with increased concentrations of Fe and Mn and decreased concentrations of target metals (e.g., Pb) in aboveground parts of plant grown in subsequent years (Figure 8.4). In the case of relatively faster processes, such as soil washing/flushing, the influence of the competing metals originating from (hydr)oxide dissolution will be significantly smaller (compared to long-term chelant-enhanced phytoextraction).

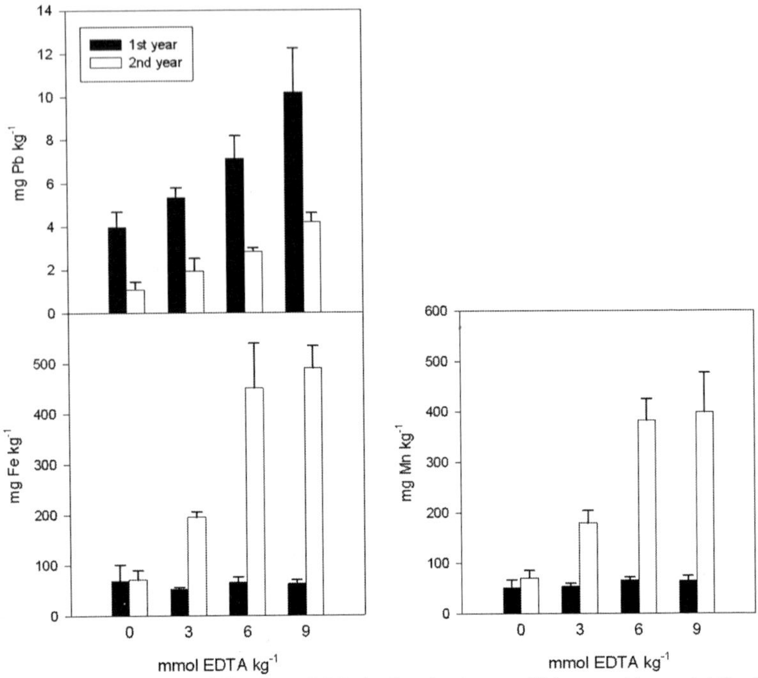

Figure 8.4 Contents of Pb, Fe and Mn in the aboveground biomass (dry weight) of corn (*Zea mays* L.) grown on a contaminated soil during a 2-year phytoextraction experiment after the addition of EDTA (mmol EDTA kg^{-1} soil) ($n = 4$) (adapted from Komárek et al., 2007b).

The competition of Fe, Mn and Al for the chelant is not the only drawback during chelant-induced washing/flushing of metals from contaminated soils. The dissolution of soil (hydr)oxides (and soil organic matter) during the process decreases the shear strength and aggregate stability of the treated soils (Tsang et al., 2007). The dissolution of amorphous (hydr)oxides will be responsible for the destabilization of the soils as crystalline phases are less susceptible to dissolution (Nowack and Sigg, 1997; Komárek et al., 2009). As mentioned before, the extraction efficiency of the chelating agents is influenced by soil properties (e.g., pH), the concentrations and availability of the targeted metals (e.g., Cu, Zn, Pb), the content and type of soil (hydr)oxides. Due to these reasons, mobilization of colloidal particles takes place and changes of physico-chemical properties of the soils (clay content, soil permeability) can be observed, leading to unwanted clogging, which present another major drawback in chelant-enhanced washing/flushing (Tsang et al., 2007).

8.3 Conclusions

Although research focused on chelant-enhanced phytoextraction is probably reaching a dead-end (Nowack et al., 2006; Komárek et al., 2010), the potential of chelating agents remains in chemical flushing/washing of metal-contaminated soils. However, there are several drawbacks of the method. One of the most important is the competition of major soil cations for the applied chelant which reduces significantly the (phyto)extraction efficiency. These competing metals include Fe, Al and Mn originating mainly from chelant/chelate-induced dissolution of soil (hydr)oxides. The dissolution increases with increasing contact (chelant-oxide) time and will be more significant during long-term chelant-enhanced phytoextraction compared to faster soil washing/flushing. Therefore, at least an approximate mineralogical composition of the treated soil needs to be known before selecting the chelants and their dosages for (pyhto)extraction purposes. For example, mineralogical methods and the simple acid oxalate extraction used for predicting the amount of amorphous and poorly crystalline Fe-, Mn- and Al-(hydr)oxides should provide useful information. Therefore, in order to overcome this drawback, large chelant excesses need to be used in some cases to counterbalance the competing effect of the major soil cations.

8.4 Acknowledgments

The author is grateful for financial support from the research projects GAČR 526/08/P042 (Czech Science Foundation) and MSM 6046070901 (Ministry of Education of the Czech Republic).

8.5 References

Bradl, H. (2004). "Adsorption of heavy metal ions on soils and soils constituents." *Journal of Colloid and Interface Science*, 277(1), 1-18.

Dzombak, D. A. and Morel, F. M. M. (1990). Surface Complexation Modeling, Hydrous Ferric Oxide. Wiley, New York, 393.

Dong, D. M., Nelson, Y. M., Lion, L. W., Shuler, M. L. and Ghiorse, W. C. (2000). "Adsorption of Pb and Cd onto metal oxides and organic material in natural surface coatings as determined by selective extractions: new evidence for the importance of Mn and Fe oxides." *Water Research*, 34(2), 427-436.

Essington, M. E. (2004). Soil and Water Chemistry, An Integrative Approach. CRC Press, Boca Raton, Florida, 534.

Evangelou, M. W. H., Ebel, M. and Schaeffer, A. (2007). "Chelate assisted phytoextraction of heavy metals from soil. Effect, mechanism, toxicity, and fate of chelating agents." *Chemosphere*, 68(6), 989-1003.

Hauser, L., Tandy, S., Schulin, R. and Nowack, B. (2005). "Column extraction of heavy metals from soils using the biodegradable chelating agent EDDS." *Environmental Science and Technology*, 39(17), 6819-6824.

Komárek, M., Tlustoš, P., Száková, J., Chrastný, V. and Ettler, V. (2007a). "The use of maize and poplar in chelant-enhanced phytoextraction of lead from contaminated agricultural soils." *Chemosphere*, 67(4), 640-651.

Komárek, M., Tlustoš, P., Száková, J., Chrastný, V. and Balík, J. (2007b). "The role of Fe- and Mn-oxides during EDTA enhanced phytoextraction of heavy metals." *Plant, Soil and Environment*, 53(5), 216-224.

Komárek., M., Tlustoš, P., Száková, J. and Chrastný, V. (2008). "The use of poplar during a two-year induced phytoextraction of metals from contaminated agricultural soils." *Environmental Pollution*, 151(1), 27-38.

Komárek, M., Vaněk, A., Száková, J., Balík, J. and Chrastný, V. (2009). "Interactions of EDDS with Fe- and Al-(hydr)oxides." *Chemosphere*, 77(1), 87-93.

Komárek, M., Vaněk, A., Mrnka, L., Sudová, R., Száková, J., Tejnecký, V. and Chrastný, V. (2010). "Potential and drawbacks of EDDS-enhanced phytoextraction of copper from contaminated soils." *Environmental Pollution*, in press, doi:10.1016/j.envpol.2010.04.002.

Koopmans, G. F., Schenkeveld, W. D. C., Song, J., Luo, Y., Japenga, J. and Temminghoff, E. J. M. (2008). "Influence of EDDS on metal speciation in soil extracts: Measurement and mechanistic multicomponent modelling." *Environmental Science and Technology*, 42(4), 1123-1130.

Kos, B. and Leštan, D. (2003). "Induced phytoextraction/soil washing of lead using biodegradable chelate and permeable barriers." *Environmental Science and Technology*, 37(3), 624-629.

Leštan, D., Luo, C. L. and Li, X. D. (2008). "The use of chelating agents in the remediation of metal-contaminated soils: A review." *Environmental Pollution*, 153(1), 3-13.

Martell, A. E., Smith, R. M. and Motekaitis, R. J. (1998). NIST SRD 46 Critically Selected Stability Constants of Metal Complexes, Version 5.0. NIST.

Meers, E., Tack, F. M. G. and Verloo, M. G. (2008). "Degradability of ethylenediaminedisuccinic acid (EDDS) in metal contaminated soils: Implications for its use in soil remediation." *Chemosphere*, 70(3), 358-363.

Nowack, B. and Sigg, L. (1996). "Adsorption of EDTA and metal–EDTA complexes onto goethite." *Journal of Colloid and Interface Science*, 177(1), 106–121.

Nowack, B. and Sigg, L. (1997). "Dissolution of Fe(III)(hydr)oxides by metal-EDTA complexes." *Geochimica Cosmochimica Acta*, 61(5), 951-963.

Nowack, B. (2002). "Environmental chemistry of aminopolycarboxylate chelating agents." *Environmental Science and Technology*, 36(19), 4009-4016.

Nowack, B., Schulin, R. and Robinson, B. H. (2006). "Critical assessment of chelant-enhanced metal phytoextraction." *Environmental Science and Technology*, 40(17), 5225-5232.

Post, J. E. (1999). "Manganese oxide minerals: Crystal structures and economic and environmental significance." *Proceedings of the National Academy of Sciences of the United States of America*, 96(7), 3447-3454.

Rubio, J. and Matijević, E. (1979). "Interactions of metal hydrous oxides with chelating agents." *Journal of Colloid and Interface Science*, 68(3), 408-421.

Sarret, G., Vangronsveld, J., Manceau, A., Musso, M., D'Haen, J., Menthonnex, J. J. and Hazemann, J. L. (2001). "Accumulation forms of Zn and Pb in

Phaseolus vulgaris in the presence and absence of EDTA." *Environmental Science and Technology*, 35(13), 2854-2859.

Shen, Z. G., Li, X. D., Wang, C. C., Chen, H. M. and Chua, H. (2002). "Lead phytoextraction from contaminated soil with high-biomass plant species." *Journal of Environmental Quality*, 31(6), 1893-1900.

Sparks, D. L. (2003). Environmental Chemistry. Academic Press, San Diego, California, 352.

Sposito, G. (2008). The Chemistry of Soils. Oxford University Press, Oxford, 329.

Szecsody, J. E., Zachara, J. M. and Bruckhart, P. L. (1994). "Adsorption-dissolution reactions affecting the distribution and stability of Co(II)EDTA in iron oxide coated sand." *Environmental Science and Technology*, 28(9), 1706-1716.

Tandy, S., Bossart, K., Mueller, R., Ritschel, J., Hauser, L., Schulin, R. and Nowack, B. (2004). "Extraction of heavy metals from soils using biodegradable chelating agents." *Environmental Science and Technology*, 38(3), 937-944.

Tandy, S., Schulin, R. and Nowack, B. (2006). "The influence of EDDS on the uptake of heavy metals in hydroponically grown sunflowers." *Chemosphere*, 62(9), 1454-1463.

Tsang, D. C. W., Zhang, W. and Lo, I. M. C. (2007). "Copper extraction effectiveness and soil dissolution issues of EDTA-flushing of artificially contaminated soils." *Chemosphere*, 68(2), 234-243.

Tsang, D. C. W., Yip, T. C. M. and Lo, I. M. C. (2009). "Kinetic interactions of EDDS with soils. 2. Metal-EDDS complexes in uncontaminated and metal-contaminated soils." *Environmental Science and Technology*, 43(3), 837-842.

Yan, D. Y. S., Yip, T. C. M., Yui, M. M. T., Tsang, D. C. W. and Lo, I. M. C. (2010). "Influence of EDDS-to-metal molar ratio, solution pH, and soil-to-solution ratio on metal extraction under EDDS deficiency." *Journal of Hazardous Materials*, 178(1-3), 890-894.

Yang, J. K. and Davis, A. P. (1999). "Competitive adsorption of Cu(II)-EDTA and Cd(II)-EDTA onto TiO_2." *Journal of Colloid and Interface Science*, 216(1), 77-85.

Yip, T. C. M., Tsang, D. C. W., Ng, K. T. W. and Lo, I. M. C. (2009a). "Kinetic interactions of EDDS with soils. 1. Metal resorption and competition under EDDS deficiency." *Environmental Science and Technology*, 43(3), 831-836.

Yip, T. C. M., Tsang, D. C. W., Ng, K. T. W. and Lo, I. M. C. (2009b). "Empirical modeling of heavy metal extraction by EDDS from single-metal and multi-metal contaminated soils." *Chemosphere*, 74(2), 301-307.

CHAPTER 9

Use of Chelating Agents in Electrochemical Remediation of Contaminated Soil

Albert T. Yeung and Ying-Ying Gu

9.1 Introduction

Soil contamination is a worldwide problem of increasing concern. Many sites are contaminated by different hazardous contaminants, both inorganic and organic. The problem is serious and fast growing, in particular in industrially developed or developing economies. The problem is persistent and long lasting, impacting public health and the ecology adversely in the long term. In fact, the long-term impacts of soil contamination are not fully understood or delineated. The nature and properties of the contaminated soil may change from site to site. However, soil can be broadly classified into two categories in accordance with its particle size distribution: (1) coarse-grained soil; and (2) fine-grained soil. Depending on: (1) the nature and properties of the contaminated soil; (2) the type, chemical state, and concentration of the contaminant; and (3) the specific environmental conditions of the contaminated site; there are many available *in-situ* or *ex-situ* remediation technologies that can achieve varying degrees of remediation (Yeung, 2009a). However, remediation of contaminated fine-grained soil is particularly difficult due to its extremely low hydraulic conductivity and very large specific area (Yeung, 2008).

The extremely low hydraulic conductivity of the fine-grained soil makes it very difficult for fluid to permeate through the soil, rendering the technology of pump-and-treat, soil flushing, etc. cost prohibitive and practically ineffective. If the hydraulic gradient imposed on the contaminated soil is too low, it will take too long to pump adequate volume of cleansing fluid through the soil to achieve the required level of remediation. If the hydraulic gradient imposed is too high, it may generate hydraulic fractures in the soil and create many uncontrolled and random paths in the subsurface for the contaminant to spread, thus aggravating the existing contamination problem. The very large specific area of the fine-grained soil provides numerous interactive reaction sites between contaminant species and soil particle surfaces. These soil-contaminant interactions are soil specific, contaminant specific, pH-dependent, dynamic, and reversible. Therefore, they cannot be relied upon as a long-term reliable contaminant containment technology. However, these interactions may immobilize the contaminant temporarily, rendering them immovable during the remediation process. As a result, remediation of contaminated fine-grained soil is not an easy task for the geoenvironmental professional to accomplish with much confidence.

Electrochemical remediation is a promising emerging technology to remediate the contaminated fine-grained soil (Reddy and Cameselle, 2009). A direct-current (dc) electric field is imposed on the contaminated soil to extract contaminants in the soil pore fluid by the combined electrokinetic mechanisms of electroosmosis, electromigration, and/or electrophoresis (Shapiro and Probstein, 1993; Yeung, 1994, 2008, 2011). However, the efficiency of electrochemical remediation, similar to all other remediation technologies, is heavily dependent on the mobility of contaminants within the soil matrix and other geochemical processes in the subsurface (Yeung, 2006a, 2009b). In general, only a small proportion of contaminants in a contaminated site would exist in the water-soluble and readily extractable form. A significant proportion of contaminants are normally sorbed on soil particle surfaces, rendering them immobile and not amenable to electrochemical remediation or other remediation technologies. As a result, the ability of electrochemical remediation in conjunction with enhancement technologies to solubilize contaminants in the soil pore fluid, rendering them mobile, becomes the key determining parameter controlling the remediation efficiency of the technology (Yeung, 2006a, 2006b).

The introduction of chelating agents to the soil pore fluid of the contaminated soil can enhance the efficiency of electrochemical remediation, as the chelating agent can significantly increase the solubility of contaminants, in particularly metals, in the soil pore fluid (Yeung, 2006a). Depending on the polarity of the chelating agent, it can be injected into the contaminated soil from the electrode reservoir by electroosmosis, electromigration, and a combination of both electrokinetic transport mechanisms. Although similar enhancement of solubilizing sorbed metal contaminants can be achieved by introduction of strong acids to the contaminated soil, the resulting destruction of soil structure and biota may make the soil no longer arable. The use of chelating agents is thus preferred as it can enhance the solubilization of sorbed contaminants at much higher soil pHs. However, the applicability of a particular chelating agent is dependent on several factors, including (1) the pH and temperature range of the contaminated soil; (2) the type of metal ions to be solubilized; (3) the potential destruction of the agricultural properties of the soil after remediation; (4) the possibility of recovery, recycling, and reusing of the used chelating agent rich in contaminant species; (5) the overall economy of the remediation process, etc. As changes in the pH and temperature of the contaminated soil are induced by the electrochemical remediation process as a function of space and time resulting from many simultaneously occurring geochemical processes (Yeung, 2009b), the selection of chelating agents to be used in electrochemical remediation is not a straightforward process. Moreover, the introduction of chelating agents to the contaminated soil also changes the electrokinetic properties of soil particle surfaces. The inter-dependency of the properties of chelating agents and the environmental conditions developed by the electrochemical remediation process further complicates the situation. On the basis of this background, a thorough review on the use of chelating agents in electrochemical remediation of contaminated soil is given in this Chapter. Every chemical name is spelt out the first time it appears and its acronym will be used afterwards for simplicity and clarity. A list of the acronyms used is compiled at the end of the Chapter.

9.2 Electrokinetic Remediation of Contaminated Soils

In the past three decades, the inadequacy of existing technologies to remediate contaminated fine-grained soils has urged the development of alternative technologies to accomplish the important task. Electrokinetics has been applied to remediate contaminated soils *in-situ* and has shown promising potential in extracting contaminants from fine-grained soils of low hydraulic conductivity and large specific area when the environmental conditions are favorable. Milestone developments and future research directions of the technology are outlined in Yeung (2011). Although developments and applications of the technology have been slowing down in the United States, they are growing very rapidly in Europe and Asia during the last decade. These rapid developments are evidenced by the organization of the Symposium on Electrokinetic Remediation (EREM) in Europe every odd year and outside Europe every even year since 2007 (Yang, 2011). The symposia were originally held in Europe biennially since, 1997. All the symposia held outside Europe to date were held in Asia, i.e., Seoul, Korea in 2008 and Kaohsiung, Taiwan in 2010, and the next outside Europe symposium has been scheduled to be held in Hokkaido, Japan in 2012 (Yeung, 2011).

The electrochemical remediation technology imposes a direct-current (dc) electric field across a wet mass of contaminated soil through electrodes emplaced in the subsurface to drive contaminants to electrodes before extracting them from the soil. Several electrokinetic phenomena arise during the electrochemical remediation process, including electroosmosis, electromigration or ionic migration, and electrophoresis which are the movement of soil pore fluid, charged contaminants in the dissolved phase, and colloidal size solid particles, respectively under the influence of the applied dc electric field, as shown in Figure 9.1. A brief discussion of these electrokinetic phenomena of direct relevancy to the use of chelating agents in electrochemical remediation of contaminated soil is given here. A detailed description of these electrokinetic phenomena is given in Yeung (1994).

9.2.1 Electroosmosis

Clay particle surfaces are normally negatively charged as a result of isomorphic substitution and presence of broken bonds (Sposito, 2008). Soil pores can be idealized as liquid-filled capillaries composed of negatively charged cylindrical interior wall surfaces. When a dc electric field is applied across a wet soil matrix, soil pore fluid is transported relative to the stationary charged solid surface from the anode (positive electrode) towards the cathode (negative electrode). This phenomenon is known as electroosmosis. If soil particle surfaces are positively charged, the soil pore fluid would flow from the cathode (negative electrode) towards the anode (positive electrode), i.e., reverse electroosmotic flow. Electroosmotic flow rate and flow direction indicate the magnitude and polarity of the zeta potential on the capillary wall or soil particle surfaces (Hunter, 1981; Yeung, 1994, 2006a; Mitchell and Soga, 2005; Sposito, 2008). Analogous to Darcy's law, the electroosmotic volume flow rate is given by

$$Q = k_e \nabla(-E) A \quad (1)$$

where Q = electroosmotic volume flow rate (m^3/s); k_e = coefficient of electroosmotic conductivity ($m^2/V\text{-}s$); E = electrical potential (V); and A = total cross-sectional area perpendicular to the flow direction. It is clearly indicated in Eq. (1) that electroosmotic flow is in the direction of the imposed electric field $\nabla(-E)$, i.e., in the direction of decreasing electrical potential when the numerical value of k_e is positive.

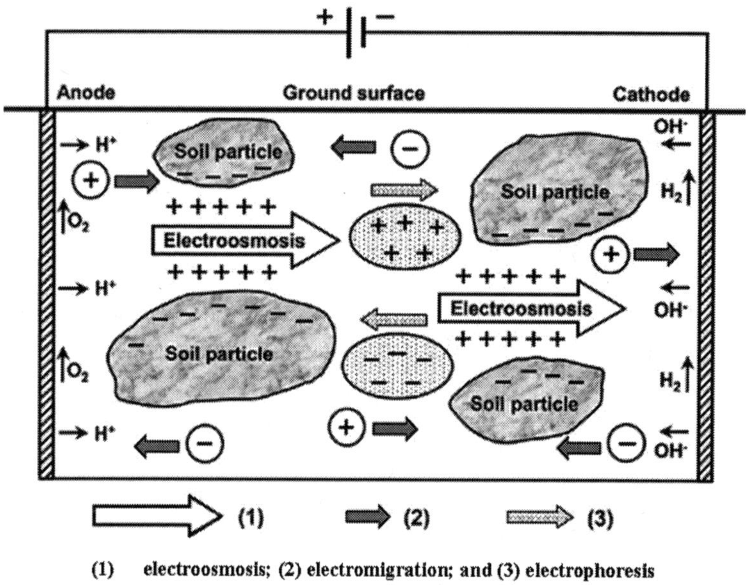

Figure 9.1 Concept of electrochemical remediation of contaminated soil (after Yeung, 2006a).

Several theories have been developed on the basis of different assumptions to describe the electroosmosis phenomenon in relation to the electrokinetic properties of the fluid and the solid surface. The Helmholtz-Smoluchowski theory is the most widely accepted model for soil (Yeung, 1994; Mitchell and Soga, 2005). It is assumed in the theory that parallel layers of opposite charges exist at the solid-fluid interface and the layers of charges can be considered as electrical condensers. Balancing the viscous frictional force on the moving fluid and the electrical force induced on the mobile ions by the dc electric field imposed yields

$$k_e = -\frac{\varepsilon \zeta n}{\eta} \tag{2}$$

where ε = permittivity of soil pore fluid (F/m); ζ = zeta potential of soil particle surfaces (V); n = porosity of soil; and η = viscosity of soil pore fluid (N-s/m^2). Detailed derivation of Eq. (2) is given by Yeung (1994). As indicated in Eq. (1), the electroosmotic volume flow rate is proportional to the coefficient of electroosmotic conductivity k_e. Therefore, the direction and magnitude of electroosmotic flow is a function of the polarity and magnitude of the zeta potential on soil particle surfaces as indicated in Eq. (2). When the numerical value of ζ is negative, i.e., when soil particle surfaces are negatively charged, k_e is positive. When the numerical value of ζ is zero, i.e., at the point of zero charge (PZC) (Sposito, 1998), k_e diminishes. When the numerical value of ζ is positive, i.e., when soil particle surfaces are positively charged, k_e becomes negative, and electroosmotic flow is from the cathode towards the anode. Therefore, any effects of the chelating agent on the zeta potential of soil particle surfaces thus have a significant impact on the direction and volume flow rate of electroosmosis, and thus the efficiency of electrochemical remediation. Typically, the zeta potential of soil particle surfaces is negative and it increases (becomes less negative) with decrease in soil pH. However, the zeta potential of soil particle surfaces does not necessarily become positive in a low pH environment, as the point of zero charge (PZC) may not exist in some soils. However, it has been demonstrated experimentally that the effect of chelating agents on zeta potential of soil particle surfaces can be significant (Popov et al., 1997, 2004; Kaya and Yukselen, 2005; Gu et al., 2009a).

It can also be observed in Eq. (2) that the coefficient of electroosmotic conductivity of soil, unlike hydraulic conductivity, is independent of the pore size distribution of soil. In fact, experimental data obtained to date indicate the values of the coefficient of electroosmotic conductivity are in the narrow range of 1×10^{-9} to 1×10^{-8} m^2/V-s for most soils (Mitchell, 1991; Yeung, 1994; Mitchell and Soga, 2005). However, the values of hydraulic conductivity of different soils span over 13 orders of magnitudes (Freeze and Cherry, 1979). An electrical gradient is thus drastically more effective than a hydraulic gradient in driving fluid through the fine-grained soil where the hydraulic conductivity is very low. As a result, electrochemical remediation is an efficient technology to remediate contaminated fine-grained soils.

9.2.2 Electromigration or Ionic Migration

When a dc electric field is imposed on a wet clay, ions and ion complexes dissolved in the soil pore fluid will be migrated towards the electrode of opposite charge relative to the soil pore fluid. The migration phenomenon is denoted as electromigration or ionic migration. It plays a very important role in the migration of ionic species during electrochemical remediation. When the electrical gradient is 1 V/m, the velocity of an ion in a stationary liquid phase is denoted as its ionic mobility (Yeung, 2009b). In dilute solution, the ionic mobility of an ion can be calculated using the Nernst-Einstein equation (Silbey et al., 2005),

$$u_{i,m} = \frac{D_i z_i F}{RT} \tag{3}$$

where $u_{i,m}$ = ionic mobility of ion or ionic complex i in free solution (m^2/V-s); D_i = diffusion coefficient of ion or ionic complex i in free solution (m^2/s); z_i = valence of ion or ionic complexes i; F = Faraday constant (96500 C/mole); R = universal gas constant (8.3145 J/K/mole); and T = absolute temperature (K). It is evident from Eq. (3) that the ionic mobility of an ion or ionic complex is of the same polarity as its valence, i.e., the ionic mobility of a cation is positive and that of an anion is negative.

The electromigration velocity of ion or ionic complex i is proportional to its ionic mobility and the electric field imposed on it, i.e.,

$$V_{i,m} = u_{i,m}\nabla(-E) \tag{4}$$

where $V_{i,m}$ = electromigration velocity of ion or ionic complex i in free solution. When the ionic mobility of an ion is positive, the ion migrates in the direction of the electric field, i.e., from the anode towards the cathode or in the direction of decreasing electrical potential. When the ionic mobility of an ion is negative, the ion migrates in the opposite direction of the electric field, i.e., from the cathode towards the anode. However, the effective ionic mobility of ion in soil is much smaller in magnitude than that in free solution as the paths for electromigration in soil are much longer and tortuous than in free solution (Yeung, 1994; Baraud et al., 1997). Assuming the validity of the Nernst-Einstein equation in soil, the effective ionic mobility of ion or ionic complex i can be estimated by (Yeung, 1990; Yeung and Mitchell, 1993)

$$u_{i,m}' = \frac{D_i' z_i F}{RT} \tag{5}$$

where $u_{i,m}'$ = effective ionic mobility of ion or ionic complex i in soil; and D_i' = effective diffusion coefficient of ion or ionic complex i in soil. There are established experimental methods and procedures to measure the effective diffusion coefficient of an ion or ionic complex in soil. As a result, Eq. (5) can be used to estimate the effective ionic mobility of the ion or ionic complex in soil. However, it has been noted by Yeung (2011) that there is no experimental evaluation on the validity of the modified Nernst-Einstein equation in soil given in Eq. (5) to date. Nonetheless, it can be observed in Eqs. (4) and (5) that the effective electromigration velocity of an ion or ionic complex under the influence of a specified dc electric field is proportional to its effective ionic mobility which is proportional to its valence and effective diffusion coefficient in soil.

Chelating agents can extract contaminants sorbed on soil particle surfaces to form soluble complexes and/or change the effective electromigration velocity of the contaminants in soil pore fluid. As a result, they can increase the mobility of the contaminants in soil and facilitate their extraction by electrochemical remediation. However, as the pH of contaminated soil during electrochemical remediation is being changed by the process itself as a function of time and space, the complexation reactions among chelating agents, contaminants, and soil particle surfaces under all these different environmental conditions have to be fully understood before appropriate chelating agents can be selected to enhance the electrochemical remediation process. Moreover, the concentration of the chelating agent has to be properly selected and an appropriate method to inject the chelating agent into the contaminated soil has to be devised.

9.2.3 Electrophoresis

Electrophoresis is the transport of fine charged particles, colloids, or bacteria suspended in the soil pore fluid towards electrodes of the opposite polarity under a dc electric field. As most clay particles and colloids carry negative charges at pHs higher than their points of zero charge (PZC) and vice versa (Sposito, 1998, 2008), they are migrated towards the anode in a high pH environment and towards the cathode in a low pH environment. As a result, contaminants bound to mobile fine particles or colloids can be migrated through soil. The mechanism is also important when surfactants are injected as the processing fluid to form mobile micelles with other species, or when contaminated slurries are to be treated by similar technologies (Acar and Alshawabkeh, 1993).

There are three different forces exerting on these particulates under a dc electric field, including: (1) electrophoretic retardation force; (2) relaxation force; and (3) hydrodynamic friction force (van Olphen, 1977). When these forces are in equilibrium, the particulates move at a constant electrophoretic velocity in free solution. When the electrical gradient is 1 V/m, the velocity of a particulate in a stationary liquid phase is denoted as its electrophoretic mobility (Yeung, 2006a).

The electrophoretic velocity of particulate i is proportional to its electrophoretic mobility and the dc electric field imposed on it, i.e.,

$$V_{i,p} = u_{i,p} \nabla(-E) \tag{6}$$

where $V_{i,p}$ = electrophoretic velocity of particulate i in free solution (m/s); $u_{i,p}$ = electrophoretic mobility of particulate i in free solution (m^2/V-s).

The electrophoretic mobility of particulates provides useful information on the net charges or surface potential of the particulates relative to that of the bulk solution (Hunter, 1981). On the basis of different assumptions on the thickness of diffuse double layer around the particulate, there are three equations relating the electrophoretic mobility of particulates to the zeta potential of particulate surfaces: (1) the Smoluchowski equation; (2) the Hückel equation; and (3) the Henry equation (Henry, 1931),

Smoluchowski equation $$u_{i,p} = \frac{\varepsilon\zeta}{\eta} \tag{7}$$

Hückel equation $$u_{i,p} = \frac{2\varepsilon\zeta}{3\eta} \tag{8}$$

Henry equation $$u_{i,p} = \frac{2\varepsilon\zeta}{3\eta} f_1(\kappa r) \tag{9}$$

where $u_{i,p}$ = electrophoretic mobility of particulate i in free solution (m^2/V-s); ε = permittivity of the solution (C^2/N-m^2); ζ = zeta potential of the particulate surface (V); η = viscosity of the solution (N-s/m^2); $1/\kappa$ = thickness of the diffuse double layer (m); r = radius of the particulate (m); and the function $f_1(\kappa r)$ depends on the size, shape, and electrical conductivity of the particulate; orientation of the particulate relative to the applied dc electric field; and type of electrolyte (Henry, 1931). It is indicated in Eqs. (7)-(9) that when ζ is negative, $u_{i,p}$ is negative. As a result, the electrophoretic velocity of particulates is in the direction of increasing electrical potential, i.e., from the cathode towards the anode, as indicated in Eq. (6). Most zeta potential measurement instruments are in fact measuring the electrophoretic mobility of particulates in free solution, and determine the zeta potential of particulate surfaces from the measured electrophoretic mobility using Eqs. (7)-(9).

The applicability of Eqs. (7)-(9) depends on the thickness of the diffuse double layer relative to the radius of the particulate. The Smoluchowski equation is applicable when κr is larger than 500, i.e., the diffuse double layer is very thin relative to the radius of the particulate. The Hückel equation is applicable when κr is less than 0.1, i.e., the diffuse double layer is very thick relative to the radius of the particulate. The Henry equation is more precise when κr takes an intermediate value (Hunter, 1981). Values of electrophoretic mobility for most clay particles are in the range of 1×10^{-8} to 3×10^{-8} m^2/V-s (Hsu, 1997). As chelating agents can change the zeta potential of suspended fine particles, colloids, or bacteria in soil pore fluid, and the thickness of the diffuse double layer around them, they can change their electrophoretic mobility as indicated in Eqs. (7)-(9).

As the effective electrophoretic mobility of particulates in soil is much smaller in magnitude than that in free solution, as the paths for electrophoresis of particulates in soil are much longer and tortuous than in free solution. The effective electrophoretic velocity of particulates in soil pore fluid can be estimated by

$$V_{i,p}' = u_{i,p}'\nabla(-E) \tag{10}$$

where $V_{i,p}'$ = effective electrophoretic velocity of particulate i in soil pore fluid (m/s); $u_{i,p}'$ = effective electrophoretic mobility of particulate i in soil pore fluid (m^2/V-s). However, there is no well-established method to estimate $u_{i,p}'$ in soil to date.

The phenomenon of electrophoresis is not only useful in the remediation of contaminated soil by injecting bacteria into the soil to enhance biodegradation of the contaminants, it has also found successful applications in other areas of waste management, such as sealing leaks in in-service liquid impoundments non-destructively using bentonite (Yeung et al., 1997a), densification of fine particles in suspensions such as slimes, sludge, and mine tailings (Sprute and Kelsh, 1976, 1979, 1980, 1982; Sprute et al., 1988; Fourie et al., 2007), sedimentation of colloidal contaminants (Sauer and Davis, 1994), dewatering of clay (Shang and Lo, 1997), etc.

9.2.4 Mechanisms of electrochemical remediation

Electrochemical remediation can extract contaminants by the combined mechanisms of electroosmosis, electromigration or ionic migration, and electrophoresis: (1) contaminants can be flushed by electroosmotic advection of soil pore fluid; (2) contaminants carrying charges are migrated towards electrodes of opposite polarity by electromigration or ionic migration; and (3) contaminants sorbed on surfaces of charged particulates and colloids are transported by electrophoresis. The concepts of these three extraction mechanisms are illustrated in Figure 9.1. The technology is applicable to various types of contaminants, including heavy metals, radionuclides, and organics (Yeung, 2006a).

The extraction efficiency of contaminants from soil by electrochemical remediation depends primarily on the mobility of contaminants in soil (Yeung et al., 1997b; Page and Page, 2002; Yeung, 2006a, 2009b). Contaminants in soil may exist in different chemical states depending on environmental conditions and types of co-existing species (Suèr and Allard, 2003). The different chemical states of contaminants can be transformed among themselves reversibly depending on environmental conditions. Only contaminants in mobile phases, such as dissolved phase in soil pore fluid, colloidal phase suspended in soil pore fluid, and/or a mobile immiscible liquid phase co-existing with soil pore fluid, can be extracted by the electrochemical remediation process. It is difficult to extract contaminants that are sorbed on soil particle surfaces or exist as a separate solid phase, such as precipitates in soil pores or on soil particle surfaces, from soil. In many cases, the use of enhancement agents is one of the possible approaches to enhance the extraction efficiency of immobile contaminants by changing the chemical characteristics of contaminants and transforming them to their mobile states.

9.3 Enhancement Agents in Electrochemical Remediation

Enhancement agents have been applied in electrochemical remediation to promote the extraction efficiency of contaminants from fine-grained soils by controlling the soil chemistry to enhance solubilization of the contaminant species and subsequent migration of the solubilized species (Ottosen et al., 2009). The enhancement agents are injected into the contaminated soil to transform contaminants sorbed soil particle surfaces into mobile states before they can be migrated by electroosmotic advection and electromigration towards the electrodes where they are collected for further treatment. In accordance with the enhancement mechanisms, enhancement agents can be categorized into: (1) surfactants and cosolvents; (2) cation solutions; (3) conditioning agents; (4) oxidizing/reducing agents; (5) complexing agents; and (6) chelating agents. As the thesis of this review Chapter is on the use of chelating agents in electrochemical remediation of contaminated soil, the first five categories of enhancement agents are briefly discussed in this section and chelating agents are discussed in detail in a separate section.

9.3.1 Surfactants or Cosolvents

Surfactants are amphiphilic compounds that can lower the surface tension of a liquid, the interfacial tension between two liquids, or the interfacial tension between a liquid and a solid. They may act as detergents, wetting agents, emulsifiers, foaming agents, and dispersants. Surfactants can be used as additives in the phase separation processes during remediation of soils contaminated by organic compounds, so as to enhance the aqueous solubility and mobility of organic contaminants (Chu and Kwan, 2003; Mulligan and Eftekhari, 2003; Fabbri et al., 2009). Moreover, surfactants have been observed by many researchers to be feasible in enhancing heavy metal extraction from soil and sludge (Doong et al., 1998).

Both synthetic surfactants and natural surfactants have been reported to be efficient in mobilizing organic contaminants, such as phenanthrene, polycyclic aromatic hydrocarbons (PAHs), 1,2-dichlorobenzene, hexachlorobenzene, ethylbenzene, trichloroethylene, petroleum hydrocarbons, gasoline, diesel oil, etc., during electrochemical remediation (Kim and Lee, 1999; Ko et al., 2000; Kolosov et al., 2001; Reddy and Saichek, 2003; Saichek and Reddy, 2003a, 2005; Sawada et al., 2003; Yuan and Weng, 2004; Yang et al., 2005; Yuan et al., 2006, 2009; Karagunduz et al., 2007; Park et al., 2007; Chang et al., 2009; Han et al., 2009; Maturi et al., 2009; Alcántara et al., 2010; Gonzini et al., 2010; Jeon et al., 2010; Pham et al., 2010).

Surfactants have been observed to be feasible in enhancing heavy metal extraction from industrial wastewater sludge. The feasibility is probably due to the existence of hydrogen bonding and electrostatic forces between surfactants and metals, and the alteration of surface properties of soil particles by surfactants (Yuan and Weng, 2006). The experimental results of Kaya and Yukselen (2005) indicate that the presence of cationic surfactants increases (become less negative) the zeta potential of soil particle surfaces significantly in an acidic environment, while that of anionic surfactants decreases (becomes more negative) the zeta potential of soil particle surfaces. The variation of zeta potential of soil particle surfaces is of great importance to the remediation process as it does not only reveal the physicochemical reactions at the interface but also influences the precipitation of heavy metals and surfactants during electrochemical remediation. However, the results of using surfactants to enhance the extraction efficiency of metal contaminants by electrochemical remediation are mixed. Some researchers reported positive results (Weng and Yuan, 2001; Yuan and Weng, 2006), while other researchers reported insignificant enhancement (Giannis et al., 2007; Srivastava et al., 2007).

Cosolvent is a second solvent added in small quantity to the primary solvent to form a mixture that may greatly enhance the solvent power of the primary solvent due to synergism. They can enhance the solubility of many organic contaminants through cosolvent effect. Several cosolvents, such as ethanol (Saichek and Reddy, 2003a; Wan et al., 2009), n-butylamine (Li et al., 2000; Reddy et al., 2006; Maturi and Reddy, 2008), n-propanol (Han et al., 2009), acetone (Li et al., 2000), and tetrahydrofuran (Li et al., 2000), have been examined for their ability to enhance the solubilization of organic compounds such as PAHs and diesel oil in soil during the electrochemical remediation process.

9.3.2 Cation Solutions

Natural clay extracts and synthetic solutions containing different concentrations of cations, such as Al^{3+}, Ca^{2+}, Mg^{2+}, K^{+}, and Na^{+}, were used by Coletta et al. (1997) as anodic purging solutions to enhance *in-situ* electrochemical remediation of Pb-contaminated kaolin clay. The polyvalent cations introduced at the anode were supposed to migrate through the soil towards the cathode by the dc electrical field imposed and replace the monovalent sodium or hydrogen ions in the soil. Replacement of monovalent ions by polyvalent ions in soil pore fluid would decrease the thickness of the diffuse double layer around kaolin clay particles, resulting in flocculation of clay particles and increase in hydraulic conductivity of the soil (Hunter, 1981; Yeung, 1992; Mitchell and Soga, 2005). Their experimental results indicate that the introduction of non-toxic cation-enhanced natural or synthetic purging solutions at the anode could decrease the duration required for electrochemical remediation and restore some of the natural soil constituents lost during the electrochemical remediation process. The contaminant extraction efficiency is dependent on the combined influences of the valence and concentration of the cation used in the anodic purging solution, and the ionic strength of the anodic purging solution.

9.3.3 Conditioning Agents

Electrolytic decomposition of electrode fluids during the electrochemical remediation process generates H^{+} ions and OH^{-} ions at the electrodes as follows,

Oxidation at the anode: $$2H_2O - 4e^{-} \rightarrow 4H^{+} + O_2 \uparrow \quad (11)$$

Reduction at the cathode: $$4H_2O + 4e^{-} \rightarrow 4OH^{-} + 2H_2 \uparrow \quad (12)$$

The H^{+} ions generated can migrate across the contaminated soil towards the cathode while the OH^{-} ions generated migrate towards the anode, thus generating a pH gradient in the soil (Acar et al., 1990). The pH gradient developed depends on the acid/base buffer capacity of the soil, i.e., the resistance of the soil to pH changes (Yeung et al., 1996, 1997b; Yeung and Hsu, 2005). A low pH environment developed in the contaminated soil can facilitate solubilization of heavy metals sorbed on soil particle surfaces to enhance the efficiency of electrochemical remediation. However, the electroosmotic flow can be reversed if the soil pH is lower than the point of zero charge (PZC) (Sposito, 1998), resulting in a slower migration velocity of cations towards the cathode and thus a lower remediation efficiency. Moreover, the low soil pH can have detrimental effects on future agricultural uses of the soil after remediation (Yeung, 2006a, 2009b). Therefore, there is a need for electrolyte conditions in the vicinity of the electrodes.

Electrolyte conditioning by introducing controlled quantities of acids or bases at the electrodes at regular time intervals can: (1) eliminate the adverse impacts of electrode reactions such as polarization of electrodes; (2) maintain the pH of electrode fluid within a narrow range favorable for efficient extraction of contaminants in soils of low acid/base buffer capacity; (3) control the electroosmotic flow direction and maintain the electroosmotic volume flow rate to facilitate electroosmotic advection of contaminants; and (4) decrease the energy consumption during electrochemical remediation by reducing the electrical conductivity of soil.

Metal cations, e.g., As^{3+}, As^{5+}, and Cr^{3+}, are stable in acidic environments while oxy-anions, e.g., $H_2AsO_4^-$, $HAsO_4^{2-}$, AsO_4^{3-}, $HCrO_4^-$, $Cr_2O_7^{2-}$, and CrO_4^{2-}, are stable in alkaline environments. The efficiency of acid and base solutions in the extraction of As, Pb, and Zn from mine tailing using the soil washing technique was observed by Yang et al. (2009). As different metal species are soluble in different ranges of pHs, conditioning agents are often utilized to enhance the efficiency of electrochemical remediation in extracting heavy metals from soils by controlling the soil pH. Success of electrochemical remediation of metal-contaminated soils depends heavily on the capability of the technology to keep the metal species in a soluble chemical state by adding acids or bases to the soil, and to elute development of environmental conditions favoring precipitation of contaminant ions and/or sorption of contaminant ions on soil particle surfaces (Page and Page, 2002).

The uses of strong acids or weak acids to enhance desorption of heavy metal cations from soil particle surfaces and migration of the desorbed cations towards the cathode have been investigated by many researchers (Acar and Alshawabkeh, 1993; Sah and Chen, 1998). The use of hydrochloric acid, sulfuric acid, or nitric acid as a neutralizing agent to prevent migration of OH^- ions into the soil from the cathode, thus preventing the precipitation of metals, have been attempted (Sah and Chen, 1998; Marceau et al., 1999; Kim et al., 2001, 2003; Zhou et al., 2003; Reddy and Chinthamreddy, 2004; Al-Shahrani and Roberts, 2005; Zhou et al., 2005a; Wang et al., 2006; Mascia et al., 2007; Buchireddy et al., 2009; Xiu and Zhang, 2009).

Weak acids, including acetic acid or acetate, citric acid, lactic acid, and phosphoric acid, have also been used successfully to depolarize the cathode reactions in electrochemical remediation of heavy metal- and radionuclide-contaminated soils (Acar and Alshawabkeh, 1993; Eykholt and Daniel, 1994; Mohamed, 1996; Puppala et al., 1997; Yang and Lin, 1998; Jackson et al., 2001; Vengris et al., 2001; Weng and Yuan, 2001; Kim et al., 2003; Zhou et al., 2003, 2004a, 2004b, 2005a, 2005b, 2006; Gent et al., 2004; Reddy and Chinthamreddy, 2004; Al-Shahrani and Roberts, 2005; Altin and Degirmenci, 2005; Kornilovich et al., 2005; Liu et al., 2005; Gidarakos and

Giannis, 2006; Jensen et al., 2007; Kim et al., 2009; Gu and Yeung, 2011). Weak acids, such as acetic acid and citric acid, have several advantages over strong acids in electrode conditioning. These advantages include: (1) they are biodegradable and environmentally friendly; (2) they possess certain acid/base buffer capacities so that they can maintain the electrolyte pH to some extent; and (3) they are complexing agents that can form soluble complexes with metals to enhance the solubilization of heavy metals sorbed on soil particle surfaces and to maintain mobility of heavy metals in soil; and (4) the concentration of ions generated by acid dissociation is very low as their pKa values are relatively high, the resulting increase in the electrical conductivity of soil and thus the power consumption are small.

The feasibility of anolyte conditioning using NaOH solution to enhance the extraction of arsenate, fluorine, Cr, and PAHs during electrochemical remediation of mine tailings or contaminated soils have been investigated (Pomès et al., 1999; Saichek and Reddy, 2003; Sawada et al., 2004; Baek et al., 2009). Weak alkaline solutions of certain acid/base buffer capacities, such as Na_2CO_3 and $NaHCO_3$, have also been introduced at the anode to neutralize the hydrogen ions generated by electrolysis of electrode fluid (Hsu, 1997; Zhou et al., 2003; Yeung and Hsu, 2005).

9.3.4 Oxidizing/Reducing Agents

Oxidizing or reducing agents can be injected into contaminated soil during the electrochemical remediation process to enhance the extraction of contaminants or to reduce the toxicity of contaminants through oxidation or reduction reactions. Le Hecho et al. (1998) accelerated the extraction of As and Cr from an industrial site soil and a kaolinite in an alkaline environment using the hydroxide ions generated at the cathode in the presence of an oxidizing agent, sodium hypochlorite NaClO. Thepsithar and Roberts (2006) enhanced the electrochemical remediation of a phenol-contaminated kaolinite using permanganate as the oxidizing agent. Their experimental results reveal that the migration of phenol (by electroosmosis) and permanganate (by electromigration) in opposite directions led to rapid oxidation of the phenol in the contaminated soil specimen and more than 90% of the phenol was extracted after 5 days of electrochemical remediation.

Hexavalent chromium Cr(VI) and trivalent chromium Cr(III) are prevailing species in the subsurface environment. Cr(VI), existing as $HCrO_4^-$, $Cr_2O_7^{2-}$, and CrO_4^{2-}, is much more mobile and toxic than Cr(III). Moreover, Cr(III) precipitates as $Cr(OH)_3$ in the pH range of 6.8 to 11.3 and forms anionic hydroxo complexes, such as $Cr(OH)_4^-$ and $Cr(OH)_3^{2-}$, at higher pHs. Reddy and Chinthamreddy (1999) investigated the transport and fate of Cr(VI) in clayey soils during electrochemical remediation enhanced by three different reducing agents, i.e., humic acid, ferrous iron, and sulfide. Their experimental results reveal that maximum reduction of Cr(VI) to Cr(III) occurs in the presence of sulfide, while the minimum occurs in the presence of humic acid. However, Cr migration in the soil specimen was slow in the presence of sulfide due to the opposite migration directions of Cr(VI) and Cr(III), sorption of Cr(VI) onto soil particle surfaces in the low pH environment in the vicinity of the anode, and sorption and precipitation of Cr(III) in the high pH regions near the cathode. Pamukcu et al. (2004) also studied the feasibility of injecting anolyte containing ferrous iron into Cr-contaminated soil to reduce the toxic and mobile Cr(VI) to the less toxic and less mobile Cr(III) by electrokinetics.

Sodium metabisulfite $Na_2S_2O_5$ is an inexpensive reducing reagent widely used in the treatment of inorganic wastes, such as heavy metals, sulfides and cyanides. It is frequently utilized to reduce the extremely toxic Cr(VI) to the less hazardous Cr(III). As a reducing agent, $Na_2S_2O_5$ provides a versatile and relatively inexpensive alternative to the more costly chelating agents (Chaiyaraksa and Sriwiriyanuphap, 2004). Gidarakos and Giannis (2006) used it as washing and purging solutions in the electrochemical remediation of the Zn-contaminated soil from an agricultural area of Chania, Crete, Greece. The $Na_2S_2O_5$ washing and purging solutions weaken the bonds between the contaminant and soil particle surfaces by the reduction-chelation mechanism between $Na_2S_2O_5$ with Zn. Zinc removal from the soil is probably enhanced by the rapid desorption/dissolution of the strong coordinated Zn-soil particle surface complexes. However, the exact nature of the reduction-chelation mechanism between $Na_2S_2O_5$ with Zn is yet to be understood (Abumaizar and Smith, 1999).

9.3.5 Complexing Agents

In this section, complexing agents are referred to the molecules or ions surrounding a central atom or ion (usually metal) of a coordination complex. A coordination complex or metal complex is a chemical species consisting of a central atom or ion bonded to surrounding molecules or ions, i.e., ligands or complexing agents. The central atom of a coordination complex is usually a metal cation, and various ligands or complexing agents may surround the central atom. The atom within a ligand that is directly bonded to the central atom or ion is called the donor atom. Coordination complexes, in contrast with chelate complexes, are composed of mono-dentate ligands which form only a single bond with the central atom. Chelating agents are organic compounds that can form much more stable chelate complexes with metal ions than complexing agents, as they are capable of forming chelates with metals through two or more atoms of the organic compounds. As chelating agents play a very important role in electrochemical remediation of contaminated soil, a detailed review of their applications in electrochemical remediation is given in a separate section of this review Chapter. However, as the applications of complexing agents in electrochemical remediation of contaminated soil are quite similar to those to chelating agents, a review of their applications is given here.

Certain complexing agents, such as I^-, Cl^-, NH_3^-, and OH^- etc., have been utilized as conditioning acids or bases during the electrochemical remediation process. These ligands can form soluble complexes with metals, such as $[HgI_4]^{2-}$, $[CuCl_2]^-$, $[CuCl_4]^{2-}$, $[Cu(NH_3)_4]^{2+}$, $[Zn(OH)_4]^{2-}$, $[Cr(OH)_4]^-$, $[Cr(OH)_3]^{2-}$, etc., to facilitate extraction of metals from contaminated soil.

Iodide ions can form anionic complexes with Hg in contaminated soil and the resulting complexes can be migrated towards the anode during electrochemical remediation. It is a stronger complexing agent for Hg than chloride. Therefore, it is more promising as an enhancement agent in the electrochemical remediation of Hg-contaminated soil (Suèr and Allard, 2003). It has been demonstrated by many researchers (Cox et al., 1996; Reddy et al., 2003; Suèr and Allard, 2003; Suèr and Lifvergren, 2003; Reddy and Ala, 2005; Shen et al., 2009) that Hg can be extracted efficiently as a soluble complex, i.e., HgI_4^{2-}, by electrochemical remediation enhanced by the injection of potassium iodide or iodine/iodide lixiviant (I_2/I^-) into Hg-contaminated soil. The formation reactions of the soluble Hg complex are as follows (Cox et al., 1996),

$$HgS + I_2 + 2I^- \rightarrow HgI_4^{2-} + S \tag{13}$$

$$Hg(l) + I_2 + 2I^- \rightarrow HgI_4^{2-} \tag{14}$$

$$HgO + 4I^- \rightarrow HgI_4^{2-} + O^{2-} \tag{15}$$

Acetic acid is an important metal complexing agent. It is preferred in electrochemical remediation not only because it can neutralize the products of cathodic reactions to avoid the formation of hydroxide ions so as to reduce the energy consumption of the electrochemical remediation process, but it can also keep the electrolyte pH within a certain range by its acid/base buffer capacity. Moreover, it is relatively cheap, biodegradable, and environmentally safe. Most metal acetates are highly soluble in water, favoring desorption of heavy metals from soil particle surfaces (Gidarakos and Giannis, 2006; Yeung, 2006a). Acetic acid has been applied intensively in electrochemical remediation as an electrode reservoir conditioning agent and purging solution to enhance extraction of heavy metals and radionuclides by many researchers (Acar and Alshawabkeh, 1993; Mohamed, 1996; Puppala et al., 1997; Yang and Lin, 1998; Vengris et al., 2001; Reddy and Chinthamreddy, 2004; Zhou et al., 2004b, 2005a; Al-Shahrani and Roberts, 2005; Altin and Degirmenci, 2005; Kornilovich et al., 2005; Gidarakos and Giannis, 2006).

Lactic acid (2-hydroxypropanoic acid) is a complexing agent that forms anionic complexes with metal ions. The performance of lactic acid as an enhancement agent was evaluated for electrokinetic remediation of Cr-, Cu-, or Cu-Zn-contaminated soils and mine tailings by Zhou et al. (2004a, 2004b, 2005a, 2005b). They considered the extraction efficiency of the metals from soils with the enhancement agent satisfactory.

Another type of complexing agent widely studied in electrochemical remediation is cyclodextrins. Cyclodextrins are cyclic oligosaccharides formed from degradation of starch by bacteria (Wang and Brusseau, 1993; Brusseau et al., 1997). Typical cyclodextrins contain glucose monomers ranging from six to eight units in a ring, creating a cone shape. They are α-cyclodextrin (6 sugar ring molecule), β-cyclodextrin (7 sugar ring molecule), and γ-cyclodextrin (8 sugar ring molecule), as shown in Figure 9.2. Cyclodextrins are non-toxic, biodegradable, and have low affinity of sorption to soil particle surfaces in a wide pH range (Maturi and Reddy, 2006), and they have been used in different soil remediation technologies (Skold et al., 2009). Moreover, they have the ability to form inclusion complexes with many substrates in aqueous solutions.

The viability of using β-cyclodextrin (β-CD), carboxymethyl-β-cyclodextrin, hydroxypropyl-β-cyclodextrin (HPCD), and methyl-β-cyclodextrin (MCD) as enhancement agents in the electrochemical remediation of organic- and heavy metal-contaminated soils and sediments have been investigated. Yuan et al. (2006) observed significant migration of hexachlorobenzene during electrochemical remediation of hexachlorobenzene-contaminated kaolin and natural clayey soil enhanced by β-CD.

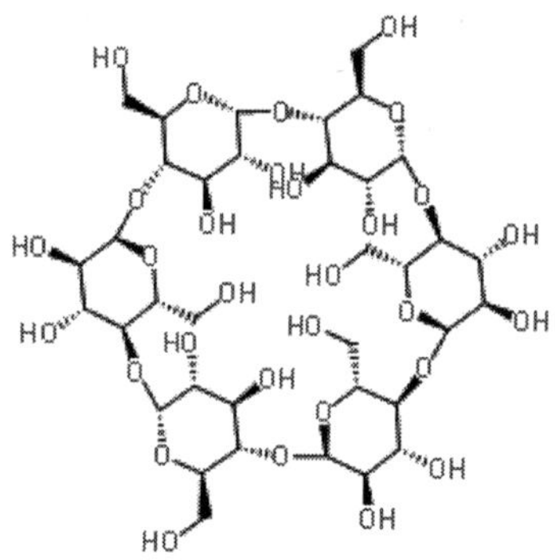

α-cyclodextrin

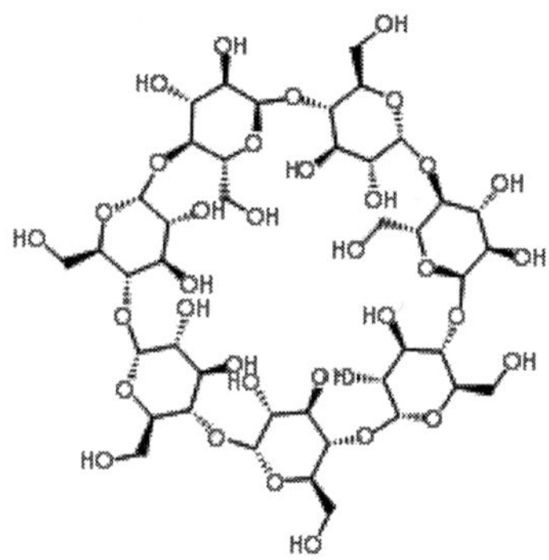

β-cyclodextrin

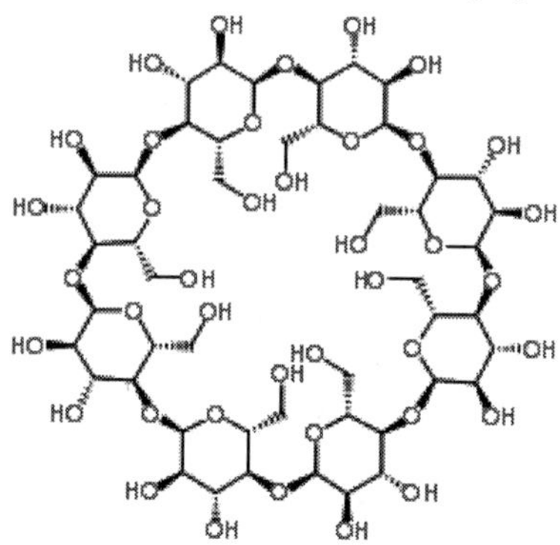

γ-cyclodextrin

Figure 9.2 Chemical structures of the three main types of cyclodextrins.

Reddy and Ala (2005) utilized HPCD as the enhancement agent in bench-scale electrochemical remediation experiments for simultaneous extraction of a variety of heavy metals from contaminated soil. However, no significant extraction of metals from soil was observed in experiments using HPCD. Maturi and Reddy (2006) studied the simultaneous extraction of PAHs and heavy metals from kaolinite by electrochemical remediation enhanced by HPCD. Their results reveal that 10% HPCD solution could cause phenanthrene to migrate towards the cathode, but further migration and extraction were retarded by reduced electric current and electroosmotic flow. Moreover, there was no significant extraction of Ni with the introduction of HPCD. Reddy et al. (2006) examined the feasibility of using HPCD to enhance the solubilization of PAHs in a contaminated manufactured gas plant soil. Their results indicate that PAHs could be effectively solubilized by HPCD-enhanced electrochemical remediation.

Li et al. (2009) conducted pilot-scale electrochemical remediation experiments enhanced by HPCD to extract hexachlorobenzene and zinc from contaminated sediments. Their results indicate that HPCD could be migrated across the sediments by electroosmosis. Significant migration of contaminants was observed but the quantity of contaminant extracted was negligible. Li et al. (2010) evaluated the performance of HPCD-enhanced electrochemical remediation of aged sediment contaminated by Zn, Ni, and hexachlorobenzene in laboratory bench-scale experiments. Their experimental results indicate that both the migration and extraction of hexachlorobenzene were significantly affected by the concentration of HPCD and cumulative electroosmotic flow volume. They also observed that the mobilization of Zn and Ni were not affected by the introduction of HPCD but were heavily dependent on sediment pH. The observation may be attributed to the fact that HPCD cannot complex with heavy metals, such as Cd, Ni, Sr, and Hg, in the presence of various organic contaminants.

Wan et al. (2009) demonstrated experimentally the increase in migration of hexachlorobenzene in sediments with increase in the concentration of MCD used as the enhancement agent, as MCD can solubilize hydrophobic organic chemicals by forming inclusion complexes between hydrophobic organic chemicals and the interior cavity of MCD.

9.4 Use of Chelating Agents in Electrochemical Remediation

Chelation is the formation or presence of two or more separate bonds between a bi-dentate or multi-dentate ligand and a single metal central atom or ion. These ligands are usually organic compounds denoted as chelants, chelators, chelating agents, or sequestering agents. Chelating agents are thus compounds that have the ability to coordinate with central metal atoms or ions at a minimum of two sites to form chelate complexes. The bi-dentate or multi-dentate coordination can typically solubilize or otherwise inactivate these metal ions, so as to reduce any adverse impacts of the metals on the system in which they are being used. The difference between chelate complexes and coordination complexes is that coordination complexes are composed of mono-dentate ligands or complexing agents, thus forming only a single bond with the central metal atom or ion, and chelate complexes are composed of bi-dentate or multi-dentate ligands that bond with metals to form stable and ring-like coordinate complexes called chelates.

The applications of chelating agents span a large number of diverse end-use markets. Important applications include pulp and paper processing; industrial wastewater treatment; household, institutional, and industrial cleaning compounds; metal finishing; agriculture; photography; rubber processing; use in food, pharmaceuticals, cosmetics, and toiletries; textile treatment; etc.

Chelating agents contain two or more ligands that can bond with metals to form stable and ring-like coordinate complexes called chelates. These chelates are usually water soluble. Therefore, metal ions can be desorbed from soil particle surfaces, and kept in a mobile chemical state in the liquid phase. As chelating agents are capable of desorbing metals from soil particle surfaces and preventing them from precipitating, their applications in the enhancement of various *in-situ* and *ex-situ* soil remediation processes, such as soil washing, bioremediation, phytoremediation, soil slurry extraction in reactors, soil heap/column leaching, etc., have been frequently studied (Chen and Hong, 1995; Chen et al., 1995; Hong et al., 1995a, 1995b, 2000; Macauley and Hong, 1995; Hong and Chen, 1996; Bordas and Bourg, 1998; Wu et al., 2003; Chen et al., 2004; Lim et al., 2004; Lestan et al., 2008; Niinae et al., 2008; Komarek et al., 2009, 2010; Mossop et al., 2009; Polettini et al., 2009; Udovic et al., 2009; Udovic and Lestan, 2009, 2010; Barrutia et al., 2010; Cestone et al., 2010; Navarro and Martinez, 2010; Qiu et al., 2010; Soleimani et al., 2010; Zaier et al., 2010).

In the last two decades, different chelating agents have also been used extensively in electrochemical remediation to enhance the extraction of different types of metals and organics from fine-grained soils of high acid/base buffer capacities.

9.4.1 *Types of Chelating Agents Used*

There are four broad categories of organic chelating agents being used or investigated in electrochemical remediation, including: (1) carboxylates; (2) organophosphonates; (3) polyamines; and (4) industrial wastewaters. Among all the chelating agents, amino-polycarboxylates, such as ethylenediaminetetraacetic acid (EDTA) and (diethylenetriamine)-pentaacetic acid (DTPA), and hydroxycarboxylates, such as citric acid, have been used most frequently in electrochemical remediation of contaminated soils. However, other categories of chelating agents, such as organophosphonates and industrial wastewaters, also have high potential to serve as effective enhancement agents for electrochemical remediation of heavy metal-contaminated soils. In fact, their performance has been evaluated experimentally to be promising by the authors of this review Chapter.

9.4.1.1 Carboxylates

Carboxylates are carboxylic acids and salts. Carboxylate chelating agents can be subdivided into several groups: (a) aminopolycarboxylates; (b) hydroxycarboxylic acids and salts; and (c) pyridinecarboxylic acids and salts; on the basis of the different functional groups in their chemical structures. Some of the common carboxylate chelating agents are shown in Figure 9.3 as illustrations (Nowack and VanBriesen, 2005).

Aminopolycarboxylates are aminopolycarboxylic acids and salts. Common examples of aminopolycarboxylic acids and salts include EDTA, DTPA, nitrilotriacetic acids and salts (NTA), ethylene glycol tetraacetic acid (EGTA), hydroxyethyl ethylenediamine triacetic acid (HEDTA), nitrilotris(methylene)triphosphonic acid (NTTA), S-carboxymethylcysteine (SCMC), N-(2-acetamido)iminodiacetic acid (ADA), trimethylenedinitrilote-tetraacetic acid (TMDTA), L-5-glutamyl-L-cysteinylglycine (GCG), etc. They contain several carboxylate groups bonded to one or more nitrogen atoms. They have the ability to form stable and water-soluble complexes with many metal ions by forming one or more heteroatomic rings. The ideal octahedral structures of metal complexes with EDTA and NTA are shown in Figure 9.4 as examples. The presence of basic secondary or tertiary amino groups and the large negative charges of aminopolycarboxylic acids contribute to the high stability of their metal complexes (Bucheli-Witschel and Egli, 2001). During the past few years, more biodegradable or greener aminopolycarboxylate chelating agents have been developed, including glutamic acid, methylglycinediacetic acid, L-aspartic acid N,N-diacetic acid tetrasodium salt, sodium diethanolglycine, 2,2'-[(2-hydroxyethyl)imino]diacetic acid, iminodisuccinic acid and salts (IDSA), ethylenediamine-N,N'-disuccinic acid and salts (EDDS), etc. These aminopolycarboxylate chelating agents are more readily biodegradable than classical aminopolycarboxylates. However, classical aminopolycarboxylates are still the most widely consumed chelating agents worldwide, accounting for 37.8% of consumption in, 2009 (Glauser et al., 2010).

Aminopolycarboxylate chelating agents are widely used in many domestic products and industrial applications to control the solubility and precipitation of metal ions. Some aminopolycarboxylate chelating agents, such as EDTA, DTPA, NTA, and EGTA, have been used in electrochemical remediation of contaminated soils as purging solutions.

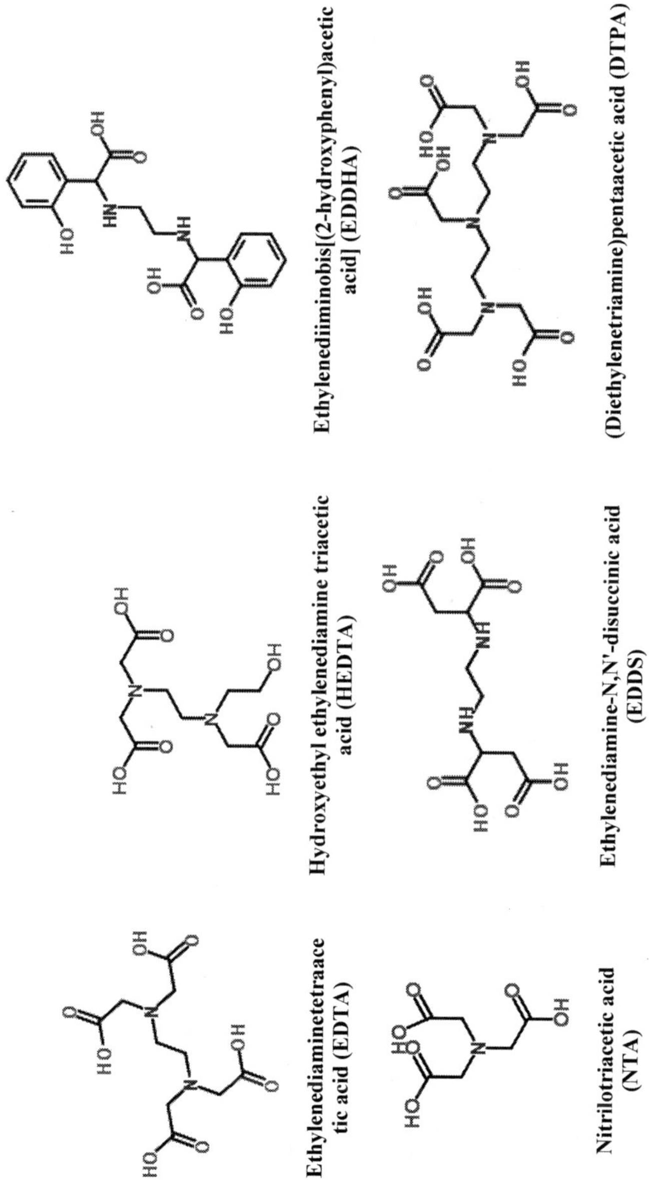

Figure 9.3 Carboxylate chelating agents.

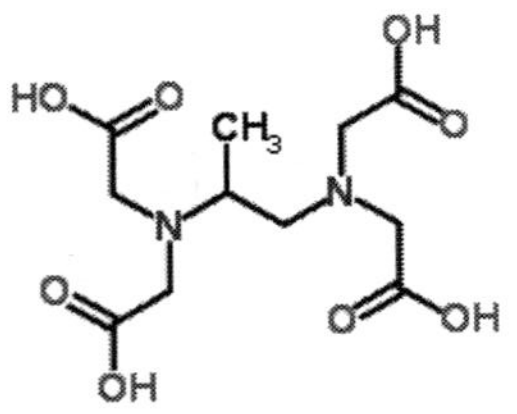

1,2-Diaminopropanetetraacetic acid (PDTA)

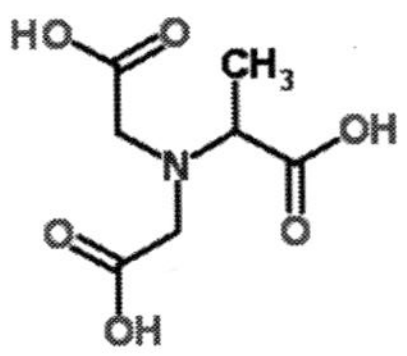

N,N-bis(carboxymethyl)alanine

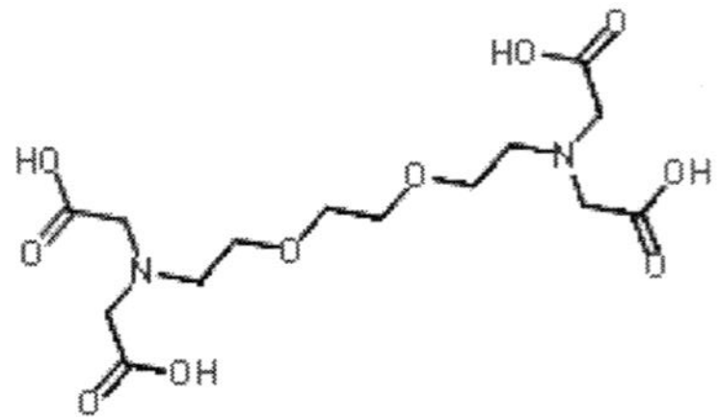

Ethylene glycol tetraacetic acid (EGTA)

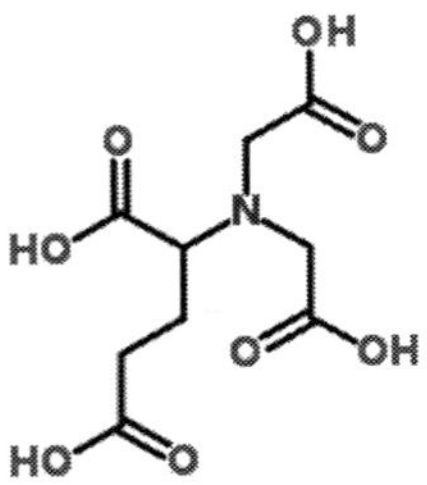

N-(2-Carboxyethyl)iminodiacetic acid

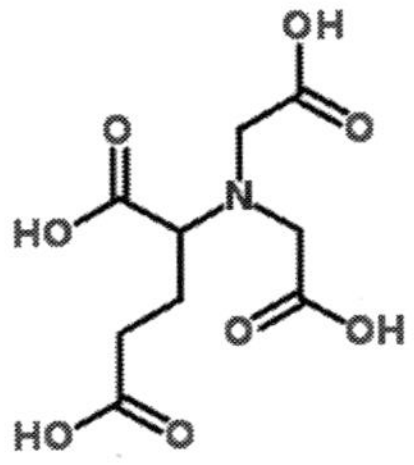

N,N-bis(carboxymethyl)glutamic acid

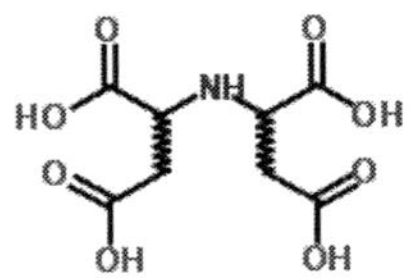

Iminodisuccinic acid (IDSA)

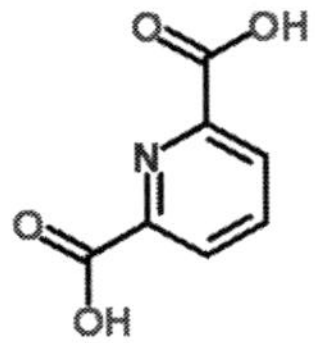

Pyridine-2,6-dicarboxylic acid (PDA)

Figure 9.3 Carboxylate chelating agents (Continued).

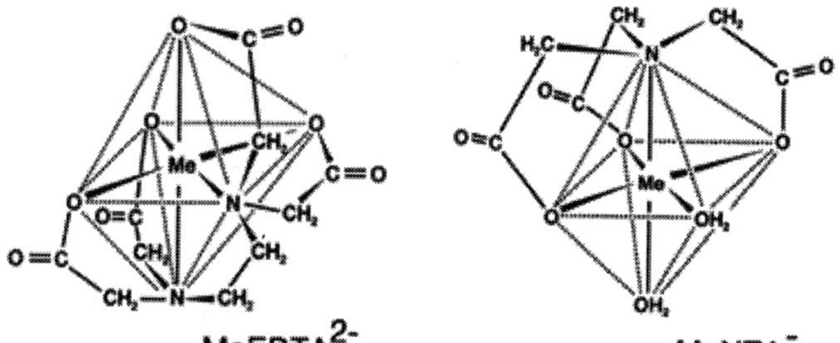

Figure 9.4 Ideal octahedral structures of metal-EDTA and metal-NTA complexes (representing work reported in Bucheli-Witschel and Egli, 2001).

Hydroxycarboxylic acids form a large group of important compounds in many branches of science and technology. They contain the −OH and −COOH functional groups and are capable of forming stable metal complexes. Many hydroxycarboxylic acids, such as citric acid and humic acid, exist in nature. Citric acid is frequently used in extraction of heavy metals in soil remediation techniques including electrochemical remediation and soil washing. Humic acid is a mixture of molecules of heterogeneous aromatic hydroxycarboxylic acids available in black or brown powder or grains. It is soluble in alkaline environments. Humic acids behave as mixtures of dibasic acids, with a pK_1 value of approximately 4 for protonation of carboxyl groups and approximately 8 for protonation of phenolate groups. The presence of carboxylate and phenolate groups gives humic acid the ability to form complexes with bi- and tri-valent metal ions. Many humic acids, such as the typical one shown in Figure 9.5, have two or more of these functional groups which enable their formation of chelate complexes with metals. Oxalic acid is a bicarboxylic acid. It is a reducing agent as well as a chelating agent, and it is widely used in cleaning or bleaching, especially for the extraction of rust.

Pyridinecarboxylic acids are organic compounds with pyridine rings and carboxyl groups. Both the carboxyl group and the nitrogen donor atom in the pyridine ring are coordination sites where metal ions can bond to. Moreover, pyridinecarboxylic acids are biodegradable under field conditions (Banerji and Regmi, 1999). Pyridine-2,6-dicarboxylic acid (PDA) was evaluated as an enhancement agent in the electrochemical remediation of Cd-contaminated soil by Gidarakos and Giannis (2006).

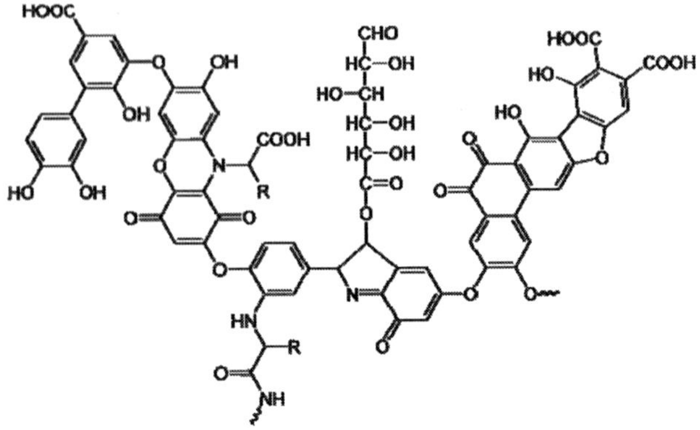

Figure 9.5 Chemical structure of a typical humic acid (representing work reported in Stevenson, 1994).

9.4.1.2 Organophosphonates

Phosphonic acids are anthropogenic complexing compounds containing one or more C–PO(OH)$_2$ or C–PO(OR)$_2$ groups, and phosphonates are the corresponding anions of phosphonic acids. Some of the phosphonate chelating agents are illustrated in Figure 9.6. Phosphonates are highly water-soluble while phosphonic acids are only sparingly soluble. Phosphonates are poorly soluble in organic solvents and not volatile.

Common phosphonates, such as (ethylenedinitrilo)-tetramethylenephosphonic acid (EDTMP), diethylenetriaminepenta(methylenephosphonic) acid (DTPMP), and (nitrilotrimethylene)triphosphonic acid (NTMP), are the structure analogues to the well-known aminopolycarboxylates of EDTA, DTPA, and NTA, respectively as shown in Figure 9.7. These organophosphonates and aminopolycarboxylic acids are the most frequently used chelating agents in the industries. They are effective chelating agents. Their primary properties of inhibiting crystal growth and scale formation and stability under various chemical conditions make them widely used in numerous technical and industrial applications as corrosion inhibitors, scale inhibitors, and laundry detergents (Nowack, 2003). The stability of their metal complexes increases with the number of phosphonic acid groups in their chemical structures. They have strong affinity for metal ions and their stability constants with different bi-valent metal ions are tabulated in Table 9.1. Phosphonates generally have the greatest affinity for metal ions when they exist as fully dissociated species in purified alkaline water. As a result, most measurements of stability constants have been made under such conditions (Gledhill and Feijtel, 1992). It can be observed from the data tabulated in Table 9.1 that phosphonates have similar or even higher affinity for bi-valent metal ions in comparison with EDTA and NTA.

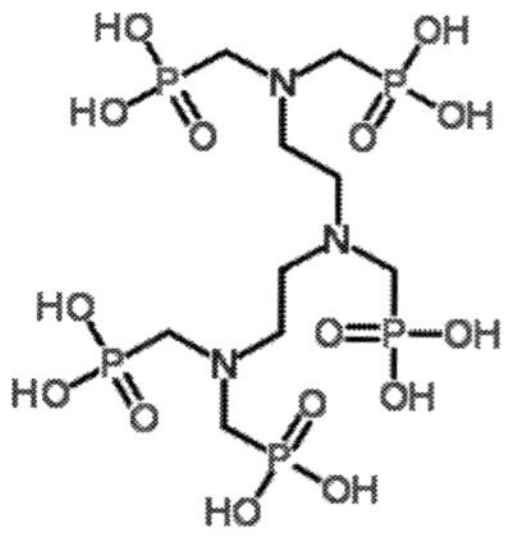

DTPMP

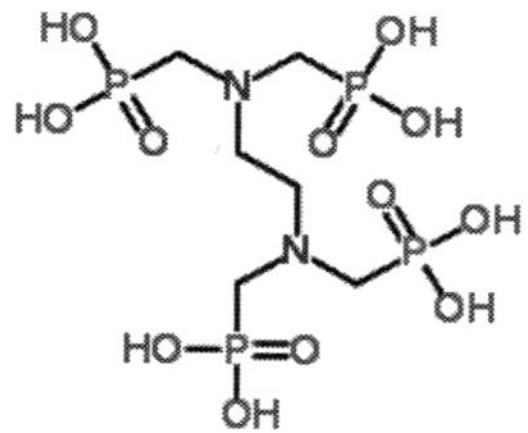

EDTMP

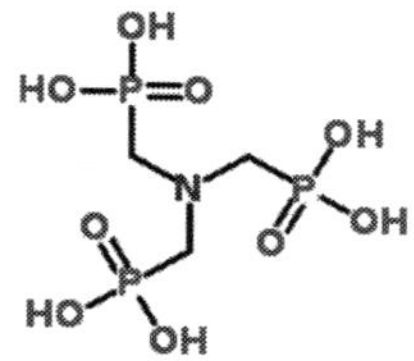

NTMP

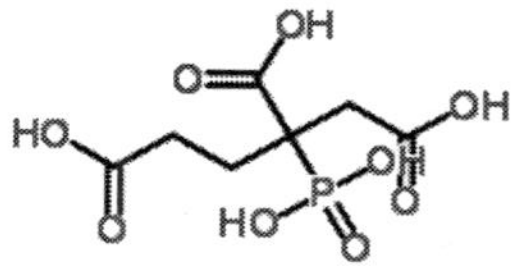

PBTC

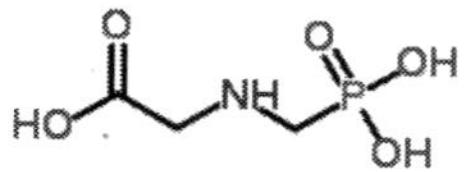

N-(Phosphonomethyl)glycine

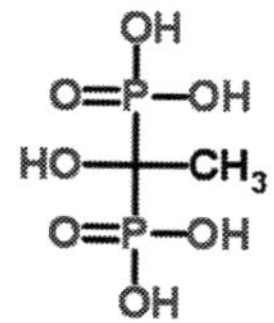

HEDP

DTPMP = Diethylenetriaminepenta(methylenephosphonic) acid; EDTMP = (ethylenedinitrilo)-tetramethylenephosphonic acid; NTMP = (Nitrilotrimethylene)triphosphonic acid; PBTC = 2-Phosphono butane-1,2,4-tricarboxylic acid; HEDP = (hydroxyethylidene)diphosphonic acid

Figure 9.6 Phosphonate chelating agents.

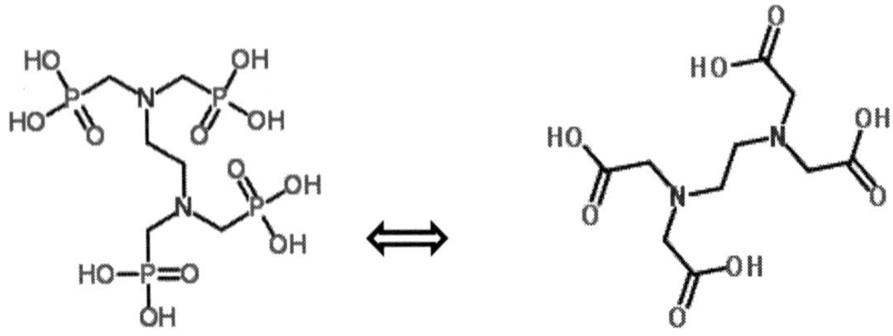

EDTMP **EDTA**

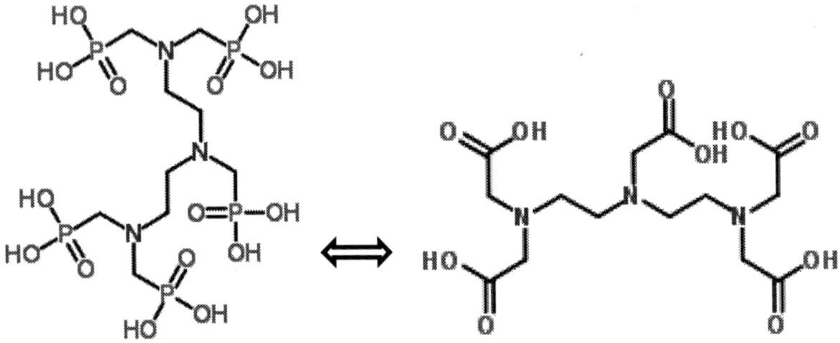

DTPMP **DTPA**

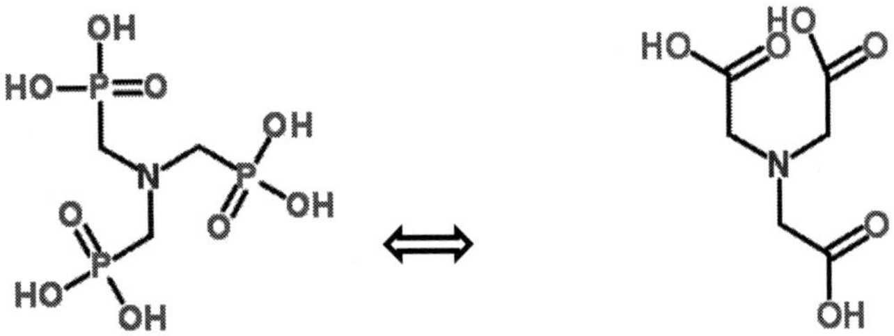

NTMP **NTA**

Figure 9.7 Structure analogues of phosphonic acids (Gu et al., 2009a; with permission from Taylor and Francis).

Table 9.1 Stability constants of phosphonate complexes with bi-valent metal ions

Log K stability constants in alkaline environments

Chelate	**Metal ion**							**Reference**
	Ca^{2+}	Mg^{2+}	Cd^{2+}	Cu^{2+}	Ni^{2+}	Pb^{2+}	Zn^{2+}	
EDTMP	21.4	10	9.6	24.3	20.2	23	21.1	Gledhill and Feijtel (1992)
DTPMP	9.7	6.6	6.7	19.5	19	8.6	19.1	
NTMP	12.7	6.7	7.6	17	15.5	16.4	14.1	
HEDP	15.8	6.2	6.8	18.7	15.8	–	16.7	
EDTA	16.4	8.8	10.6	18.7	18.5	17.9	16.4	Martell and Smith (1974)
NTA	9.8	5.5	6.4	13	11.5	11.3	10.7	

Note: Data for EDTA and NTA are included for comparison.

9.4.1.3 Polyamine

Polyamines, such as ethylenediamine (EDA), 1,3-diaminopropane, hexamethylenediamine, are organic compounds having two or more primary amino groups $-NH_2$ as shown in Figure 9.8. Polyamines are essential for the growth and functioning of normal cells. However, their functions in living organisms are not fully understood (Igarashi and Kashiwagi, 2000). Nonetheless, they are primarily used in the field of medicine (Wang and Casero, 2006; Pegg and Casero, 2011). They also have certain metal chelating ability contributed by the nitrogen donor atoms.

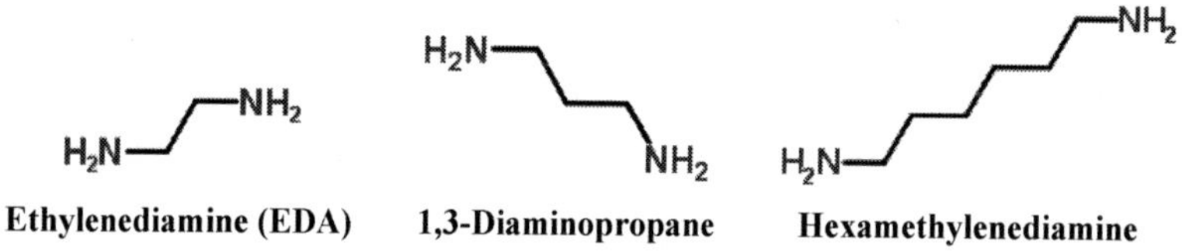

Figure 9.8 Polyamine chelating agents.

9.4.1.4 Non-toxic Industrial Wastewaters

Different industrial wastewaters are being generated from different industrial manufacturing processes in large quantities on a daily basis. Proper and economic handling and disposal of these industrial wastewaters is already a challenging environmental management problem by itself. Depending on the nature of the manufacturing process, the industrial wastewaters may contain different chelating agents as by-products of the manufacturing process. Some non-toxic industrial wastewaters, such as citric acid industrial wastewater (CAIW) and mono-sodium glutamate wastewater, contain organic chelating agents and other non-toxic impurities, and can be used in lieu of commercially available chelating agents in electrochemical remediation. For example, laboratory synthesized CAIW was used as the enhancement agent for electrochemical remediation of a Cd-contaminated natural clay of high acid/base buffer capacity for economic reasons (Gu, 2011; Gu and Yeung, 2011). Reuse of these industrial wastewaters in soil remediation can put a waste product into a productive use.

9.4.2 Use of Carboxylate Chelating Agents in Electrochemical Remediation

Carboxylate chelating agents, such as EDTA, DTPA, PDA, EGTA, NTA, and some weak acids, have been utilized to enhance electrochemical remediation of contaminated soils in many studies. Some of these studies are reviewed in this section.

EDTA is probably one of the most frequently utilized chelating agents for extracting potentially toxic heavy metals from soil due to its strong chelating ability with most toxic metals and universal availability. It is a tetraprotic acid and can be abbreviated as H_4EDTA. Each $EDTA^{4-}$ ion can bond to a metal ion at six different sites since each of the four acetate groups and the two nitrogen atoms have free electron pairs available for coordination bonds, leading to the high stability of metal-EDTA complexes. The configuration of metal-EDTA complexes is shown in Figure 9.9 and the ideal octahedral structure of metal-EDTA complexes is shown in Figure 9.4.

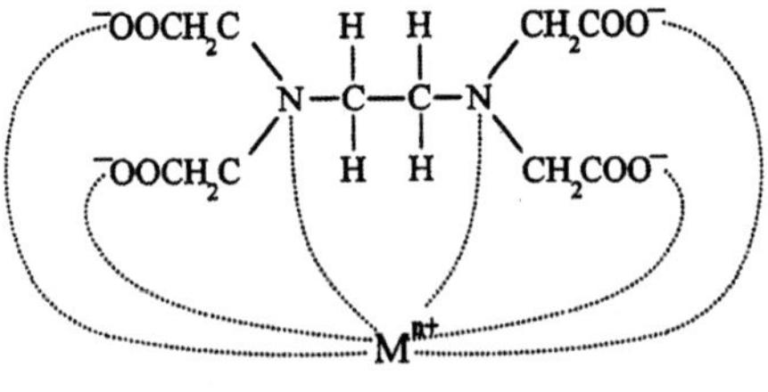

Figure 9.9 Configuration of metal-EDTA complexes (after Yeung et al., 1996).

EDTA can modify the sorption characteristics of Pb and Cd significantly by promoting their desorption from soil particle surfaces and solubilization into the liquid phase at high pHs (Yeung et al., 1996; Yeung and Hsu, 2005). The effects of EDTA on desorption of sorbed Cd from particle surfaces of Milwhite kaolinite, a natural clay of high acid/base buffer capacity, are illustrated in Figure 9.10. Experimental parameters used to produce the results are tabulated in Table 9.2. The results indicate the significant positive impact of EDTA in solubilizing the Cd sorbed on soil particle surfaces when the soil pH is higher than approximately 5. Moreover, the proportion of desorbed Cd increases with EDTA concentration.

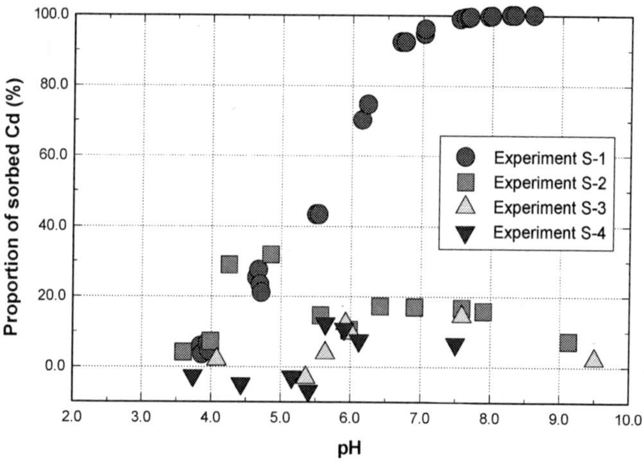

Figure 9.10 Effects of EDTA concentration on Cd sorption on Milwhite kaolinite (after Yeung and Hsu, 2005).

Table 9.2 Experimental parameters for the results presented in Figure 9.10 (after Yeung and Hsu, 2005)

Experiment Number	Concentration of Milwhite kaolinite (kg/L)	Initial concentration of Cd (mg/L)	Background electrolyte	Initial concentration of EDTA (M)
S-1	0.1	26.26	None	0
S-2	0.1	24.74	None	0.01
S-3	0.1	24.74	None	0.05
S-4	0.1	24.74	None	0.1

Yeung et al. (1996) and Yeung and Hsu (2005) also performed bench-scale electrochemical remediation experiments to investigate the effect of injecting EDTA into the contaminated soil on the extraction efficiency of Pb and Cd from kaolinites. Their results indicate that approximately 90% of Pb spiked into the soil could be migrated towards the anode and accumulated within approximately 15% of the total volume of soil for much easier removal after remediation. Similarly, the injection of EDTA from the cathode by a reverse electroosmotic flow could mobilize the Cd spiked in the specimen effectively. When the electroosmotic flow was from the anode towards the cathode, mobilization of Cd was less significant. However, accumulation of Cd near the anode was still observed. Similar phenomenon of heavy metal accumulation in soil specimens was also observed in EDTA-enhanced electrochemical remediation experiments performed by many other researchers (Reddy and Chinthamreddy, 2004; Reddy and Ala, 2005; Gidarakos and Giannis, 2006; Kimura et al., 2007; Gu et al., 2009b). Kimura et al. (2007) reported the accumulation of $Cu(OH)_2$ and Cu-EDTA complex at the pH junction due to isoelectric focusing of EDTA and Me-EDTA as a result of acid dissociation between the acidic and alkaline zones as shown in Figure 9.11.

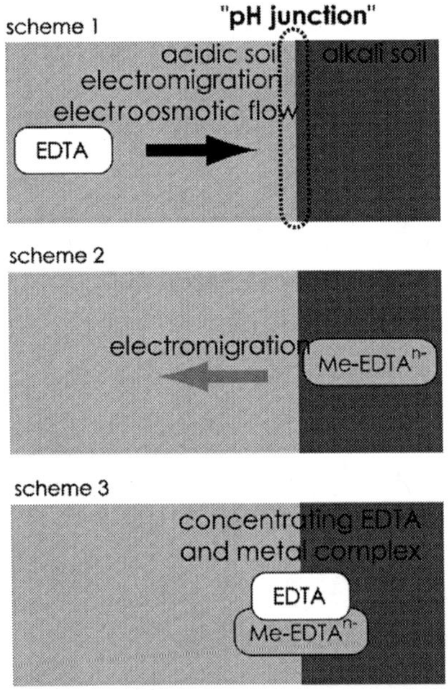

Figure 9.11 Mechanism of Me-EDTA concentration at the pH junction (Kimura et al., 2007; with permission from Elsevier).

There are many studies on EDTA-enhanced electrochemical remediation of metal- or radionuclide-contaminated soils, sludges, sediments, mine tailings, concrete, and cement mortar. Wong et al. (1997) investigated the feasibility of extracting spiked Pb and Zn from sandy soil by injecting EDTA from the cathode. The injected EDTA solubilized metal precipitates, and the resulting soluble complexes were then migrated to the anode and extracted with high efficiencies even in the presence of carbonates and sulfates. Weng and Yuan (2001) reported that the extraction efficiency of total Cr(III) using EDTA as the processing fluid was approximately 10-20% higher than that of tap water, and up to 80-90% of Cr(III) was accumulated near the anode. Zhou et al. (2003, 2005a) remediated a Cu-contaminated kaolinite and mine tailings by conditioning catholyte pH with citric acid and EDTA. Their results reveal that a significant proportion of exchangeable Cu species in the kaolinite specimen can be extracted, indicating the strong complexation ability of EDTA. Significant enhancement in the extraction of Cu from mine tailings using EDTA together with HAc-NaAc as the catholyte was also observed. Amrate et al. (2005) demonstrated experimentally the efficient migration of Pb in a contaminated soil of high acid/base buffer capacity collected near a battery plant towards the anode by EDTA-enhanced electrochemical remediation in the presence of calcite. During the electrochemical remediation process, simultaneous recovery of EDTA and Pb from effluents was performed by inserting cation exchange membranes in an extra compartment (Amrate et al., 2006). The effect of electrochemical remediation on metal fractionation in anaerobic granular sludge spiked with Cu was studied by Virkutyte et al. (2005). It was observed that the introduction of EDTA did not enhance the mobility of Cu from the organic/sulfides and residual fractions. However, it reduced the total Cu content of the sludge. The significant improvement by EDTA on the overall performance of electrochemical remediation of natural contaminated sediment was also demonstrated experimentally by De Gioannis et al. (2008).

However, major disadvantages of EDTA includes (Barona et al., 2001; Gidarakos and Giannis, 2006): (a) it may form stable complexes with a variety of metals in natural soils, such as Ca, Al, Fe, and Mn, and the resulting complexes may sorb onto soil particle surfaces and no longer be available for extraction when their concentrations are low; (b) it is relatively expensive, making the soil remediation excessively costly given tons of soils to be remediated; (c) the non-selective nature of EDTA has limited its use in the remediation of metal-contaminated sites; (d) EDTA resists biodegradation and can be sorbed on soil particle surfaces, therefore, soils remediated by EDTA or similar chelating agents may not be suitable for future agricultural use (Wasay et al., 1998, 2001).

DTPA is an aminopolycarboxylic acid consisting of a diethylenetriamine backbone and five carboxymethyl groups. It has been considered intensively for use in remediation of metal-contaminated soils as its metal chelating ability is stronger than EDTA (Lee and Kao, 2004; Khodadoust et al., 2004, 2005; Giannis et al., 2009). Reddy and Ala (2005) investigated the use of DTPA-enhanced electrochemical remediation to extract heavy metals from soil. Their results reveal that a substantial electroosmotic flow could be induced in soil treated with DTPA and Pb was migrated from the cathode towards the middle of the soil specimen as a result of the formation of anionic lead-DTPA complexes.

EDTA, PDA, EGTA, and NTA have also been utilized as washing solutions prior to electrochemical remediation. Triay et al. (1992) demonstrated experimentally that the extraction efficiency of Pu was increased by more than 23% after the Pu-contaminated soil had been pre-mixed with EDTA. Kornilovich et al. (2005) investigated possible enhancement of electrochemical remediation of clay contaminated with radionuclides, i.e., ^{137}Cs, ^{90}Sr, and U, and Co, by wetting the contaminated soil with EDTA, NTA, and acetic acid prior to electrochemical remediation. Their results reveal that considerably better remediation efficiency was achieved by the use of chelating agents. In another study, Na_2H_2EDTA solution was added to the granular sludge contaminated with $Cu(NO_3)_2$ at the molar ratio of Cu^{2+}:$EDTA^{4-}$ of 1.2:1 to enhance extraction of Cu (Virkutyte et al., 2006). It was observed that the introduction of EDTA at pH 12.5 did not induce any migration of Cu or Fe species due to the formation of insoluble complexes. However, at pH 7.7, both Cu and Fe were migrated exclusively towards the anode by the introduction of EDTA, as a result of the formation of $CuEDTA^{2-}$ and $Fe(III)EDTA^{-}$ complexes. Gidarakos and Giannis (2006) used chelating agents, including EDTA, PDA, and citric acid, as purging solution and/or washing solution to extract Cd and Zn from soil and achieved satisfactory extraction efficiencies. NTA, DTPA, and EGTA of different concentrations were used as washing solutions at different pHs to enhance the extraction of Cd, Pb, and Cu from contaminated soil by electrochemical remediation (Giannis et al., 2009). Their results reveal that the extraction efficiency for Cd is in the order of NTA > EGTA > DTPA, while that for Pb and Cu is in the order of DTPA > NTA > EGTA.

Weak acids have been utilized in many studies of enhanced electrochemical remediation to depolarize the cathode reactions. Some of the weak acids, such as citric acid, humic acid, and oxalic acid, are also chelating agents which can form soluble complexes with metals and prevent their precipitation that may occur upstream of the cathode when chelating agents are not used. In comparison with synthetic chelating agents, such as EDTA, these organic weak acids are biodegradable and thus environmentally safe. However, their major drawback lies in their much weaker chelating abilities for metal ions than those of synthetic chelating agents.

Citric acid is an environmentally safe organic acid which has certain acid/base buffer capacity and can neutralize the product of the electrolysis reactions at the cathode. It can form mono-nuclear, bi-nuclear, or poly-nuclear, and bi-, tri-, and multi-dentate complexes, depending on the type of metal ion available (Gidarakos and Giannis, 2006). It has several advantages as an enhancement agent in soil remediation, such as relatively low cost and low affinity for alkaline and alkaline earth metals (Bassi et al., 2000). It has thus been used intensively as conditioning solutions as well as purging solutions to extract heavy metals and radionuclides from contaminated soils, concrete and waste printed circuit boards by electrochemical remediation. Introduction of citric acid can slightly enhance the mobilization of Cu precipitate near the cathode (Eykholt and Daniel, 1994). The experimental results of Yang and Lin (1998) reveal that the extraction efficiency of Pb from a silt loam soil was higher when citric acid was used as the anode reservoir fluid than when acetic acid and EDTA were used. The results obtained by Weng and Yuan (2001) indicate that an extraction efficiency of approximately 30-40% of total Cr(III) from clay soil was achieved when citric acid was used. However, the extraction efficiency was only approximately, 20% when tap water was used. Kim et al. (2003) reported that the use of citric acid was not efficient in removing U from kaolinite because the direction of electromigration of the negatively charged citrate chelates was opposite to that of advective electroosmotic flow. Reddy and Chinthamreddy (2004) used citric acid to enhance the electrochemical remediation of Ni, Cd, and Cr from a glacial till soil. Their experimental results reveal that a high Cr extraction efficiency of approximately 82% was achieved when citric acid was used as the enhancement agent, but the extraction of Ni and Cd was insignificant. Al-Shahrani and Roberts (2005) reported that the extraction efficiency of Cs enhanced by citric acid was lower than that by stronger nitric and sulfuric acid. They considered that citrate might be able to desorb Cs ions from soil particle surfaces by complexing with Cs ions. However, the resulting complex might be uncharged and thus would not be subject to electromigration. Gent et al. (2004) conducted laboratory bench- and *in-situ* field-scale citric acid-enhanced electrochemical remediation experiments to extract Cr and Cd from contaminated soil. The results of their bench-scale experiments indicate that citric acid could improve the extraction efficiency, especially in the vicinity of the cathode. Citric acid was used as the conditioning solution and chelating agent at the cathode in the field tests. The results indicate that Cr was extracted from 78% of the soil to below background level and Cd was extracted from 70 % of the soil. The results by Ravera et al. (2006) indicate that the extraction of Cu was not significantly enhanced by the introduction of citrates due to the dominant presence of clay minerals and organic matter in the soil used in the study.

Humic acid is a complex mixture of many different acids containing carboxyl and phenolate groups produced by biodegradation of dead organic matter and is widely distributed in the environment. It contains both hydrophilic and hydrophobic moieties, thus serving as surfactants in the dissolution of many hydrophobic organic chemicals. The presence of carboxylate and phenolate functional groups gives humic acid the ability to form complexes with metal ions. Many humic acids have two or more of these functional groups so that they can form very stable chelate complexes with many metal cations such as Fe^{3+}, Al^{3+}, Pb^{2+}, Ca^{2+}, Mn^{2+}, and Mg^{2+} (Gu et al., 1994). They have also been used as the chelating agent to enhance mobilization of heavy metals from soil, sediments, and mine tailings (Wang and Mulligan, 2006, 2009). The application of humic acid as a surfactant in electrochemical remediation of Cd-contaminated soil was evaluated by Giannis et al. (2007). The mechanism of enhancement by humic acid is the formation of weak complexes with Cd in an acidic environment.

PDA is a base containing a nitrogen donor atom and two carboxyl groups. Therefore, it can bond to a metal ion at three coordinating sites. It is able to maintain heavy metals in the liquid phase at pH 7 due to its strong complexing ability (Hong and Chen, 1996). Gidarakos and Giannis (2006) evaluated the effects of using PDA as the purging solution to extract Cd from soil by electrochemical remediation. Their experimental results reveal that more than 95% of Cd could be extracted in a very short time.

9.4.3 Use of Organophosphonate Chelating Agents in Electrochemical Remediation

Efficient re-mobilization of Cu, Cd, and Pb from river sediments was observed by Bordas and Bourg (1998) when the concentrations of EDTA and NTMP used were higher than 0.1 mM in a neutral environment. The solubilizing power of NTMP is only slightly lower than that of EDTA. Nonetheless, NTMP is already very efficient in desorbing Cu from the sediment. Nowack and Stone (1999) reported that the sorption of Cu onto goethite was enhanced in the presence of (hydroxyethylidene)diphosphonic acid (HEDP), NTMP, EDTMP, and DTPMP at pHs below the Cu sorption edge of approximately 5.3. The sorption is attributed to the electrostatic sorption of phosphonates onto the net positive charged surfaces of goethite. However, the sorption decreases substantially in the presence of NTMP, EDTMP, and DTPMP at pHs higher than 5.3 due to the formation of strong Cu-phosphonate complex. In excess of phosphonates, there is a predominance of anionic complexed species in the pH range of non-acidic natural waters. Therefore, there is little chance for such species to be sorbed electrostatically onto natural solids (Deluchat et al., 1997). Hong and Jiang (2005) also studied the effectiveness of NTMP and many other chelating agents for extraction of Pb from contaminated soil. Their results indicate that the complexing effectiveness of NTMP is slightly lower than that of EDTA, reaching the extraction efficiency of Pb of approximately 75% when the NTMP concentration is 50 mM. However, it is much more selective for the target Pb than other chelating agents in the presence of large amounts of Ca ions in the background. Gu et al. (2009a) investigated the effects of four frequently used phosphonates, i.e., HEDP, NTMP, EDTMP, and DTPMP, on desorption of Cd from a natural clay of high acid/base buffer capacity from Shanghai, China in comparison with those of EDTA. Their results indicate that the phosphonates could effectively desorb Cd from the soil. The proportion of Cd desorbed from soil using 0.1 M NTMP in comparison with that of 0.1 M EDTA are shown in Figure 9.12. The results are in general agreement with those of Hong and Jiang (2005).

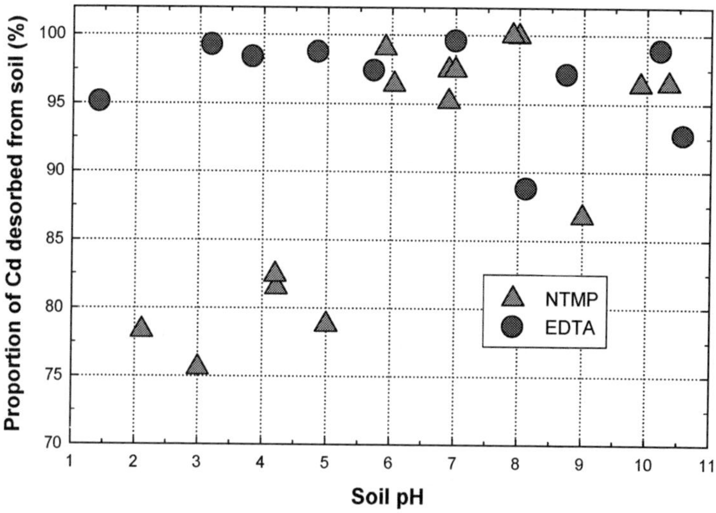

Figure 9.12 Desorption of Cd from the soil using NTMP and EDTA.

Chelating agents, such as phosphonates, EDTA, NTMP, citric acid, and succinic acid, can decrease (becomes more negative) the zeta potential of soil particle surfaces as a function of the specific complex formation ability of the chelating agent (Popov et al., 1997, 2004). Gu et al. (2009a) evaluated the effects of different chelating agents, including HEDP, NTMP, EDTMP, DTPMP, and EDTA, on the zeta potential of a natural clay soil of high acid/base buffer capacity with and without the presence of contaminant, i.e., Cd, in the pH range of 2-11. Their results indicate that introduction of chelating agents to the system decreased (became more negative) the zeta potentials of soil particle surfaces as shown in Figures 9.13 and 9.14, thus enhancing forward electroosmotic flow. The zeta potential of soil particle surfaces is an important factor affecting the extractability of contaminants form soils, as it controls the direction and magnitude of advective electroosmotic flow of dissolved species (Yeung et al., 1997b; Yeung, 2006a, 2009b; Gu et al., 2009a). The enhancement of electroosmotic flow during electrochemical remediation by the introduction of HEDP, EDTA, and citric acid have also been investigated experimentally by other researchers (Popov et al., 1999, 2001; Kolosov et al., 2001). The experimental data obtained also indicate that chelating agents significantly decrease (become more negative) the zeta potential of the clayey soil and increase the forward electroosmotic flow rate, i.e., from the anode towards the cathode. The phenomenon facilitates the extraction of non-ionic organic contaminants such as chlorobenzene, trichloroethylene, and phenol from soil by advective electroosmotic flow.

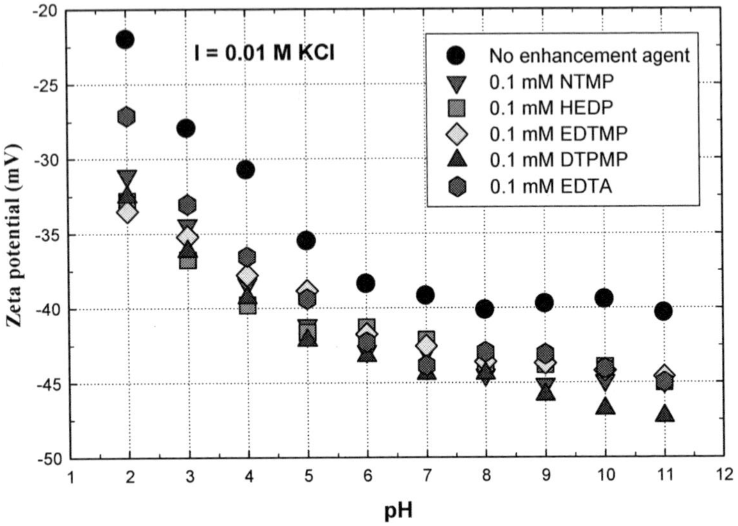

Figure 9.13 Effects of EDTA and phosphonates on zeta potential of a natural clay without contaminant.

Although phosphonates have been investigated as enhancement agents to promote the electroosmotic flow and the migration of organic contaminants in soil, few studies have been conducted on the use of phosphonates as chelating agents to enhance electrochemical remediation of contaminated soil to date. Popov et al. (2005) investigated the effects of several chelating agents, including three phosphonates, i.e., HEDP, NTMP, and DTPMP, and EDTA and citric acid in the enhancement of extraction of uranium U(VI) from the contaminated soil of Oak Ridge K-25 site through batch experiments and bench-scale electrochemical remediation experiments. Their experimental results of batch experiments indicate that the three phosphonates and citric acid were the most effective chelating agents. DTPMP and citric acid were thus selected for their enhanced electrochemical remediation experiments. The DTPMP experiment was abandoned due to the drying of soil near the anode along with a drastic decrease of electric current. However, the introduction of citric acid as an enhancement agent has greatly enhanced the extraction of U from the heavily contaminated Oak Ridge K-25 soil.

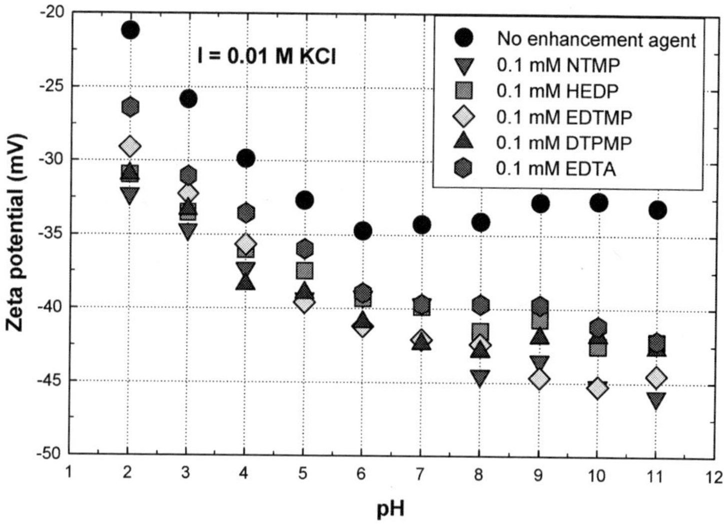

Figure 9.14 Effects of EDTA and phosphonates on zeta potential of a natural clay in the presence of 0.1 mM Cd as contaminant.

Gu (2011) studied the feasibility of using 0.1 M NTMP, EDTMP, and EDTA as the enhancement agent for the extraction of Cd from a natural soil of high acid/base buffer capacity. Hong Kong tap water was used in her control experiment. The resulting distributions of Cd after 5-6 days of electrochemical remediation are shown in Figure 9.15. The introduction of the chelating agent during the electrochemical remediation process could enhance the extraction of Cd significantly as shown in Figure 9.15(a)-(c). However, Hong Kong tap water could not mobilize the spiked Cd in the specimen as shown in Figure 9.15(d). EDTA is a much more efficient chelating agent in migrating Cd from the cathode towards the anode, resulting in accumulation of Cd near the anodes. However, no significant accumulation of Cd was observed at the anodes when phosphonates were used as chelating agents. The extraction efficiencies of Cd enhanced by phosphonates were thus even higher than that by EDTA.

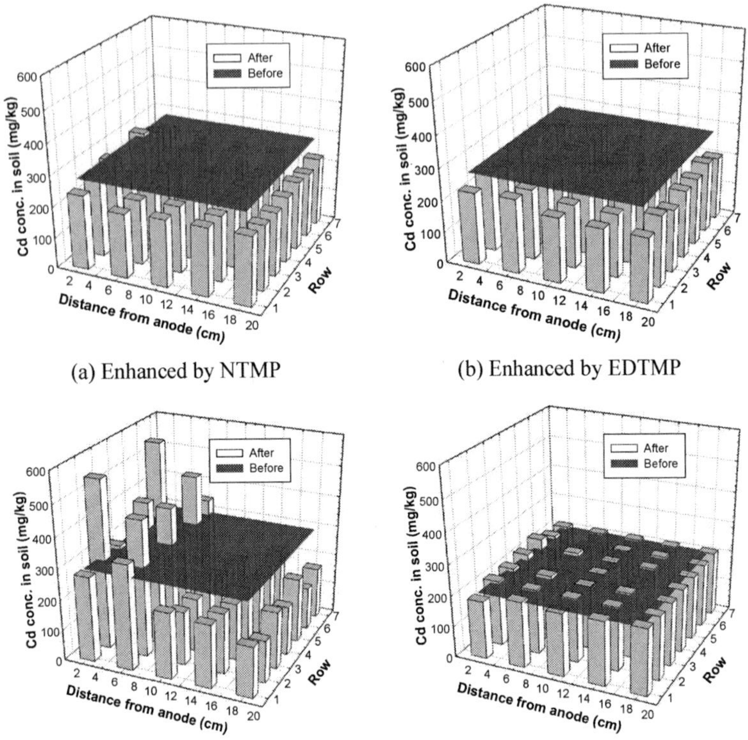

(a) Enhanced by NTMP

(b) Enhanced by EDTMP

(c) Enhanced by EDTA

(d) Enhanced by Hong Kong tap water

Figure 9.15 Cd distribution before and after enhanced electrochemical remediation.

9.4.4 Use of Polyamine Chelating Agents in Electrochemical Remediation

EDA is a strong water-soluble alkaline amine of the chemical formula $C_2H_4(NH_2)_2$. Each EDA molecule provides two nitrogen donor atoms for the formation of cationic complexes with metal ions. It was studied by Wittle and Pamukcu (1993) as the chelating agent in the enhanced electrochemical remediation of Co-contaminated soil. Their experimental results indicate that the introduction of EDA from the anode chamber prevented Co from precipitating and being sorbed onto clay particle surfaces. However, the ionic mobility of the complexes was low and the extraction was primarily provided by advective electroosmotic flow.

9.4.5 Use of Industrial Wastewater Chelating Agents in Electrochemical Remediation

CAIW was proposed by Gu and Yeung (2011) as a potentially cost-effective chelating agent for electrochemical remediation of Cd-contaminated soil of high acid/base buffer capacity. It is a waste liquid of low pH of approximately 4 containing large quantities of citric acid, acetic acid, and other ionic impurities (Wilson et al., 1998). The chemical properties of the laboratory synthesized CAIW to replicate real-life CAIW are tabulated in Table 9.3. Being weak acids and complexing agents, citric acid and acetic acid have been demonstrated to be efficient in enhancing the extraction of metals from soils by electrochemical remediation by many researchers (Acar and Alshawabkeh, 1993; Mohamed, 1996; Puppala et al., 1997; Yang and Lin, 1998; Weng and Yuan, 2001; Vengris et al., 2001; Kim et al., 2003; Gent et al., 2004; Reddy and Chinthamreddy, 2004; Zhou et al., 2003, 2004b, 2005a; Al-Shahrani and Roberts, 2005; Altin and Degirmenci, 2005; Kornilovich et al., 2005; Gidarakos and Giannis, 2006). It has been reported that CAIW can enhance the bioavailability of different Cu forms in contaminated soil (Chen et al., 2004). As CAIW is a recalcitrant wastewater, successful application of CAIW as an enhancing agent in lieu of commercially available citric acid in the electrochemical remediation process will put a waste product into productive use.

Table 9.3 Chemical properties of laboratory synthesized CAIW

Property	Value
pH	3.87
Citric acid (mg/L)	3000
Acetic acid (mg/L)	5500
SO_4^{2-} (mg/L)	2001.1
Ca^{2+} (mg/L)	504.5
Cl^- (mg/L)	984.4
NH_4^+ (mg/L)	45.1
Na^+ (mg/L)	958.9

Gu and Yeung (2011) studied the feasibility of desorbing Cd from a natural clay of high acid/base buffer capacity using synthesized CAIW. Their results indicate that CAIW could effectively enhance desorption of Cd from soil particle surfaces although the proportion of desorbed Cd decreased significantly with increase in soil pH as shown in Figure 9.16. The dominant Cd desorption mechanism of CAIW is the complexion of Cd with the citric acid and acetic acid in CAIW. CAIW can be a promising enhancement agent for electrochemical remediation of Cd-contaminated soils of high acid/base buffer capacity when the environment conditions are favorable.

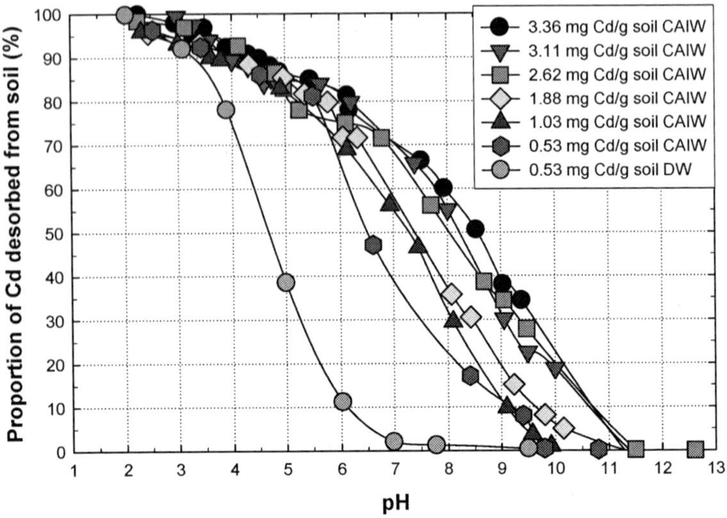

Figure 9.16 Desorption of Cd by CAIW.

Gu (2011) utilized laboratory synthesized CAIW to enhance electrochemical remediation of a Cd-contaminated natural clay. Hong Kong tap water was used in her control experiment. The results of her bench-scale electrochemical remediation experiments indicate that synthesized CAIW could efficiently enhance the solubilization and migration of Cd in the soil specimen while Hong Kong tap water could hardly mobilize the Cd sorbed on soil particle surfaces after approximately 5 days of electrochemical remediation as shown in Figure 9.17. A non-uniform distribution of Cd was observed in the specimen after the CAIW-enhanced electrochemical remediation as shown in Figure 9.17(a). In the vicinity of the anodes, concentrations of Cd in Rows 2, 4 and 6 where electrodes were located decreased from 144 mg Cd/kg soil to approximately 95 mg Cd/kg soil while those in Rows 1, 3, 5, and 7 increased dramatically to approximately 180 mg Cd/kg soil, indicating the existence of localized electrical gradients between anodes and 3-dimensional effects when rod electrodes were used at regular intervals.

Although Cd was observed to be mobilized in the specimen after approximately 5 days of synthesized CAIW-enhanced electrochemical remediation, the extraction efficiency was still very low as the high acid/base buffer capacity of the soil kept the soil pH higher than 7 in most parts of the specimen during the remediation process. Another CAIW-enhanced electrochemical experiment was conducted for 514 hours to evaluate the effects of synthesized CAIW on extraction efficiency of Cd from the soil over a longer remediation period. A high Cd extraction efficiency of 84.7% was achieved by synthesized CAIW-enhanced electrokinetic remediation when the final soil pH was in the range of 4.3-4.7. It can be concluded that synthesized CAIW is a promising and cost-effective enhancement agent for electrochemical remediation.

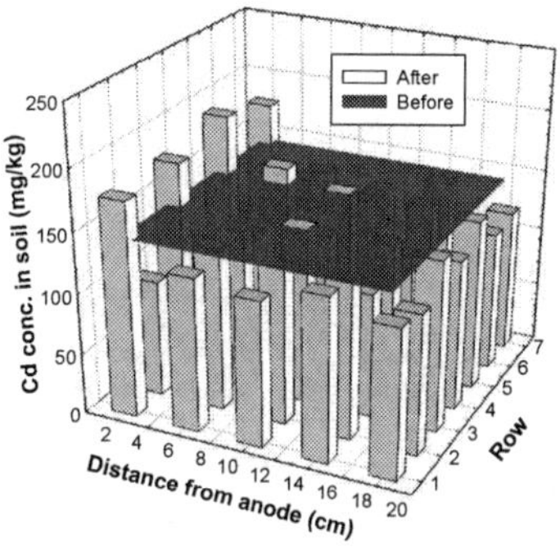

(a) Enhanced by CAIW

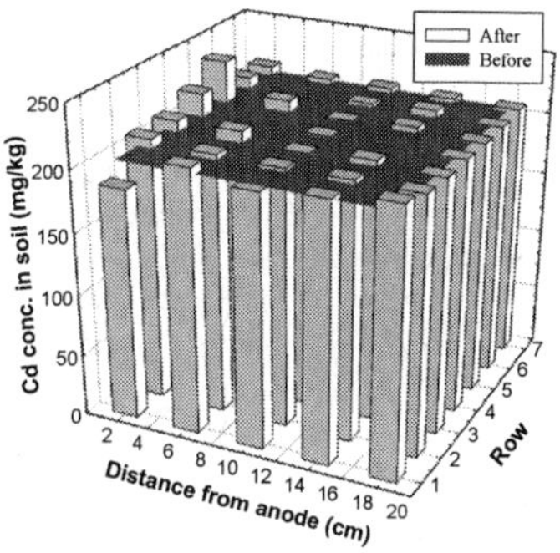

(b) Enhanced by Hong Kong tap water

Figure 9.17 Cd distribution before and after electrochemical remediation.

9.5 Selection of Chelating Agents in Electrochemical Remediation

There are numerous chelating agents on the market available for electrochemical remediation of contaminated soils. Selection of the most suitable chelating agent for the target contaminant under the given environmental conditions and operation constraints is of paramount importance to the success of the electrochemical remediation process. In general, chelating agents to be used in electrochemical remediation have to satisfy these criteria (Hong and Jiang, 2005; Yeung, 2006a):

(1) The chelating agent should have strong extraction strength for the target contaminant, and it should form stable and soluble complexes instead of insoluble salts with the target contaminant within the range of soil pHs during the electrochemical remediation process. The complexation constant K for the formation of the metal-ligand chelate complex can be used as an indication of the stability of the chelate complex composed of the metal ion M and the ligand L. It is defined as:

$$K = \frac{[ML]}{[M][L]} \tag{16}$$

where [M] = concentration of the metal ion; [L] = concentration of the ligand; and [ML] = concentration of the metal-ligand chelate complex. It should be noted that the constant is conditional and different values may be obtained at different pHs. In general, the higher is the value of K or pK, the more stable is the chelate complex, and the higher is the extraction strength of the chelating agent. The complexation constant is also a measure of the preference of a chelating agent for a particular metal ion.

(2) The chelating agent should have extraction selectivity towards the target contaminant. The efficiency of the chelating agent in extracting the target contaminant from the particular soil should be investigated by laboratory-scale batch experiments prior to the performance of bench-scale and/or field-scale experiments on the electrochemical remediation process. Hong et al. (2000) developed the concept of selectivity ratio to quantify the selectivity of a chelating agent for target contaminants. The selectivity ratio is taken to be the ratio of the average complexation constants, i.e., pKs, of the chelating agent for 6 common metal contaminants, i.e., Pb, Cu, Cd, Zn, Ni, and Hg, to that for Ca. The validity of the concept has been established by their experimental results using different chelating agents and different metal contaminants. The concept can also be applied to a particular metal contaminant instead of all six common metal contaminants (Hong and Jiang, 2005).

(3) The chelating agent should have a higher affinity for the target contaminant than soil particle surfaces and it should have a low affinity for soil particle surfaces. Sorption of chelating agents on soil particle surfaces does not only increase the cost of chemicals for the process, but the sorbed chelating agents may also become groundwater pollutants when they are desorbed from soil and form soluble complexes with heavy metals in the groundwater when there are changes in the environmental conditions of the subsurface.

(4) It should form soluble chelates that have low sorption affinity for soil particle surfaces. Only soluble species can be extracted from the contaminated soil by electrochemical remediation. Precipitation of metal chelates and/or sorption of the metal chelates on soil particle surfaces still have the potential to generate secondary pollutants to the environment when there are changes in the environmental conditions of the subsurface, as these soil-contaminant interactions are dynamic and reversible.

(5) It should not solubilize soil minerals excessively, induce any harmful species in the soil pore fluid, or generate toxic residues in the remediated soil during the process of physicochemical or electrochemical reactions, or biodegradation.

(6) It should have the potential to be recovered and reused for several cycles of electrochemical remediation. The recovery and reuse of the used chelating agent can greatly reduce the cost of chemicals, reduce the quantities of waste products to be disposed of, and thus increase the sustainability of the enhanced electrochemical remediation technology. Hong et al. (1995a, 1995b)

and Macauley and Hong (1995) investigated the use of SCMC, ADA, and PDA to extract spiked Cu, Zn, and Pb from soils, respectively. Extracted metals were recovered as oxide, hydroxide, and/or carbonate precipitates by increasing the solution pH. Hong and Chen (1996) also investigated the use of PDA to extract spiked Cd from soil and to recover the extracted Cd by increasing the solution pH. Their results indicate that the chelating agents could extract and release the metals selectively and reversibly, depending on the pH of the system. However, the chelating agent must be relatively stable with respect to biodegradation in the remediation process and during the remediation duration, so that it can be used for at least several consecutive remediation cycles (Hong and Jiang, 2005).

(7) It should be of low toxicity and have a low potential to deteriorate the environment and/or the ecology.

(8) It should be cost-effective.

The proper choice of chelating agents depends on the type of contaminants, soil properties, and environmental conditions, as the soil-contaminant-chelating agent interactions are soil specific, contaminant specific, chelating agent specific, dynamic, and reversible. An improper choice of chelating agent may aggravate the existing contamination problem and make the remediation process much more difficult and costly (Yeung et al., 1996; Yeung, 2006a).

9.6 Effects of Electrochemical Reactions during Electrochemical Remediation on Performance of Chelating Agents

Electrochemical remediation of contaminated soils involves the application of a dc electric field across the soil and utilization of the resulting electrokinetic flow processes, geochemical processes, and electrochemical reactions to remediate the contaminated soil (Yeung, 2009b). Electrolysis of pore fluids at the electrodes is the primary electrochemical reaction during electrochemical remediation, rendering the soil-fluid-contaminant system an electrochemical system (Yeung, 2006b). When a dc field is applied across a wet soil, the products of electrolysis depend on the redox or electrode potentials of the chemicals in the system and the chemical properties of the electrodes (Antropov, 1972). When graphite or most commonly used metals are applied as the electrodes, hydrogen ions and oxygen gas are generated at the anode, and hydroxide ions and hydrogen gas are generated at the cathode as a result of the electrolytic decomposition of water, as described in Eqs. (11) and (12). As hydrogen ions are migrated towards the cathode, an acid front is formed and migrated from the anode towards the cathode. Similarly, an alkaline front is formed at the cathode and migrated towards the anode. As a result, the soil pH varies temporally and spatially during the electrochemical remediation process.

Soil pH is an important parameter influencing the stability of chelate complexes. The logarithm of the conditional complexation constants pKs for various phosphonate-metal ion complexes as a function of pH are tabulated in Table 9.4. It can be observed from the data that phosphonates generally possess greater affinity for metal ions at higher pHs. Moreover, the water solubility of chelate complexes may also changes with pH. Insoluble complexes may form under certain conditions and precipitates on soil particle surfaces or in the soil pore fluid.

The acid/base buffer capacity of soil is also an important factor affecting the performance of chelating agents during electrochemical remediation of contaminated soil. During the electrochemical remediation of soil of low acid/base buffer capacity, such as Georgia kaolinite, soil pH can be decreased to as low as 2 at the anode and increased to approximately 12 at the cathode due to the electrolysis of water at the electrodes and subsequent migration of the acid and alkaline fronts into the soil (Scott, 1994). However, natural soils may have high acid/base buffer capacities which resist soil pH changes. During electrochemical remediation, the high acid/base buffer capacity of the soil may lead to an increase in soil pH at the cathode but practically no decrease of soil pH at the anode without any enhancement as shown in Figure 9.18. As the acid/base buffer capacity of soil affects the variation of soil pH significantly, it should be carefully evaluated prior to the application of electrochemical remediation to extract contaminants from soil (Yeung et al., 1996, 1997b; Yeung and Hsu, 2005).

Table 9.4 Logarithm of conditional complexation constants of Cd-phosphonates complexes pKs as a function of pH (data from Gledhill and Feijtel, 1992)

Chelates \ pH	3	4	5	6	7	8	9	10	11	12
Cd-EDTMP	–	2.2	5.0	8.3	11.8	13.9	15.4	16.5	16.7	16.4
Cd-DTPMP	–	–	–	–	–	4.8	6.0	6.9	7.9	7.4
Cd-NTMP	–	1.9	4.0	5.8	7.5	9.5	11.4	10.0	9.5	7.8
Cd-HEDP	2.6	3.0	4.5	6.1	8.2	10.5	12.1	10.7	–	–

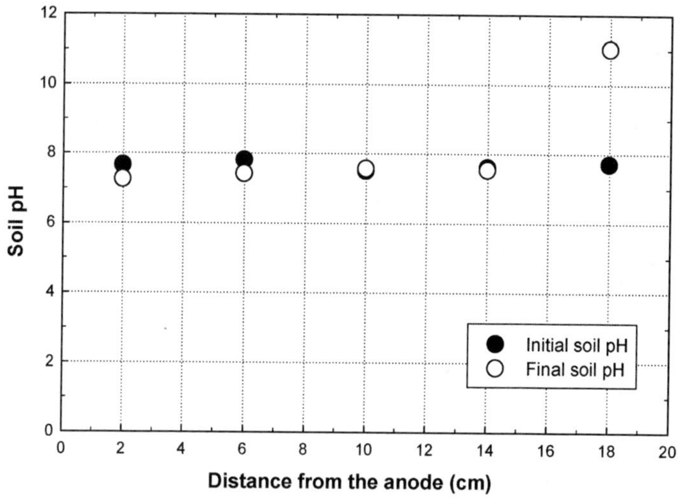

Figure 9.18 Initial and final pH of soil of high acid/base buffer capacity after 115 hours of electrochemical remediation without enhancement (after Gu, 2011).

Changes in soil pH induced by electrochemical reactions may lead to changes in zeta potential of soil particle surfaces. The change in zeta potential of soil particle surfaces influences the direction and magnitude of electroosmotic flow dramatically. The decrease in soil pH generally increases (becomes less negative) the zeta potential of soil particle surfaces and reduces the electroosmotic flow rate. As a result, the migration of positively charged contaminant species and zero charged contaminants, such as organic compounds, towards the cathode are retarded. When the soil pH is lower than the point of zero charge (PZC), the polarity of the zeta potential of soil particle surfaces reverses from negative to positive, resulting in a reverse electroosmotic flow from the cathode towards the anode (Yeung et al., 1996; Yeung and Hsu, 2005; Yeung, 2009b; Gu et al., 2009b). The reverse electroosmotic flow may promote the migration of contaminant species with negative charges, such as most metal-chelating agent complexes, towards the anode. The introduction of chelating agents to soil would decrease (become more negative) the zeta potential and increase the electroosmotic flow rate during electrochemical remediation. The phenomenon affects the performance of chelating agents in enhancing electrochemical remediation of contaminated soil.

The soil temperature is increased by a few to several tens of a degree Celsius, especially at the electrodes (Penn and Savvidou, 1997), during electrokinetic remediation. The increase in temperature may impact the performance of chelating agents by changing the chelating reaction rates and complexation constants of metal complexes.

During electrochemical remediation, complicated electrochemical reactions, electrokinetic flow processes, and geochemical processes, including electroosmotic flow, electromigration, electrophoresis, electrolysis of soil pore water at the electrodes and subsequent migration of hydrogen and hydroxide ions into the soil; change of soil pH; gas generation at electrodes; development of non-uniform electric field; occurrence of reverse electroosmotic flow; changes in electrokinetic properties of soil; hydrolysis; phase change of contaminants; soil-contaminant interactions such as sorption/desorption of contaminants onto/from soil particle surfaces; formation of complexes of contaminants; precipitation of contaminants; etc.; occur simultaneously (Yeung, 2009b). These reactions are contaminant-specific, soil-specific, chelating agent-specific, dynamic, reversible, and interactive among soil, contaminant, and chelating agent. As a result, they may have positive or negative impacts on the performance of chelating agents and extractability of contaminants. The importance of the effects of electrochemical reactions on the performance of chelating agents during electrochemical remediation of contaminated soil cannot be over-emphasized.

9.7 Limitations of the Use of Chelating Agents in Enhancing Electrochemical Remediation

Although very promising results on contaminant extraction efficiencies have been reported for electrochemical remediation enhanced by chelating agents, the use of chelating agents in electrochemical remediation has several limitations pending improvement of the technology:

(1) Some synthetic chelating agents are slowly or hardly biodegradable. As a result, the residual chelating agent may remain in soil for a long time after completion of the remediation process. These residual chelating agents and their metal complexes may cause secondary groundwater pollution (Fischer et al., 1998; Neilson et al., 2003).

(2) Strong chelating agents, such as EDTA and DTPA, have low selectivity for target contaminants and they complex strongly with many metals in soils, including alkaline earth and metal cations. The use of these chelating agents may result in many disturbing physical, chemical, and biological impacts on the soil environment (Bassi et al., 2000), and/or detrimental effects for future agricultural uses of the remediated soil (Wasay et al., 1998, 2001).

(3) As the complexation constant for the complexation reactions is conditional and pH-dependent, the chelating ability of the chelating agent is influenced by the variation of pH during the electrochemical remediation process. In fact, re-sorption of heavy metals onto soil particle surfaces and/or re-precipitation of heavy metals may occur as a result of the change in soil pH during the electrochemical remediation process. Therefore, the sorption characteristics of the metal contaminant in the presence of the chelating agent as a function of pH must be carefully evaluated as suggested by Yeung (2009b). The pH

should be maintained in a range favorable for metal complexion throughout the remediation process, if possible.

(4) The cost of the chelating agent can be a significant factor in the soil remediation operation. The synthetic chelating agents are relatively expensive although they have stronger chelating abilities than natural or quasi-natural chelating agents. Although the extracted metals can be recovered by simply increasing the pH of the system and the chelating agents can be reused in some cases (Hong et al., 1995a, 1995b; Macauley and Hong, 1995; Hong and Chen, 1996), the recovery of the chelating agent, such as EDTA, may require the use of other expensive chemicals and/or demanding technology (Allen and Chen, 1993; Amrate et al., 2006; Lestan et al., 2008). Effective recovery technologies for used chelating agents are still needed to be developed. The use of industrial wastewaters as enhancement agents for electrochemical remediation may be an economical alternative. However, the variation in quality and chemical composition of the industrial wastewater collected from an industrial plant with time should be taken into account carefully, not to mention such variations from industrial plant to industrial plant.

9.8 Future Research Directions

Although many successful, promising, and encouraging chelating agent-enhanced electrochemical remediation studies on different contaminants and of different scales have been conducted, there are still many aspects of the electrochemical processes not fully understood. Therefore, these future research directions in chelating agent-enhanced electrochemical remediation are suggested:

9.8.1 Interactions among Soil, Contaminant, and Chelating Agents

Although the use of chelating agents in electrochemical remediation has been experimentally studied in laboratory bench-scale and/or field pilot-scale experiments by many researchers, the interactions among soil, contaminants, and chelating agents as a function of environmental conditions, such as sorption/desorption characteristics, soil acid/base buffer capacity, and zeta potential of soil particle surfaces, etc. are not fully understand. As these interactions are soil-specific, contaminant-specific, chelating agent-specific, dynamic, reversible, and pH-dependent, they have to be carefully evaluated in the laboratory before full-scale applications of electrochemical remediation of contaminated soil enhanced by chelating agents can take place in the field. A better understanding of these interactions does not only improve the extraction efficiency of electrochemical remediation, but also the extraction efficiency of many other soil remediation technologies.

9.8.2 Sequential Extraction of Multiple Contaminants Using Multiple Chelating Agents

Many sites are contaminated by multiple contaminants, including heavy metals and organic compounds. As many chelating agents are selective and contaminant specific, it is difficult to extract all the contaminants simultaneously using a single enhancement agent. Sequential extraction of multiple contaminants using different chelating agents may be a promising approach in electrochemical remediation. However, the sequence of chelating agents to be applied and the interactions among chelating agents, if any, have to be fully understood to remediate sites contaminated by multiple contaminants by electrochemical remediation and to make the sites useful afterwards.

9.8.3 Recovery of Chelating Agents and Contaminants

Considering the large amounts of different chelating agents required for the remediation of tons of soil contaminated by different heavy metals and organic compounds, efficient and cost-effective recovery methods for used chelating agents and contaminants have to be developed to make the electrochemical remediation technology sustainable.

It has been demonstrated by Allen and Chan (1993) that the liquid stream containing the metal and EDTA resulting from remediation of toxic heavy metal-contaminated soils is amenable to further treatment by electrolysis so that the metal can be separated from EDTA. The extracted metal can be removed for reuse or treated for final disposal by conventional technologies, and the recovered EDTA can be used again for soil remediation. The results of experimental studies on extracting metal contaminants by chelating agents of Hong et al. (1995a, 1995b), Macauley and Hong (1995), and Hong and Chen (1996) indicate that the extracted metals could be recovered by simply increasing the pH of the system as oxide, hydroxide, and/or carbonate precipitates and the chelating agents could be reused in some cases for some specific combinations of chelating agents and metal contaminants. Pociecha and Lestan (2009) studied the feasibility of using an advanced electrochemical oxidation process to treat and reuse the EDTA washing solution used for Cu leaching in a closed loop. A boron-doped diamond anode was used for the generation of hydroxyl radicals and oxidative decomposition of EDTA-metal complexes at a constant electric current density of 40 mA/cm^2. The released Cu was removed from the solution mostly as electro-deposit on the cathode. The feasibility of electrochemical treatment of the EDTA washing solution using Al anode after extraction of Pb (Pociecha and Lestan, 2009) and Cu (Voglar and Lestan, 2010) for reuse in the soil remediation process has been studied. The results indicate the recovered EDTA still possessed the metal extraction potential through several steps of soil extraction and washing solution treatment. However, there was loss of EDTA to soil sorption. Hasegawa et al. (2010) used a selective immobilized marcocyclic material, commonly known as a molecular recognition technology gel, to separate metal ions from wastewater containing large quantities of aminopolycarboxylate chelating agents. Amrate et al.

(2006) used cation-exchange membranes to recover Pb and EDTA in their experimental study on electrochemical remediation of Pb-contaminated soil.

However, some of these recovery technologies are still too expensive for recovery of large quantities of chelating agents. Much research is still needed for the recovery different types of chelating agents and different contaminants so as to minimize the quantities of waste products required for final disposal and the total disposal cost.

9.8.4 Two-dimensional and Three-dimensional Experiments on Electrochemical Remediation

Most laboratory bench-scale or pilot-scale field experiments performed to date are one-dimensional and the electric field imposed on the contaminated soil during electrochemical remediation has been assumed to be uniform. The assumption may be practically valid if plate electrodes are used and boundary effects are neglected. However, it is impractical, if not impossible, to use plate electrodes in real-life field applications. It has been demonstrated experimentally by Gu (2011) as shown in Figures 9.15 and 9.17 that the electric field imposed on the laboratory contaminated soil specimen is not uniform if rod electrodes are used. Therefore, it is evident that 2-dimensional and 3-dimensional electrochemical remediation experiments should be conducted in the laboratory to better simulate field conditions under which non-uniform electric fields are generated. The experimental parameters, including soil pH, electrical conductivity of the soil, zeta potential of soil particle surfaces, redox potential of soil, contaminant distribution, electrical current, and voltage distribution as a function of time and space, should be analyzed to better understand the real-life electrochemical remediation process when the electric field imposed on contaminated soil is non-uniform.

9.8.5 Use of Natural/Natural-equivalent Chelating Agents or Non-toxic Industrial Wastewaters as Chelating Agents

Natural or natural-equivalent chelating agents, such as biosurfactants and amino acids which are naturally produced and biodegradable, are preferred for their environmental compatibility, biodegradability, and relatively low cost. Some biosurfactants, such as rhamnoplipids, surfactin, and sophorolipid, have been demonstrated to be effective in the extraction of heavy metals when used in different soil remediation technologies (Mulligan et al., 2001; Shi et al., 2004; Singh and Cameotra, 2004; Mulligan, 2009). The efficiency of biomass residues containing various amino acids in the extraction of heavy metals from contaminated soils has also been evaluated. The results indicate that they are able to extract many heavy metals, including Cd, Zn, Cu, Ni, Cr and Pb, from soils (Fischer et al., 1994, 1997, 1998; Leidmann et al., 1994, 1995). Some industrial wastewaters from citric acid industries and amino acid industries, etc. are non-toxic and they contain large quantities of organic acids which can function as chelating agents (Gu and Yeung, 2011). These wastewaters may be potential cost-effective sources of chelating agents to enhance electrochemical remediation of metal-contaminated soils. Moreover, effective use of these wastewaters can put them into a productive use.

9.8.6 Large-scale Experiments and Field-scale Applications on Heterogeneous Natural Soils

Laboratory bench-scale electrochemical remediation experiments may not be able to replicate field conditions due to different environmental conditions and scale effects. Moreover, there are many practical aspects of electrochemical remediation have to be addressed in real-life engineering practice (Alshawabkeh et al., 1999; Pazos et al., 2009). Large-scale experiments and field-scale applications of electrochemical remediation enhanced by chelating agents should be performed on contaminated natural soils which are generally heterogeneous and anisotropic. Detailed characterization of the site and delineation of the contamination have to be performed prior to *in-situ* electrochemical remediation to facilitate interpretation of the field data obtained.

9.8.7 Modeling of the Fate and Transport of Contaminants in Chelating Agent-Assisted Electrochemical Remediation Processes

There are many numerical models describing various electrochemical remediation processes with varying degrees of success (Yeung, 1990; Alshawabkeh and Acar, 1992, 1996; Eykholt, 1992, 1997; Datla, 1994; Eykholt and Daniel, 1994; Hicks and Tondorf, 1994; Jacobs et al., 1994; Choi and Lui, 1995; Yeung and Datla, 1995; Denisov et al., 1996; Menon, 1996; Yu and Neretnieks, 1996, 1997; Haran et al., 1997; Liu and Lui, 1997; Ribeiro and Mexia, 1997; Kim, 1998; Jennings and Mansharamani, 1999; Narasimhan and Ranjan, 2000; Shiba et al., 2000; Mattson et al., 2002a, 2002b; Rahner et al., 2002; Rohrs et al., 2002; Kim et al., 2003; Musso, 2003; Oyanader et al., 2003; Vereda-Alonso et al., 2004; Al-Hamdan and Reddy, 2005, 2008; Teutli-Leon et al., 2005; Ribeiro et al., 2005; Lima et al., 2010). However, there are very few models describing the complexing actions of chelating agents during electrochemical remediation. Amrate and Akretche (2005) developed a model for EDTA-enhanced electrochemical remediation of Pb-contaminated soil. More sophisticated models should be developed to describe the fate and transport of contaminants during chelating agent-enhanced electrochemical remediation on the basis of a better understanding of the soil-contaminant-chelating agent interactions and electrochemical phenomena. In addition to all the electrochemical reactions, these models should also take the kinetics of the chelating reactions into account.

9.9 Summary

Electrochemical remediation is a promising soil remediation technology, especially for the fine-grained soil of low hydraulic conductivity and large specific area. During the electrochemical remediation process, electroosmosis, electromigration or ionic migration, and electrophoresis may occur, and contaminants may be extracted from soil by these electrokinetic extraction mechanisms. Moreover, many other electrochemical reactions are occurring simultaneously and these electrochemical reactions may impact the extraction efficiency of electrochemical remediation positively or negatively.

It is always essential to inject various enhancement agents into the soil during the remediation process to solubilize the contaminants sorbed on soil particle surfaces, to mobilize the contaminants in soil, or to reduce the toxicity of contaminants. In general, the enhancement agents used in electrochemical remediation include surfactants or cosolvents, cation solutions, conditioning agents, reducing/oxidizing agents, complexing agents, and chelating agents.

Chelating agents are special types of complexing agents that form stable, cyclic or ring-like complexes with metal ions through bonding between two or more atoms of the chelating agent and the central metal ion. Several types of chelating agents, including aminopolycarboxylates, hydroxycarboxylic acids and salts, pyridinecarboxylic acids and salts, organophosphonates, polyamines, and industrial wastewaters, have been utilized to enhance electrochemical remediation of contaminated soil. Chelating agents used in electrochemical remediation should have strong extraction ability and selectivity for target contaminants, have low affinity for soil particle surfaces, be environmentally safe, have high potential for recovery and reuse, and be of low cost, etc.

The importance of the effects of electrochemical reactions during electrochemical remediation on the performance of chelating agents cannot be over-emphasized. Soil pH is an important factor that influences the metal complexing ability of chelating agents. Electrolysis of water at the electrodes induces temporal and spatial variation of soil pH. Acid/base buffer capacity of soil resists the change of soil pH and influences the chelating ability of the chelating agent indirectly. The zeta potential of soil particle surfaces is affected by the introduction of chelating agents. As the zeta potential of soil particle surfaces influences the direction and magnitude of advective electroosmotic flow, the chelating agent thus impacts the extraction efficiency of electrochemical remediation in addition to its complexing actions. Moreover, the increase of soil temperature during electrochemical remediation affects the performance of chelating agents by increasing the chelating reaction rates and changing the complexation constants.

The soil-contaminant-chelating agent interactions are not fully understood. The chelating agents now in use have some limitations, such as slow biodegradability, non-selectivity, and high cost. However, these limitations may be overcome by further development of the technology. More numerical and experimental studies are thus needed to better understand the chelating agent-enhanced electrochemical process and to select appropriate chelating agents and operational parameters to achieve satisfactory extraction of contaminants from soil by chelating agent-enhanced electrochemical remediation at a reasonable cost.

9.10 Acknowledgments

It is impossible to mention, or even to be conscious of all the people whose work, talk, and publications have affected one's thinking or implanted ideas which have later flourished. Therefore, the authors would thank all the researchers and practitioners working diligently on different aspects of electrochemical remediation and chelating agents, who have contributed selflessly to the literature for the advancement of these technologies to improve the quality of our environment. Without the treasure of literature available, any attempt to conduct a review of the subject will definitely be an impossible mission.

The financial support for the doctoral study of the second author provided by The University of Hong Kong is gratefully acknowledged. The research support provided by Department of Civil Engineering, The University of Hong Kong and the technical support provided by the Technical Support Centre of The University of Hong Kong for the relevant research conducted by the authors are also acknowledged. The sabbatical leave of the first author in California supported by the Faculty of Engineering of The University of Hong Kong to finish the review Chapter is appreciated. Last but the least, the invitation and encouragement of the editors are highly appreciated. Without these support and encouragement, completion of the review Chapter will never be realized.

9.11 References

Abumaizar, R. J., and Smith, E. H. (1999). "Heavy metal contaminants removal by soil washing." *Journal of Hazardous Materials*, 70(1-2), 71-86.

Acar, Y. B., and Alshawabkeh, A. N. (1993). "Principles of electrokinetic remediation." *Environmental Science & Technology*, 27(13), 2638-2647.

Acar Y. B., Gale R. J., Putnam G. A., Hamed J., and Wong R. L. (1990). "Electrochemical processing of soils - Theory of pH gradient development by diffusion, migration, and linear convection." *Journal of Environmental Science & Health Part A*, 25(6), 687-714.

Alcántara, M. T., Gómez, J., Pazos, M., and Sanromán, M. A. (2010). "Electrokinetic remediation of PAH mixtures from kaolin." *Journal of Hazardous Materials*, 179(1-3), 1156-1160.

Al-Hamdan, A. Z., and Reddy, K. R. (2005). "Surface speciation modeling of heavy metals in kaolin: Implications for electrokinetic soil remediation processes."

Adsorption - Journal of the International Adsorption Society, 11(5-6), 529-546.

Al-Hamdan, A. Z., and Reddy, K. R. (2008). "Electrokinetic remediation modeling incorporating geochemical effects." *Journal of Geotechnical & Geoenvironmental Engineering*, ASCE, 134(1), 91-105.

Al-Shahrani, S. S., and Roberts, E. P. L. (2005). "Electrokinetic removal of cesium from kaolin." *Journal of Hazardous Materials*, 122(1-2), 91-101.

Allen, H. E., and Chen, P. H. (1993). "Remediation of metal-contaminated soil by EDTA incorporating electrochemical recovery of metal and EDTA." *Environmental Progress*, 12(4), 284-293.

Alshawabkeh, A., and Acar, Y. B. (1992). "Removal of contaminants from soils by electrokinetics: A theoretical treatise." *Journal of Environmental Science & Health Part A*, 27(7), 1835-1861.

Alshawabkeh, A. N., and Acar, Y. B. (1996). "Electrokinetic remediation. 2. Theoretical model." *Journal of Geotechnical Engineering*, ASCE, 122(3), 186-196.

Alshawabkeh, A. N., Yeung, A. T., and Bricka, M. R. (1999). "Practical aspects of in-situ electrokinetic extraction." *Journal of Environmental Engineering*, ASCE, 125(1), 27-35.

Altin, A., and Degirmenci, M. (2005). "Lead (II) removal from natural soils by enhanced electrokinetic remediation." *Science of the Total Environment*, 337(1-3), 1-10.

Amrate, S., and Akretche, D. E. (2005). "Modeling EDTA enhanced electrokinetic remediation of lead contaminated soils." *Chemosphere*, 60(10), 1376-1383.

Amrate, S., Akretche, D. E., Innocent, C., and Seta, P. (2005). "Removal of Pb from a calcareous soil during EDTA-enhanced electrokinetic extraction." *Science of the Total Environment*, 349(1-3), 56-66.

Amrate, S., Akretche, D. E., Innocent, C., and Seta, P. (2006). "Use of cation-exchange membranes for simultaneous recovery of lead and EDTA during electrokinetic extraction." *Desalination*, 193(1-3), 405-410.

Antropov, L. (1972). *Theoretical electrochemistry*. Mir, Moscow.

Baek, K., Kim, D. H., Park, S. W., Ryu, B. G., Bajargal, T., and Yang, J.S. (2009). "Electrolyte conditioning-enhanced electrokinetic remediation of arsenic-contaminated mine tailing." *Journal of Hazardous Materials*, 161(1), 457-462.

Banerji, S. K., and Regmi, T. P. (1998). "Biodegradation of the chelator 2,6-pyridine dicarboxylic acid (PDA) used for soil metal extraction." *Waste Management*, 18(5), 331-338.

Baraud, F., Tellier, S., and Astruc, M. (1997). "Ion velocity in soil solution during electrokinetic remediation." *Journal of Hazardous Materials*, 56(3), 315-332.

Barona, A., Aranguiz, I., and Elias, A. (2001). "Metal associations in soils before and after EDTA extractive decontamination: Implications for the effectiveness of further cleanup procedures." *Environmental Pollution*, 113(1), 79-85.

Barrutia, O., Garbisu, C., Hernandez-Allica, J., Garcia-Plazaola, J. I., and Becerril, J. M. (2010). "Differences in EDTA-assisted metal phytoextraction between metallicolous and non-metallicolous accessions of Rumex acetosa L." *Environmental Pollution*, 158(5), 1710-1715.

Bassi, R., Prasher, S. O., and Simpson, B. K. (2000). "Extraction of metals from a contaminated sandy soil using citric acid." *Environmental Progress*, 19(4), 275-282.

Bordas, F., and Bourg, A. C. M. (1998). "Effect of complexing agents (EDTA and ATMP) on the remobilization of heavy metals from a polluted river sediment." *Aquatic Geochemistry*, 4(2), 201-214.

Brusseau, M. L., Wang, X. J., and Wang, W. Z. (1997). "Simultaneous elution of heavy metals and organic compounds from soil by cyclodextrin." *Environmental Science & Technology*, 31(4), 1087-1092.

Bucheli-Witschel, M., and Egli, T. (2001). "Environmental fate and microbial degradation of aminopolycarboxylic acids." *Fems Microbiology Reviews*, 25(1), 69-106.

Buchireddy, P. R., Bricka, R. M., and Gent, D. B. (2009). "Electrokinetic remediation of wood preservative contaminated soil containing copper, chromium, and arsenic." *Journal of Hazardous Materials*, 162(1), 490-497.

Cestone, B., Quartacci, M. F., and Navari-Izzo, F. (2010). "Uptake and translocation of CuEDDS complexes by Brassica carinata." *Environmental Science & Technology*, 44(16), 6403-6408.

Chaiyaraksa, C., and Sriwiriyanuphap, N. (2004). "Batch washing of cadmium from soil and sludge by a mixture of $Na_2S_2O_5$ and Na_2EDTA." *Chemosphere*, 56(11), 1129-1135.

Chang, J. H., Qiang, Z. M., Huang, C. P., and Ellis, A .V. (2009). "Phenanthrene removal in unsaturated soils treated by electrokinetics with different surfactants-Triton X-100 and rhamnolipid." *Colloids & Surfaces A - Physicochemical & Engineering Aspects*, 348(1-3), 157-163.

Chen T. C., and Hong A. (1995). "Chelating extraction of lead and copper from an authentic contaminated soil using n-(2-acetamido)iminodiacetic acid and s-carboxymethyl-l-cysteine." *Journal of Hazardous Materials*, 41(2-3), 147-160.

Chen, T. C., Macauley, E., and Hong, A. (1995). "Selection and test of effective chelators for removal of heavy metals from contaminated." *Canadian Journal of Civil Engineering*, 22(6), 1185-1197.

Chen, Y. X., Shi, J. Y., Zhang, W. D., Lin, Q., and Tian, G. M. (2004). "EDTA and industrial waste water improving the bioavailability of different Cu forms in contaminated soil." *Plant & Soil*, 261(1-2), 117-125.

Choi, Y. S., and Lui, R. (1995). "A mathematical-model for the electrokinetic remediation of contaminated soil." *Journal of Hazardous Materials*, 44(1), 61-75.

Chu, W., and Kwan, C. Y. (2003). "Remediation of contaminated soil by solvent/surfactant system." *Chemosphere*, 53(1), 9-15.

Coletta, T. F., Bruell, C. J., Ryan, D. K., and Inyang, H. I. (1997). "Cation-enhanced removal of lead from kaolinitc by electrokinetics." *Journal of Environmental Engineering*, ASCE, 123(12), 1227-1233.

Cox, C. D., Shoesmith, M. A., and Ghosh, M. M. (1996). "Electrokinetic remediation of mercury-contaminated soils using iodine/iodide lixiviant." *Environmental Science & Technology*, 30(6), 1933-1938.

Datla, S. (1994). *A unified theory on electro-kinetic extraction of contaminants*. M.S.

thesis, Department of Civil Engineering, Texas A&M University, College Station, Texas.

De Gioannis, G., Muntoni, A., Polettini, A., and Pomi, R. (2008). "Enhanced electrokinetic treatment of different marine sediments contaminated by heavy metals." *Journal of Environmental Science & Health Part A*, 43(8), 852-865.

Deluchat, V., Bollinger, J. C., Serpaud B., and Caullet C. (1997). "Divalent cations speciation with three phosphonate ligands in the pH-range of natural waters." *Talanta*, 44(5), 897-907.

Denisov, G., Hicks, R. E., and Probstein, R. F. (1996). "On the kinetics of charged contaminant removal from soils using electric fields." *Journal of Colloid & Interface Science*, 178(1), 309-323.

Doong, R. A., Wu, Y. W., and Lei, W. G. (1998). "Surfactant enhanced remediation of cadmium contaminated soils." *Water Science & Technology*, 37(8), 65-71.

Eykholt, G. R. (1992). *Driving and complicating features of electrokinetic treatment of soils*. Ph.D. dissertation, Department of Civil Engineering, University of Texas at Austin, Texas.

Eykholt, G. R. (1997). "Development of pore pressures by nonuniform electroosmosis in clays." *Journal of Hazardous Materials*, 55(1-3), 171-186.

Eykholt, G. R., and Daniel, D. E. (1994). "Impact of system chemistry on electroosmosis in contaminated soil". *Journal of Geotechnical Engineering*, ASCE, 120(5), 797-815.

Fabbri, D., Crime, A., Davezza, M., Medana, C., Baiocchi, C., Prevot, A. B., and Pramauro, E. (2009). "Surfactant-assisted removal of swep residues from soil and photocatalytic treatment of the washing wastes." *Applied Catalysis B - Environmental*, 92(3-4), 318-325.

Fischer, K., Bipp, H. P., Riemschneider, P., Leidmann, P., Bieniek, D., and Kettrup, A. (1998). "Utilization of biomass residues for the remediation of metal-polluted soils." *Environmental Science & Technology*, 32(14), 2154-2161.

Fischer, K., Blay, K., Kotalik, J., Riemschneider, P., Klotz, D., and Kettrup, A. (1997). "Extraction of a copper contaminated soil material by the percolation of an amino acid containing residue hydrolysate .2. Time course of amino acid elution and input/output balance of amino acids." *Zeitschrift Fur Pflanzenernahrung Und Bodenkunde,* 160(5), 511-518.

Fischer, K., Riemschneider, P., Bieniek, D., and Kettrup, A. (1994). "Food engineering residues - amino-acid-composition of hydrolysates and application for the decontamination of metal polluted soils." *Fresenius Journal of Analytical Chemistry*, 350(7-9), 520-527.

Fourie, A. B., Johns, D. G., and Jones, C. J. F. P. (2007). "Dewatering of mine tailings using electrokinetic geosynthetics." *Canadian Geotechnical Journal*, 44(2), 160-172.

Freeze, R. A., and Cherry, J. A. (1979). *Groundwater*. Prentice-Hall, Englewood Cliffs, N.J.

Gent, D. B., Bricka, R. M., Alshawabkeh, A. N., Larson, S. L., Fabian, G., and Granade, S. (2004). "Bench- and field-scale evaluation of chromium and cadmium extraction by electrokinetics." *Journal of Hazardous Materials*, 110(1-3), 53-62.

Giannis, A., Gidarakos, E., and Skouta, A. (2007). "Application of sodium dodecyl sulfate and humic acid as surfactants on electrokinetic remediation of cadmium-contaminated soil." *Desalination*, 211(1-3), 249-260.

Giannis, A., Nikolaou, A., Pentari, D., and Gidarakos, E. (2009). "Chelating agent-assisted electrokinetic removal of cadmium, lead and copper from contaminated soils." *Environmental Pollution*, 157(12), 3379-3386.

Gidarakos, E., and Giannis, A. (2006). "Chelate agents enhanced electrokinetic remediation for removal cadmium and zinc by conditioning catholyte pH." *Water, Air, & Soil Pollution*, 172(1-4), 295-312.

Glauser, J., Blagoev, M., and Kishi, A. (2010). "Chelating agents." *Chemical economics handbook*. SRI Consulting, Menlo Park, California (available at http://www.sriconsulting.com/CEH/Public/Reports/515.5000/).

Gledhill, W. E., and Feijtel, T. C. J. (1992). "Environmental properties and safety assessment of organic phosphonates used for detergent and water treatment applications." *The handbook of environmental chemistry*. Vol. 3, Part F., O. Hutzinger, Editor, Berlin, Springer-Verlag, 261-285.

Gonzini, O., Plaza, A., Di Palma, L., and Lobo, M. C. (2010). "Electrokinetic remediation of gasoil contaminated soil enhanced by rhamnolipid." *Journal of Applied Electrochemistry*, 40(6), 1239-1248.

Gu, B. H., Schmitt, J., Chen, Z. H., Liang, L. Y., and McCarthy, J. F. (1994). "Adsorption and desorption of natural organic matter on iron oxide: Mechanisms and models." *Environmental Science & Technology*, 28(1), 38-46.

Gu, Y.-Y. (2011). *Electrokinetic remediation of cadmium-contaminated natural clay of high buffer capacity*. Ph.D. thesis, Department of Civil Engineering, University of Hong Kong, Hong Kong.

Gu, Y.-Y., and Yeung, A.T. (2011). "Desorption of cadmium from a natural shanghai clay using citric acid industrial wastewater." *Journal of Hazardous Materials*, in press.

Gu, Y.-Y., Yeung, A. T., Koenig, A., and Li, H.-J. (2009a). "Effects of chelating agents on zeta potential of cadmium-contaminated natural clay." *Separation Science & Technology*, 44(10), 2203-2222.

Gu, Y.-Y., Yeung, A. T., and Li H.-J. (2009b). "EDTA-enhanced electrokinetic extraction of cadmium from a natural clay of high buffer capacity." *Advances in Environmental Geotechnics*, Proc., International Symposium on Geoenvironmental Engineering, Hangzhou, Y. Chen, X. Tang, and L. Zhan, Editors, Zhejiang University Press, Hangzhou, 790-798.

Han, H., Lee, Y. J., Kim, S. H., and Yang, J. W. (2009). "Electrokinetic remediation of soil contaminated with diesel oil using EDTA-cosolvent solutions." *Separation Science & Technology*, 44(10), 2437-2454.

Haran, B. S., Popov, B. N., Zheng, G., and White, R. E. (1997). "Mathematical modeling of hexavalent chromium decontamination from low surface charged soils." *Journal of Hazardous Materials*, 55(1-3), 93-107.

Hasegawa, H., Rahman, I. M. M., Kinoshita, S., Maki, T., and Furusho, Y. (2010). "Non-destructive separation of metal ions from wastewater containing excess aminopolycarboxylate chelant in solution with an ion-selective immobilized macrocyclic material." *Chemosphere*, 79(2), 193-198.

Henry, D. C. (1931). "The cataphoresis of suspended particle. Part I: The equation of cataphoresis." *Proceedings of the Royal Society of London Series A*, 133(821), 106-129.

Hicks, R. E., and Tondorf, S. (1994). "Electrorestoration of metal contaminated soils." *Environmental Science & Technology*, 28(12), 2203-2210.

Hong, A. P. K., and Chen, T. C. (1996). "Chelating extraction and recovery of cadmium from soil using pyridine-2,6-dicarboxylic acid". *Water, Air, & Soil Pollution*, 86, 335–346.

Hong, A., Chen, T. C., and Okey, R. W. (1995a). "Chelating extraction of copper from soil using s-carboxymethylcysteine." *Water Environment Research*, 67(6), 971-978.

Hong, A. P., Chen, T. C., and Okey, R. W. (1995b). "Chelating extraction of zinc from soil using N-(2-acetamido)iminodiacetic acid." *Emerging technologies in hazardous waste management V*. D.W. Tedder, and F.G. Pohland, Editors, ACS Symposium Series 607, American Chemical Society, Washington, D.C., 210-223.

Hong, P. K. A., and Jiang, W. (2005). "Factors in the selection of chelating agents for extraction of lead from contaminated soil: Effectiveness, selectivity, and recoverability." *Biogeochemistry of chelating agents*. B. Nowack, and J.M. VanBriesen, Editors, ACS Symposium Series 910, American Chemical Society, Washington, D.C., 421-432.

Hong, P. K. A., Li, C., Jiang, W. M., Chen, T. C., and Peters, R. W. (2000). "Chelating agents for extraction of heavy metals from soil - Complexing power, selectivity, and recoverability." *Emerging technologies in hazardous waste management 8*. D.W. Tedder, and F.G. Pohland, Editors, Kluwer Academic/Plenum Publishers, New York, N.Y., 9-20.

Hsu, C. (1997). *Electrokinetic remediation of heavy metal contaminated soils*. Ph.D. thesis, Department of Civil Engineering, Texas A&M University, College Station, Texas.

Hunter, R. J. (1981). *Zeta potential in colloid science: Principles and applications*. Academic press, London.

Igarashi, K., and Kashiwagi, K. (2000). "Polyamines: Mysterious modulators of cellular functions." *Biochemical & Biophysical Research Communications*, 271(3), 559-564.

Jackson, W. A., Jeon, M. A., Pardue, J. H., and Anderson, T. (2001). Enhanced natural attenuation of perchlorate in soils using electrokinetic injection. *Bioremediation of inorganic compounds*. Proc., 6th International In Situ and On-Site Bioremediation Symposium, San Diego, Vol. 9, 273-280.

Jacobs, R. A., Senguin, M. Z., Hicks, R. E., and Probstein, R. F. (1994). "Model and experiments on soil remediation by electric fields." *Journal of Environmental Science & Health Part A*, 29(9), 1933-1955.

Jennings, A. A., and Mansharamani, P. (1999). "Modeling electrokinetically-enhanced aggregate remediation." *Environmental Modelling & Software*, 14(6), 625-634.

Jensen, P. E., Ahring, B. K., and Ottosen, L. M. (2007). "Organic acid enhanced electrodialytic extraction of lead from contaminated soil fines in suspension." *Journal of Chemical Technology & Biotechnology*, 82(10), 920-928.

Jeon, C. S., Yang, J. S., Kim, K. J., and Baek, K. (2010). "Electrokinetic removal of petroleum hydrocarbon from residual clayey soil following a washing process." *Clean-Soil Air Water*, 38(2), 189-193.

Karagunduz, A., Gezer, A., and Karasuloglu, G. (2007). "Surfactant enhanced electrokinetic remediation of DDT from soils." *Science of the Total Environment*, 385(1-3), 1-11.

Kaya, A., and Yukselen, Y. (2005). "Zeta potential of soils with surfactants and its relevance to electrokinetic remediation." *Journal of Hazardous Materials*, 120(1-3), 119-126.

Khodadoust, A. P., Reddy, K. R., and Maturi, K. (2004). "Removal of nickel and phenanthrene from kaolin soil using different extractants." *Environmental Engineering Science*, 21(6), 691-704.

Khodadoust, A. P., Reddy, K. R., and Maturi, K. (2005). "Effect of different extraction agents on metal and organic contaminant removal from a field soil." *Journal of Hazardous Materials*, 117(1), 15-24.

Kim, G. (1998). *Two-dimensional numerical modeling of electrokinetic extraction of contaminants from fine-grained soils*. PhD dissertation, Department of Civil Engineering, Texas A&M University, College Station, Texas.

Kim, G. N., Yang, B .I., Choi, W. K., and Lee, K. W. (2009). "Development of vertical electrokinetic-flushing decontamination technology to remove Co-60 and Cs-137 from a Korean nuclear facility site." *Separation & Purification Technology*, 68(2), 222-226.

Kim, J., and Lee, K. (1999). "Effects of electric field directions on surfactant enhanced electrokinetic remediation of diesel-contaminated sand column." *Journal of Environmental Science & Health Part A*, 34(4), 863-877.

Kim, S. O., Kim, J. J., Yun, S. T., and Kim, K. W. (2003). "Numerical and experimental studies on cadmium(II) transport in kaolinite clay under electrical fields." *Water, Air, & Soil Pollution*, 150(1-4), 135-162.

Kim, S. O., Moon, S. H., and Kim, K. W. (2001). "Removal of heavy metals from soils using enhanced electrokinetic soil processing." *Water, Air, & Soil Pollution*, 125(1-4), 259-272.

Kimura, T., Takase, K. I., and Tanaka, S. (2007). "Concentration of copper and a copper-EDTA complex at the pH junction formed in soil by an electrokinetic remediation process." *Journal of Hazardous Materials*, 143(3), 668-672.

Ko, S. O., Schlautman, M. A., and Carraway, E. R. (2000). "Cyclodextrin-enhanced electrokinetic removal of phenanthrene from a model clay soil." *Environmental Science & Technology*, 34(8), 1535-1541.

Kolosov, A. Y., Popov, K. I., Shabanova, N. A., Artemeva, A. A., Kogut, B. M., Frid, A. S., Zelvenskii, V. Y., and Urinovich, E. M. (2001). "Electrokinetic removal of hydrophobic organic compounds from soil." *Russian Journal of Applied Chemistry*, 74(4), 631-635.

Komarek, M., Vanek, A., Mrnka, L., Sudova, R., Szakova, J., Tejnecky, V., and Chrastny, V. (2010). "Potential and drawbacks of EDDS-enhanced phytoextraction of copper from contaminated soils." *Environmental Pollution*, 158(7), 2428-2438.

Komarek, M., Vanek, A., Szakova, J., Balik, J., and Chrastny, V. (2009). "Interactions of EDDS with Fe- and Al-(hydr)oxides." *Chemosphere*, 77(1), 87-93.

Kornilovich, B., Mishchuk, N., Abbruzzese, K., Pshinko, G., and Klishchenko, R. (2005). "Enhanced electrokinetic remediation of metals-contaminated clay." *Colloids & Surfaces A – Physicochemical & Engineering Aspects*, 265(1-3), 114-123.

Le Hecho, I., Tellier, S., and Astruc, M. (1998). "Industrial site soils contaminated with arsenic or chromium: Evaluation of the electrokinetic method." *Environmental Technology*, 19(11), 1095-1102.

Lee, C. S., and Kao, M. M. (2004). "Effects of extracting reagents and metal speciation on the removal of heavy metal contaminated soils by chemical extraction." *Journal of Environmental Science & Health Part A*, 39(5), 1233-1249.

Leidmann, P., Fischer, K., Bieniek, D., and Kettrup, A. (1995). "Chemical characterization of silage effluents and their influence on soil bound heavy-metals." *International Journal of Environmental Analytical Chemistry*, 59(2-4), 303-316.

Leidmann, P., Fischer, K., Bieniek, D., Nusslein, F., and Kettrup, A. (1994). "Removal of heavy-metals from polluted soil with grass-silage juice." *Chemosphere*, 28(2), 383-390.

Lestan, D., Luo, C. L., and Li, X. D. (2008). "The use of chelating agents in the remediation of metal-contaminated soils: A review." *Environmental Pollution*, 153(1), 3-13.

Li, A., Cheung, K. A., and Reddy, K. R. (2000). "Cosolvent-enhanced electrokinetic remediation of soils contaminated with phenanthrene." *Journal of Environmental Engineering*, ASCE, 126(6), 527-533.

Li, T. P., Yuan, S. H., Wan, J. Z., Lin, L., Long, H. Y., Wu, X. F., and Lu, X. H. (2009). "Pilot-scale electrokinetic movement of HCB and Zn in real contaminated sediments enhanced with hydroxypropyl-beta-cyclodextrin." *Chemosphere*, 76(9), 1226-1232.

Li, T. P., Yuan, S. H., Wan, J. Z., and Lu, X. H. (2010). "Hydroxypropyl-beta-cyclodextrin enhanced electrokinetic remediation of sediment contaminated with HCB and heavy metals." *Journal of Hazardous Materials*, 176(1-3), 306-312.

Lim, J. M., Salido, A. L., and Butcher, D. J. (2004). "Phytoremediation of lead using Indian mustard (Brassica juncea) with EDTA and electrodics." *Microchemical Journal*, 76(1-2), 3-9.

Lima, A. T., Ribeiro, A. B., Rodriguez-Maroto, J. M., Mateus, E. P., Castro, A. M., and Ottosen, L. M. (2010). "Experimental and modeling of the electrodialytic and dialytic treatment of a fly ash containing Cd, Cu and Pb." *Journal of Applied Electrochemistry*, 40(9), 1689-1697.

Liu, B. Y., and Lui, R. (1997). "Electrokinetic remediation of contaminated soil." *Nonlinear Analysis - Theory Methods & Applications*, 30(6), 3391-3398.

Liu, S. H., Wang, H. P., Yu, M. Y., Huang, Y. J., and Wang, H. C. (2005). "Speciation of copper in a contaminated soil during H3PO4-assisted EKR." *Journal of Electron Spectroscopy & Related Phenomena*, 144, 311-314.

Macauley, E., and Hong, A. (1995). "Chelation extraction of lead from soil using pyridine-2,6-dicarboxylic acid." *Journal of Hazardous Materials*, 40(3), 257-270.

Marceau, P., Broquet, P., and Baticle, P. (1999). "Electrokinetic remediation of cadmium-spiked clayey medium. Pilot test." *Earth & Planetary Sciences*, 328(1), 37-43 (In French).

Martell, A. E., and Smith, R. M. (1974). *Critical stability constants, Vol. 1: Amino acids*. Plenum Press, New York, N.Y.

Mascia, M., Palmas, S., Polcaro, A. M., Vacca, A., and Muntoni, A. (2007). "Experimental study and mathematical model on remediation of Cd spiked kaolinite by electrokinetics." *Electrochimica Acta*, 52(10), 3360-3365.

Mattson, E. D., Bowman, R. S., and Lindgren, E. R. (2002a). "Electrokinetic ion transport through unsaturated soil: 1. Theory, model development, and testing." *Journal of Contaminant Hydrology*, 54(1-2), 99-120.

Mattson, E. D., Bowman, R. S., and Lindgren, E. R. (2002b). "Electrokinetic ion transport through unsaturated soil: 2. Application to a heterogeneous field site." *Journal of Contaminant Hydrology*, 54(1-2), 121-140.

Maturi, K., and Reddy, K. R. (2006). "Simultaneous removal of organic compounds and heavy metals from soils by electrokinetic remediation with a modified cyclodextrin." *Chemosphere*, 63(6), 1022-1031.

Maturi, K., and Reddy, K. R. (2008). "Cosolvent-enhanced desorption and transport of heavy metals and organic contaminants in soils during electrokinetic remediation." *Water, Air, & Soil Pollution*, 189(1-4), 199-211.

Maturi, K., Reddy, K. R., and Cameselle, C. (2009). "Surfactant-enhanced electrokinetic remediation of mixed contamination in low permeability soil." *Separation Science & Technology*, 44(10), 2385-2409.

Menon, R. M. (1996). *Numerical modeling and experimental study on electro-kinetic extraction*. Ph.D. dissertation, Department of Civil Engineering, Texas A&M University, College Station, Texas.

Mitchell, J. K. (1991). "Conduction phenomena - From theory to geotechnical practice." *Géotechnique*, 41(3), 299-340.

Mitchell, J. K., and Soga, K. (2005). *Fundamentals of soil behavior*, 3rd Edition. John Wiley & Sons, Hoboken, N.J.

Mohamed, A. M. O. (1996). "Remediation of heavy metal contaminated soils via integrated electrochemical processes." *Waste Management*, 16(8), 741-747.

Mossop, K. F., Davidson, C. M., Ure, A. M., Shand, C. A., and Hillier, S. J. (2009). "Effect of EDTA on the fractionation and uptake by Taraxacum officinale of potentially toxic elements in soil from former chemical manufacturing sites." *Plant & Soil*, 320(1-2), 117-129.

Mulligan, C. N. (2009). "Recent advances in the environmental applications of biosurfactants." *Current Opinion in Colloid & Interface Science*, 14(5), 372-378.

Mulligan, C. N., and Eftekhari, F. (2003). "Remediation with surfactant foam of PCP-contaminated soil." *Engineering Geology*, 70(3-4), 269-279.

Mulligan, C. N., Yong, R. N., and Gibbs, B. F. (2001). "Surfactant-enhanced remediation of contaminated soil: A review." *Engineering Geology*, 60(1-4), 371-380.

Musso, G. (2003). "Transport phenomena in electrokinetic soil remediation." *Mathematical & Computer Modelling*, 37(5-6), 589-594.

Narasimhan, B., and Ranjan, R. S. (2000). "Electrokinetic barrier to prevent subsurface contaminant migration: Theoretical model development and validation." *Journal of Contaminant Hydrology*, 42(1), 1-17.

Navarro, A., and Martinez, F. (2010). "The use of soil-flushing to remediate metal contamination in a smelting slag dumping area: Column and pilot-scale experiments." *Engineering Geology*, 115(1-2), 16-27.

Neilson, J. W., Artiola, J. F., and Maier, R. M. (2003). "Characterization of lead removal from contaminated soils by nontoxic soil-washing agents." *Journal of Environmental Quality*, 32(3), 899-908.

Niinae, M., Nishigaki, K., and Aoki, K. (2008). "Removal of lead from contaminated soils with chelating agents." *Materials Transactions*, 49(10), 2377-2382.

Nowack, B. (2003). "Environmental chemistry of phosphonates." *Water Research*, 37(11), 2533-2546.

Nowack, B., and Stone, A. T. (1999). "The influence of metal ions on the adsorption of phosphonates onto goethite." *Environmental Science & Technology*, 33(20), 3627-3633.

Nowack, B., and VanBriesen, J. W. (2005). "Chelating agents in the environment." *Biogeochemistry of chelating agents*, B. Nowack, and J.M. VanBriesen, Editors, ACS Symposium Series 910, American Chemical Society, Washington, D.C., 1-18.

Ottosen, L. M., Hansen, H. K., and Jensen, P. E. (2009). "Electrokinetic removal of heavy metals." *Electrochemical remediation technologies for polluted soils, sediments and groundwater*. K.R. Reddy, and C. Cameselle, Editors, John Wiley & Sons, Hoboken, N.J., 97-126.

Oyanader, M. A., Arce, P., and Dzurik, A. (2003). "Avoiding pitfalls in electrokinetic remediation: Robust design and operation criteria based on first principles for maximizing performance in a rectangular geometry." *Electrophoresis*, 24(19-20), 3457-3466.

Page, M. M., and Page, C. L. (2002). "Electroremediation of contaminated soils." *Journal of Environmental Engineering*, ASCE, 128(3), 208-219.

Pamukcu, S., Weeks, A., and Wittle, J. K. (2004). "Enhanced reduction of Cr(VI) by direct electric current in a contaminated clay." *Environmental Science & Technology*, 38(4), 1236-1241.

Park, J. Y., Lee, H. H., Kim, S. J., Lee, Y. J., and Yang, J. W. (2007). "Surfactant-enhanced electrokinetic removal of phenanthrene from kaolinite." *Journal of Hazardous Materials*, 140(1-2), 230-236.

Pazos, M., Alcantara, M. T., Cameselle, C., and Sanroman, M. A. (2009). "Evaluation of electrokinetic technique for industrial waste decontamination." *Separation Science & Technology*, 44(10), 2304-2321.

Pegg, A. E., and Casero, R. A., Jr. (2011). Editors. *Polyamines: Methods and protocols*. Humana Press, New York, N.Y.

Penn, M., and Savvidou, C. (1997). "Temperature effects on electrokinetic remediation of contaminated soil." *Engineering geology and the environment*. P.G. Marinos, G.C. Koukis, G.C. Tsiambaos, and G.C. Stournaras, Editors, Balkema, Rotterdam, 2081-2086.

Pham, T. D., Shrestha, R. A., and Sillanpaa, M. (2010). "Removal of hexachlorobenzene and phenanthrene from clayey soil by surfactant- and ultrasound-assisted electrokinetics." *Journal of Environmental Engineering*, ASCE, 136(7), 739-742.

Pociecha, M., and Lestan, D. (2009). "EDTA leaching of Cu contaminated soil using electrochemical treatment of the washing solution." *Journal of Hazardous Materials*, 165(1-3), 533-539.

Pociecha, M., and Lestan, D. (2010). "Electrochemical EDTA recycling with sacrificial Al anode for remediation of Pb contaminated soil." *Environmental Pollution*, 158(8), 2710-2715.

Polettini, A., Pomi, R., and Calcagnoli, G. (2009). "Assisted washing for heavy metal and metalloid removal from contaminated dredged materials." *Water, Air, & Soil Pollution*, 196(1-4), 183-198.

Pomès, V., Fernandez, A., Costarramone. N., Grano, B., and Houi, D. (1999). "Fluorine migration in a soil bed submitted to an electric field: influence of electric potential on fluorine removal." *Colloids & Surfaces A: Physicochemical & Engineering Aspects*, 159(2-3), 481-490.

Popov, K., Kolosov, A., Ermakov, Y., Yachmenev, V., Yusipovich, A., Shabanova, N., Kogut, B. and Frid, A. (2004). "Enhancement of clay zeta-potential by chelating agents." *Colloids & Surfaces A: Physicochemical & Engineering Aspects*, 244(1-3), 25-29.

Popov, K., Kolosov, A., Yachmenev, V. G., Shabanova, N., Artemyeva, A., Frid, A., Kogut, B., Vesnovskii, S., and Sukharenko, V. (2001). "A laboratory-scale study of applied voltage and chelating agent on the electrokinetic separation of phenol from soil." *Separation Science & Technology*, 36(13), 2971-2982.

Popov, K., Yachmenev, V., and Barinov, A. (2005) "Enhancement of the electrokinetic remediation of soil contaminated with U(VI) by chelating agents." *Biogeochemistry of chelating agents*, B. Nowack, and J.M. VanBriesen, Editors, ACS Symposium Series 910, American Chemical Society, Washington, D.C., 398-420.

Popov, K., Yachmenev, V., Kolosov, A., and Shabanova, N. (1999). "Effect of soil electroosmotic flow enhancement by chelating reagents." *Colloids & Surfaces A: Physicochemical & Engineering Aspects*, 160(2), 135-140.

Popov, K. I., Shabanova, N. A., Artemeva, A. A., Urinovich, E. M., and Tulaeva, Y. V. (1997). "Influence of chelating agents on the electrokinetic potential of the clay fraction of soddy podzolic soils." *Colloid Journal*, 59(2), 212-214.

Puppala, S. K., Alshawabkeh, A. N., Acar, Y. B., Gale, R. J., and Bricka, M. (1997). "Enhanced electrokinetic remediation of high sorption capacity soil." *Journal of Hazardous Materials*, 55(1-3), 203-220.

Qiu, R. L., Zou, Z. L., Zhao, Z. H., Zhang, W. H., Zhang, T., Dong, H. Y., and Wei, X. G. (2010). "Removal of trace and major metals by soil washing with Na(2)EDTA and oxalate." *Journal of Soils & Sediments*, 10(1), 45-53.

Rahner, D., Ludwig, G., and Rohrs, J. (2002). "Electrochemically induced reactions in soils - A new approach to the in-situ remediation of contaminated soils? Part 1: The microconductor principle." *Electrochimica Acta*, 47(9), 1395-1403.

Ravera, M., Ciccarelli, C., Gastaldi, D., Rinaudo, C., Castelli, C., and Osella, D. (2006). "An experiment in the electrokinetic removal of copper from soil contaminated by the brass industry." *Chemosphere*, 63(6), 950-955.

Reddy, K. R., and Ala, P. R. (2005). "Electrokinetic remediation of metal-contaminated field soil." *Separation Science & Technology*, 40(8), 1701-1720.

Reddy, K. R., Ala, P. R., Sharma, S., and Kumar, S. N. (2006). "Enhanced electrokinetic remediation of contaminated manufactured gas plant soil." *Engineering Geology*, 85(1-2), 132-146.

Reddy, K. R., and Cameselle, C. (2009). Editors. *Electrochemical remediation technologies for polluted soils, sediments and groundwater*. John Wiley & Sons, Hoboken, N.J.

Reddy, K. R., Chaparro, C., and Saichek, R. E. (2003). "Removal of mercury from clayey soils using electrokinetics." *Journal of Environmental Science & Health Part A*, 38(2), 307-338.

Reddy, K. R., and Chinthamreddy, S. (1999). "Electrokinetic remediation of heavy metal-contaminated soils under reducing environments." *Waste Management*, 19(4), 269-282.

Reddy, K. R., and Chinthamreddy, S. (2004). "Enhanced electrokinetic remediation of heavy metals in glacial till soils using different electrolyte solutions." *Journal of Environmental Engineering*, ASCE, 130(4), 442-455.

Reddy, K. R., and Saichek, R. E. (2003). "Effect of soil type on electrokinetic removal of phenanthrene using surfactants and cosolvents." *Journal of Environmental Engineering*, ASCE, 129(4), 336-346.

Ribeiro, A. B., and Mexia, J. T. (1997). "A dynamic model for the electrokinetic removal of copper from a polluted soil." *Journal of Hazardous Materials*, 56(3), 257-271.

Ribeiro, A. B., Rodriguez-Maroto, J. M., Mateus, E .P., and Gomes, H. (2005). "Removal of organic contaminants from soils by an electrokinetic process: the case of atrazine. Experimental and modeling." *Chemosphere*, 59(9), 1229-1239.

Rohrs, J., Ludwig, G., and Rahner, D. (2002). "Electrochemically induced reactions in soils - A new approach to the in-situ remediation of contaminated soils? Part 2: remediation experiments with a natural soil containing highly chlorinated hydrocarbons." *Electrochimica Acta*, 47(9), 1405-1414.

Sah, J. G., and Che, J. Y. (1998). "Study of the electrokinetic process on Cd and Pb spiked soils." *Journal of Hazardous Materials*, 58(1-3), 301-315.

Saichek, R. E., and Reddy, K. R. (2003a). "Effects of system variables on surfactant enhanced electrokinetic removal of polycyclic aromatic hydrocarbons from clayey soils." *Environmental Technology*, 24(4), 503-515.

Saichek, R. E., and Reddy, K. R. (2003b). "Effect of pH control at the anode for the electrokinetic removal of phenanthrene from kaolin soil." *Chemosphere*, 51(4), 273-287.

Saichek, R. E., and Reddy, K. R. (2005). "Surfactant-enhanced electrokinetic remediation of polycyclic aromatic hydrocarbons in heterogeneous subsurface environments." *Journal of Environmental Engineering & Science*, 4(5), 327-339.

Sauer, J. E., and Davis, E. J. (1994). "Electrokinetically enhanced sedimentation of colloidal contaminants." *Environmental Science & Technology*, 28(4), 737-745.

Sawada, A., Mori, K., Tanaka, S., Fukushima, M., and Tatsumi, K. (2004). "Removal of Cr(VI) from contaminated soil by electrokinetic remediation." *Waste Management*, 24(5), 483-490.

Sawada, A., Tanaka, S., Fukushima, M., and Tatsumi, K. (2003). "Electrokinetic remediation of clayey soils containing copper(II)-oxinate using humic acid as a surfactant." *Journal of Hazardous Materials,* 96(2-3), 145–154.

Scott, T. B. (1994). *Physicochemical phenomena of electro-kinetic extraction of inorganic contaminants from kaolinite*. M.S. thesis, Department of Civil Engineering Texas A&M University, College Station, Texas.

Shang, J. Q., and Lo, K. Y. (1997). "Electrokinetic dewatering of a phosphate clay." *Journal of Hazardous Materials*, 55(1-3), 117-133.

Shapiro, A. P., and Probstein, R. F. (1993). "Removal of contaminants from saturated clay by electroosmosis." *Environmental Science & Technology*, 27(2), 283-291.

Shen, Z. M., Zhang, J. D., Qu, L. Y., Dong, Z. Q., Zheng, S. S., and Wang, W. H. (2009). "A modified EK method with an I-/I-2 lixiviant assisted and approaching cathodes to remedy mercury contaminated field soils." *Environmental Geology*, 57(6), 1399-1407.

Shi, J. G., Yuan, Z. X., Zhang, S. F., Huang, G. H., Dai, F., Li, J. B., Chakma, A., Qin, X. S., and Liu, H. L. (2004). "Removal of cadmium and lead from sediment by rhamnolipid." *Transactions of Nonferrous Metals Society of China*, 14, 66-70.

Shiba, S., Hino, S., Hirata, Y., and Seno, T. (2000). "Removal of heavy metal from soil and groundwater by in-situ electrokinetic remediation." *Water Science & Technology*, 42(7-8), 335-343.

Silbey, R. J., Alberty, R. A., and Bawendi, M. G. (2005). *Physical chemistry*, 4th Edition. Wiley, Hoboken, N.J.

Singh, P., and Cameotra, S. S. (2004). "Enhancement of metal bioremediation by use of microbial surfactants." *Biochemical & Biophysical Research Communications,* 319(2), 291-297.

Skold, M. E., Thyne, G. D., Drexler, J. W., and McCray, J. E. (2009). "Solubility enhancement of seven metal contaminants using carboxymethyl-beta-cyclodextrin (CMCD)." *Journal of Contaminant Hydrology*, 107(3-4), 108-113.

Soleimani, M., Hajabbasi, M. A., Afyuni, M., Akbar, S., Jensen, J. K., Holm, P. E., and Borggaard, O. K. (2010). "Comparison of natural humic substances and synthetic ethylenediaminetetraacetic acid and nitrilotriacetic acid as washing agents of a heavy metal-polluted soil." *Journal of Environmental Quality*, 39(3), 855-862.

Sposito, G. (1998). "On points of zero charge." *Environmental Science & Technology*, 32(19), 2815-2819.

Sposito, G. (2008). *The surface chemistry of soils*, 2nd Edition. Oxford University Press, New York, N.Y.

Sprute, R. H., and Kelsh, D. J. (1976). *Dewatering and densification of coal waste by direct current - Laboratory tests*. Report of investigations 8197, Bureau of Mines. Department of the Interior, Washington, D.C.

Sprute, R. H., and Kelsh, D. J. (1979). *Slimes consolidation at the Henderson mine*. Report of investigations 8441, Bureau of Mines, Department of the Interior, Washington, D.C.

Sprute, R. H., and Kelsh, D. J. (1980). "Dewatering fine-particle suspensions with direct current." *Proc. Int. Symp. Fine Particle Process*, Las Vegas.

Sprute, R. H., and Kelsh, D. J. (1982). *Electrokinetic densification of solids in a coal mine sediment pond: A feasibility study. Part 1. Laboratory and field tests*. Report of Investigations 8666, Bureau of Mines, U.S. Dept. of the Interior, Avondale, Maryland.

Sprute, R. H., Kelsh, D. J., and Thompson, S. L. (1988). *Electrokinetic densification of solids in a coal mine sediment pond: A feasibility study. Part 2. Design of an operational system*. Report of Investigations 9137, Bureau of Mines, U.S. Dept. of the Interior, Avondale, Maryland.

Srivastava, R. K., Tiwari, R. P., and Bala Ramudu, P. (2007). "Removal of cadmium using electrokinetics – Effect of enhancement agents." *Proc., 1st Sri Lankan Geotechnical Society International Conference on Soil and Rock Engineering*, Colombo.

Stevenson, F. J. (1994). *Humus chemistry: Genesis, composition, reactions*, 2nd Edition. John Wiley & Sons, New York, N.Y.

Suèr, P., and Allard, B. (2003). "Mercury transport and speciation during electrokinetic soil remediation." *Water, Air, & Soil Pollution*, 143(1-4), 99-109.

Suèr, P., and Lifvergren, T. (2003). "Mercury-contaminated soil remediation by iodide and electroreclamation." *Journal of Environmental Engineering*, ASCE, 129(5), 441-446.

Teutli-Leon, M. M., Oropeza, M. T., Gonzalez, I., and Soria, A. (2005). "Mathematical Modeling of a galvanostatic soil electroremediation process." *AICHE Journal*, 51(6), 1822-1833.

Thepsithar, P., and Roberts, E. P. L. (2006). "Removal of phenol from contaminated kaolin using electrokinetically enhanced in situ chemical oxidation." *Environmental Science & Technology*, 40(19), 6098-6103.

Triay, I., Miller G., and Mitchell, A. (1992). "Electrochemical removal of actinides." *Proc. Electrokinetics Workshop*, Atlanta, U.S. Department of Energy.

Udovic, M., Drobne, D., and Lestan, D. (2009). "Bioaccumulation in Porcellio scaber (Crustacea, Isopoda) as a measure of the EDTA remediation efficiency of metal-polluted soil." *Environmental Pollution*, 157(10), 2822-2829.

Udovic, M., and Lestan, D. (2009). "Pb, Zn and Cd mobility, availability and fractionation in aged soil remediated by EDTA leaching." *Chemosphere*, 74(10), 1367-1373.

Udovic, M., and Lestan, D. (2010). "Fractionation and bioavailability of Cu in soil remediated by EDTA leaching and processed by earthworms (Lumbricus terrestris L.)." *Environmental Science & Pollution Research*, 17(3), 561-570.

Van Olphen, H. (1977). *An introduction to clay colloid chemistry*, 2nd edition. John Wiley & Sons, New York, N.Y.

Vengris, T., Binkiene, R., and Sveikauskaite, A. (2001). "Electrokinetic remediation of lead-, zinc- and cadmium-contaminated soil." *Journal of Chemical Technology & Biotechnology*, 76(11), 1165-1170.

Vereda-Alonso, C., Rodriguez-Maroto, J. M., Garcia-Delgado, R. A., Gómez-Lahoz, C., and García-Herruzo, F. (2004). "Two-dimensional model for soil electrokinetic remediation of heavy metals - Application to a copper spiked kaolin." *Chemosphere*, 54(7), 895-903.

Virkutyte, J., van Hullebusch, E., Sillanpaa, M., and Lens, P. (2005). "Copper and trace element fractionation in electrokinetically treated methanogenic anaerobic granular sludge." *Environmental Pollution*, 138(3), 517-528.

Voglar, D., and Lestan, D. (2010). "Electrochemical separation and reuse of EDTA after extraction of Cu contaminated soil." *Journal of Hazardous Materials*, 180(1-3), 152-157.

Wan, J. Z., Yuan, S. H., Chen, J., Li, T. P., Lin, L., and Lu, X. H. (2009). "Solubility-enhanced electrokinetic movement of hexachlorobenzene in sediments: A comparison of cosolvent and cyclodextrin." *Journal of Hazardous Materials*, 166(1), 221-226.

Wang, J. Y., Huang, X. J., Kao, J. C. M., and Stabnikova, O. (2006). "Removal of heavy metals from kaolin using an upward electrokinetic soil remedial (UESR) technology." *Journal of Hazardous Materials*, 136(3), 532-541.

Wang, J.-Y., and Casero, R. A., Jr. (2006). Editors. *Polyamines cell signaling: Physiology, pharmacology, and cancer research*. Humana Press, New York, N.Y.

Wang, S. L., and Mulligan, C. N. (2006). "Effect of natural organic matter on arsenic release from soils and sediments into groundwater." *Environmental Geochemistry & Health*, 28(3), 197-214.

Wang, S. L., and Mulligan, C. N. (2009). "Enhanced mobilization of arsenic and heavy metals from mine tailings by humic acid." *Chemosphere*, 74(2), 274-279.

Wang, X. J., and Brusseau, M. L. (1993). "Solubilization of some low-polarity organic compounds by hydroxypropyl-β-cyclodextrin." *Environmental Science & Technology*, 27(13), 2821-2825.

Wasay, S. A., Barrington, S. F., and Tokunaga, S. (1998). "Remediation of soils polluted by heavy metals using salts of organic acids and chelating agents." *Environmental Technology*, 19(4), 369-379.

Wasay, S. A., Barrington, S., and Tokunaga, S. (2001). "Organic acids for the in situ remediation of soils polluted by heavy metals: Soil flushing in columns." *Water, Air, & Soil Pollution*, 127(1-4), 301-314.

Weng, C. H., and Yuan, C. (2001). "Removal of Cr(III) from clay soils by electrokinetics." *Environmental Geochemistry & Health*, 23(3), 281-285.

Wilson, D. R., Page, I. C., Cocci, A. A., and Landine, R. C. (1998). "Case history - Two stage, low-rate anaerobic treatment facility for South American alcochemical citric acid wastewater." *Water Science & Technology*, 38(4-5), 45-52.

Wittle, J. K., and Pamukcu, S. (1993). *Electrokinetic treatment of contaminated soils, sludges, and lagoons*. Final Report, Contract Ns 02112406, Office of Research and Development, Technology Development, Environmental Restoration and Waste Management, U.S. Department of Energy, Washington, D.C.

Wong, J. S. H., Hicks, R. E., and Probstein, R. F. (1997). "EDTA-enhanced electroremediation of metal-contaminated soils." *Journal of Hazardous Materials*, 55(1-3), 61-79.

Wu, L. H., Luo, Y. M., Christie, P., and Wong, M. H. (2003). "Effects of EDTA and low molecular weight organic acids on soil solution properties of a heavy metal polluted soil." *Chemosphere*, 50(6), 819-822.

Yang, G. C. C. (2011). "Guest editorial (Preface)." *Separation & Purification Technology*, in press.

Yang, G. C. C., and Lin, S. L. (1998). "Removal of lead from a silt loam soil by electrokinetic remediation." *Journal of Hazardous Materials*, 58(1-3), 285-299.

Yang, J. S., Lee, J. Y., Baek, K., Kwon, T. S., and Choi, J. (2009). "Extraction behavior of As, Pb, and Zn from mine tailings with acid and base solutions." *Journal of Hazardous Materials*, 171(1-3), 443-451.

Yang, J. W., Lee, Y. J., Park, J. Y., Kim, S. J., and Lee, J. Y. (2005). "Application of APG and Calfax 16L-35 on surfactant-enhanced electrokinetic removal of phenanthrene from kaolinite." *Engineering Geology*, 77(3-4), 243-251.

Yeung, A. T. (1990). "Coupled flow equations for water, electricity, and ionic contaminants through clayey soils under hydraulic, electrical and chemical gradients." *Journal of Non-Equilibrium Thermodynamics*, 15(3), 247-267.

Yeung, A. T. (1992). "Diffuse double-layer equations in SI units." *Journal of Geotechnical Engineering*, ASCE, 118(12), 2000-2005.

Yeung, A. T. (1994). "Electrokinetic flow processes in porous media and their applications." *Advances in porous media*. M.Y. Corapcioglu, Editor, Elsevier, Amsterdam, Vol. 2, 309-395.

Yeung, A. T. (2006a). "Contaminant extractability by electrokinetics." *Environmental Engineering Science*, 23(1), 202-224.

Yeung, A. T. (2006b). "Fundamental aspects of prolonged electrokinetic flows in kaolinites." *Geomechanics & Geoengineering: An International Journal*, 1(1), 13-25.

Yeung, A. T. (2008). "Electrokinetics for soil remediation." *Environmental Geotechnology and Global Sustainable Development, 2008*. A.T. Yeung, and I.M.C. Lo, Editors, Advanced Technovation Limited, Hong Kong, 16-25.

Yeung, A. T. (2009a). "Remediation technologies for contaminated sites." *Advances in Environmental Geotechnics*. Proc., International Symposium on

Geoenvironmental Engineering, Hangzhou, Y. Chen, X. Tang, and L. Zhan, Editors, Zhejiang University Press, 328-369.

Yeung, A. T. (2009b). "Geochemical processes affecting electrochemical remediation." *Electrochemical remediation technologies for polluted soils, sediments and groundwater*. K.R. Reddy, and C. Cameselle, Editors, John Wiley & Sons, Hoboken, N.J., 65-94.

Yeung, A. T. (2011). "Milestone developments, myths, and future directions of electrokinetic remediation." *Separation & Purification Technology*, in press.

Yeung, A. T., Chung, M., Corapcioglu, M. Y., and Stallard, W. M. (1997a). "Impoundment liner repair by electrophoresis of clay." *Journal of Environmental Engineering*, ASCE, 123(10), 993-1001.

Yeung, A. T., and Datla, S. (1995). "Fundamental formulation of electro-kinetic extraction of contaminants from soil." *Canadian Geotechnical Journal*, 32(4), 569-583.

Yeung, A. T., and Hsu, C. (2005). "Electrokinetic remediation of cadmium-contaminated clay." *Journal of Environmental Engineering*, ASCE, 131(2), 298-304.

Yeung, A. T., Hsu, C., and Menon, R. M. (1996). "EDTA-enhanced electrokinetic extraction of lead." *Journal of Geotechnical Engineering*, ASCE 122(8), 666-673.

Yeung, A. T., Hsu, C., and Menon, R. M. (1997b). "Physicochemical soil-contaminant interactions during electrokinetic extraction." *Journal of Hazardous Materials*, 55(1-3), 221-237.

Yeung A. T., and Mitchell, J. K. (1993). "Coupled fluid, electrical and chemical flows in soil." *Géotechnique*, 43(1), 121-134.

Yu, J. W., and Neretnieks, I. (1996). "Modelling of transport and reaction processes in a porous medium in an electrical field." *Chemical Engineering Science*, 51(19), 4355-4368.

Yu, J. W., and Neretnieks, I. (1997). "Theoretical evaluation of a technique for electrokinetic decontamination of soils." *Journal of Contaminant Hydrology*, 26(1-4), 291-299.

Yuan, C., Hung, C. H., and Huang, W. L. (2009). "Enhancement with carbon nanotube barrier on 1,2-dichlorobenzene removal from soil by surfactant-assisted electrokinetic (SAEK) process - The effect of processing fluid." *Separation Science & Technology*, 44(10), 2284-2303.

Yuan, C., and Weng, C. H. (2004). "Remediating ethylbenzene-contaminated clayey soil by a surfactant-aided electrokinetic (SAEK) process." *Chemosphere*, 57(3), 225-232.

Yuan, C., and Weng, C. H. (2006). "Electrokinetic enhancement removal of heavy metals from industrial wastewater sludge." *Chemosphere*, 65(1), 88-96.

Yuan, S. H., Tian, M., and Lu, X. H. (2006). "Electrokinetic movement of hexachlorobenzene in clayed soils enhanced by Tween 80 and beta-cyclodextrin." *Journal of Hazardous Materials*, 137(2), 1218-1225.

Zaier, H., Ghnaya, T., Ben Rejeb, K., Lakhdar, A., Rejeb, S., and Jemal, F. (2010). "Effects of EDTA on phytoextraction of heavy metals (Zn, Mn and Pb) from sludge-amended soil with Brassica napus." *Bioresource Technology*, 101(11),

3978-3983.
Zhou, D. M., Alshawabkeh, A., Deng, C. F., Cang, L., and Si, Y. B. (2004a). "Electrokinetic removal of chromium and copper from contaminated soils by lactic acid enhancement in the catholyte." *Journal of Environmental Sciences - China*, 16(4), 529-532.
Zhou, D. M., Cang, L., Alshawabkeh, A. N., Wang, Y. J., and Hao, X. Z. (2006). "Pilot-scale electrokinetic treatment of a Cu contaminated red soil." *Chemosphere*, 63(6), 964-971.
Zhou, D. M., Deng, C. F., Alshawabkeh, A. N., and Cang, L. (2005a). "Effects of catholyte conditioning on electrokinetic extraction of copper from mine tailings." *Environment International*, 31(6), 885-890.
Zhou, D. M., Deng, C. F., and Cang, L. (2004b). "Electrokinetic remediation of a Cu contaminated red soil by conditioning catholyte pH with different enhancing chemical reagents." *Chemosphere*, 56(3), 265-273.
Zhou, D. M., Deng, C. F., Cang, L., and Alshawabkeh, A. N. (2005b). "Electrokinetic remediation of a Cu-Zn contaminated red soil by controlling the voltage and conditioning catholyte pH." *Chemosphere*, 61(4), 519-527.
Zhou, D. M., Zorn, R., and Kurt, C. (2003). "Electrochemical remediation of copper contaminated kaolinite by conditioning anolyte and catholyte pH simultaneously." *Journal of Environmental Sciences - China*, 15(3), 396-400.

9.12 Acronyms

dc	=	direct-current
ADA	=	N-(2-acetamido)iminodiacetic acid
CAIW	=	citric acid industrial wastewater
CCA	=	chromated copper arsenate
DTPA	=	(diethylenetriamine)pentaacetic acid
DTPMP	=	diethylenetriaminepenta(methylenephosphonic) acid
EDA	=	ethylenediamine
EDDS	=	ethylenediamine-N,N'-disuccinic acid
EDTA	=	ethylenediaminetetraacetic acid
EDTMP	=	(ethylenedinitrilo)-tetramethylenephosphonic acid
EGTA	=	ethylene glycol tetraacetic acid
GCG	=	L-5-glutamyl-L-cysteinylglycine
HEDP	=	(hydroxyethylidene)diphosphonic acid
HEDTA	=	hydroxyethyl ethylenediamine triacetic acid
HPCD	=	hydroxypropyl-β-cyclodextrin
IDSA	=	iminodisuccinic acid

MCD	=	methyl-β-cyclodextrin
NTA	=	nitrilotriacetic acid
NTTA	=	nitrilotris(methylene)triphosphonic acid
NTMP	=	(nitrilotrimethylene)triphosphonic acid
PAH	=	polycyclic aromatic hydrocarbon
PBTC	=	2-phosphono butane-1,2,4-tricarboxylic acid
PDA	=	pyridine-2,6-dicarboxylic acid
SCMC	=	S-carboxymethylcysteine
TMDTA	=	trimethylenedinitrilote-tetraacetic acid
β-CD	=	β-cyclodextrin

Editor Biographies

Dr. Daniel Tsang is a Senior Lecturer in the Department of Civil and Natural Resources Engineering at the University of Canterbury in New Zealand. He received his Ph.D. degree from Hong Kong University of Science and Technology in 2006. Dr. Tsang was a post-doctoral fellow at Imperial College London in 2007-2008 and a visiting scholar at Stanford University in 2011. His area of teaching and research is environmental science and engineering. Dr. Tsang aims to develop economical and sustainable technologies for contaminated soil/sediment management and industrial wastewater treatment. His current research focuses on the applications of chelating agents, surfactants, natural organic matter, activated carbon, zero-valent iron, and low-cost materials for contaminant extraction or stabilization. He is also interested in the interactions of emerging contaminants in wastewater and sustainability assessment of environmental technologies. He has published over 50 refereed journal and conference papers, and served as a peer reviewer for 15 journals. Dr. Tsang is a Chartered Environmentalist in the UK and Management Committee of three Specialist Groups of International Water Association.

Prof. Irene M. C. Lo is currently Professor in the Department of Civil and Environmental Engineering, and Director of Environmental Engineering Program at The Hong Kong University of Science and Technology (HKUST). Prof. Lo is a Fellow of the Hong Kong Institution of Engineers and a Fellow of the American Society of Civil Engineers (ASCE). She is also a Guest Professor of Jilin University and Harbin Institute of Technology. She has been a Visiting Associate Professor of Technical University of Denmark and the University of Wisconsin at Madison. She received both her M.S. and Ph.D. in Civil Engineering from the University of Texas at Austin, USA. She was bestowed the 2008 EWRI Best Practice-Oriented Paper Award by ASCE in recognition of her outstanding applied research achievement. Prof. Lo was the recipient of the 2004 ASCE James Croes Medal and the 2007 ASCE Samuel Arnold Greeley Award. In 2006, she received HKUST's highest accolade in teaching, the Michael G Gale Medal for Distinguished Teaching. Prof. Lo has edited 4 books and published more than 250 journal and conference papers in the areas of remediation technologies for contaminated soils and groundwater, magnetic nano- and microparticles for environmental pollution control, carbon footprint analysis and life cycle assessment of remediation technology, waste containment facilities, and pollutant migration in soils.

Prof. Rao Y. Surampalli received his M.S. and Ph.D. degrees in Environmental Engineering from Oklahoma State University and Iowa State University, respectively. He is a Registered Professional Engineer in the branches of Civil and Environmental Engineering, and also a Board Certified Environmental Engineer (BCEE) of the American Academy of Environmental Engineers (AAEE) and Diplomate of the American Academy of Water Resources Engineers (DWRE). He has authored more than 500 technical publications, including twelve (12) patents, eleven (11) books, 77 book chapters, 203 refereed (peer-reviewed) journal articles, presented at more than 200 national and international conferences, edited 13 refereed conference proceedings, and given over 80 plenary, keynote or invited presentations worldwide. Dr. Surampalli is a

Distinguished Member of American Society Civil Engineers, Fellow of the American Association for the Advancement of Science, and a Member of the European Academy of Sciences and Arts. He also is Vice-Chair of Editorial Board of Water Environment Research Journal and Editor of ASCE's Journal of Hazardous, Toxic and Radioactive Waste.

Index

Page numbers followed by *t* indicate a table. Those followed by *f* indicate a figure.